THE
DICTIONARY for
HUMAN FACTORS
ERGONOMICS

James H. Stramler, Jr.

CRC Press
Boca Raton   Ann Arbor   London   Tokyo

Library of Congress Cataloging-in-Publication Data

Stramler, Jim
    The dictionary for human factors/ergonomics / Jim Stramler.
        p.   cm.
    ISBN 0-8493-4236-8
    1. Human engineering—Dictionaries.   I. Title.
TA166.S77 1993
620.8'2'03—dc20

                                                                92-2348
                                                                    CIP

This book represents information obtained from authentic and highly regarded sources. Reprinted material is quoted with permission, and sources are indicated. A wide variety of references are listed. Every reasonable effort has been made to give reliable data and information, but the author and the publisher cannot assume responsibility for the validity of all materials or for the consequences of their use.

Direct all inquiries to CRC Press, Inc., 2000 Corporate Blvd., N. W., Boca Raton, Florida, 33431.

© 1993 by CRC Press, Inc.

International Standard Book Number 0-8493-4236-8

Library of Congress Card Number 92-2348
Printed in the United States 1 2 3 4 5 6 7 8 9 0
Printed on acid-free paper

TO MY WIFE DIANNE

CONTENTS

INTRODUCTION

Terminology is probably the most important factor in learning or working in any field. It forms the basis for communication, discussion, and evolution of the discipline.

The most practical way to establish some commonalities, consistencies, and standardization in terminology is by knowing and having readily available a list of the terms already in use. *The Dictionary for Human Factors/Ergonomics* makes readily available a large number of the terms used in the human factors field. Hopefully, it will also eliminate at least some confusion about the current usage of terms by providing comparisons and distinctions.

Human factors is a field whose primary function is to help people in virtually all phases of life. Many of the concepts are much more complex and difficult to clarify and study than the concepts of other scientific and engineering fields. The lack of such a book as this has probably hindered the development of the field, because it is harder to gain a complete understanding without a basic reference for its terminology.

The need for a book like this is of especially great importance now that various attempts are appearing to develop and implement legislation dealing directly with human factors issues. Even though a court has just overturned the San Franciso ordinance emphasizing human factors in using VDTs at the time of this writing, other cities and states are looking into such legal avenues. Additional laws written in the future will be upheld because they are needed. The only way any sense or fairness will come out of such legislation is to standardize the terminology in the field.

In developing this dictionary, I have tried to collect the basic terms used in all areas of human factors. This dictionary is by far the most comprehensive such book available in the field. It is also the first such work to include significant numbers of terms from the areas of human-computer interaction, speech and language, robotics, and modeling.

Many terms are present because I believed they were required for completeness within this work. I saw little or no justification for making a reader seek out a medical dictionary or other reference book to obtain a term and its definition which could reasonably be included in this book, and which may have a slightly different usage in human factors than in other fields.

While some terms may not be of direct use to human factors' people in their particular work, they will provide a more general or historical background. Since most people working in this field enter it by way of experimental or some other area of psychology (*Human Factors Society, Directory and Yearbook*), I have provided a number of physical science terms applicable in some of the human factors areas to help those less familiar with the physical sciences understand the relationships. Such terms should also be useful in helping human factors people communicate with people in other scientific and engineering fields.

The Dictionary for Human Factors/Ergonomics is more than just a collection of terms and their definitions. It is an attempt to achieve at least the beginnings of an integration and internal consistency for terminology across the different human factors areas. That quality is clearly not achievable by smaller, independent works within the various areas. The book also provides a basic framework within which people can make comments and elaborate on the definitions for their own specific needs.

Within the field of human factors, this dictionary should be useful for the experienced worker, the student who is new to the field, and everyone between these extremes. I also see the book proving useful to industrial engineers, safety personnel, psychologists, management, environmentalists, lawyers, and many other groups.

I realize the present version of this dictionary is only the beginning. Some terms could still use further elaboration. Not all terms used by human factors professionals are compiled here yet. I still occasionally find an additional term I have overlooked and which should obviously be included. I have collected many other terms which I have found used only once or without clear definition in the literature, or which I have not been able to resolve to my own satisfaction by the submission deadline. These will be researched further and placed in future editions. Certainly new terms will continue to evolve as technology advances and as we learn more about humans working in different environments. These, too, will be incorporated as they are developed.

My desire is that, in time, this book or one of its descendents will serve as a terminological base or definitional standard in human factors. However, even if the book does nothing more than stimulate or enhance communication and discussion, it will have served a very useful purpose.

While doing research for this dictionary, I discovered a statement in the preface of an AGARD publication on human performance which bemoaned the communication problems among scientists, not with national languages, but with scientific language (AGARD, 1989). Within such an important field as human factors, all students, educators, and workers need to speak and write the same scientific language. With the advent of computer databases and search routines, a relatively simple matter such as the precise spelling of terms becomes of great importance.

The Dictionary for Human Factors/Ergonomics is intended to provide a major step toward eliminating language problems within the field. It should also make those outside the field more aware of the kinds of things human factors does, and the benefits it can provide, thus promoting the field.

To help fulfill my goal of achieving standardization, I would certainly appreciate any efforts of those working in human factors or other fields who wish to provide additional/alternative terms which they believe should be included in future editions. In doing so, please provide two or more independent published citations of a term's use. Unfortunately, I cannot promise acknowledgment of those who submit terms.

I realized early in my human factors career how valuable such a document as this would be. I would forget the detailed meaning of some terms through disuse, would confuse the meanings of similar terms, or would encounter a term whose definition I had never learned. Following a thorough literature search, I discovered that there really was no adequate comprehensive general dictionary of human factors or ergonomics.

Based on my conversations with fellow workers and occasional mention in research or other papers, many have recognized the need for such a collection of terms.

Writing *The Dictionary for Human Factors/Ergonomics* has been a tremendous learning experience for me. At the same time, writing this book has also been an extremely difficult, frustrating, and time-consuming job — much more so than I ever imagined when I began working on it. Aside from the general problem of reducing some concepts to words which have useful meaning, there were a number of significant problems with the current terminologies in use. I think I now know why no one has previously undertaken such a book as this.

I devoted a great deal of thought to the title of this book. Should I call it a dictionary of human factors? A dictionary of ergonomics? A dictionary of human factors and ergonomics? I clearly preferred human factors in the title because that, to me, represents a broader approach to the interaction of humans with the various aspects of life. It more easily allows for the inclusion of human interaction in the home, as well as recreational and other activities. Ergonomics, on the other hand, though a more well-recognized term now, is derived from Greek terms meaning "the study of work", and its applications were originally and still are more in that area.

I finally decided on *The Dictionary for Human Factors/Ergonomics* for two primary reasons: (a) to incorporate both terms, taking advantage of the name recognition ergonomics carries, and (b) because the terms are in the process of becoming virtually synonymous. I believe the wisdom of this choice is supported by the recent effort of the Human Factors Society to undergo a similar name change. At the time of this writing, however, that process is still ongoing.

No other field I know of is as comprehensive as human factors. In such a comprehensive and multidisciplinary field, it is very difficult to have expertise in all areas. I must acknowledge the receipt of helpful information from many colleagues in developing this book. I have made special efforts to seek a greater understanding in those areas where I had less expertise. I hope any errors of either omission or commission are minimal, though some may be present in a first work of this magnitude.

I am grateful for CRC Press offering to publish the dictionary at this time. I need some time now to step back and regain some perspective before proceeding. I would like to thank Jeffrey Woldstad, Karl Kroemer, and John Guignard for their efforts in reviewing parts or all of the book. Their suggestions resulted in many improvements.

HOW TO USE THIS BOOK

There are several guidelines I would like to offer those using this book.

A "term" may involve more than one word.

All terms are arranged in one continuous alphabetical sequence. The book is not divided into subject matter areas. A few numerical terms are placed at the beginning. I didn't want to force a user to have to guess where the verbal spelling of the term was. These numerical terms generally carry a reference to a verbal term.

The general format or plan is to have the user look up the full term or phrase as normally used, not a key word followed by some modifiers — for example, the user should look for "wet bulb temperature," not "temperature, wet bulb." There will often be a pointer to the other term. In some cases, however, these modifiers are necessary to avoid possible confusion or major separation of closely-related terms. The latter is especially true for some of the anthropometric measures so that similar measures which are distinguished by sitting, standing, or other variation could be grouped alphabetically.

There are instances where a term has more than one usage/definition. When this occurs, I have listed the terms and their definitions separately, rather than combining them under one listing. This approach provided me with much greater flexibility in tracking and cataloguing the terms. There is certainly a precedence for doing this, and I hope that by mentioning such organization here, this structure will not be an inconvenience. I have added superscripts following these terms to note an occurrence of multiple wage/definition.

For those instances where multiple terms have the same meaning, the dictionary is arranged so that the definition resides with the preferred term. Synonyms are listed with this preferred term.

I have provided a track or cross-reference to related terms with "*see*" or "*see also*" and via the use of such abbreviations as "*syn,*" "*opp,*" or "*ant.*" The meanings of these abbreviations are listed under Internal Abbreviations.

Words such as "man," "his," and others are used in the scientific context, and refer generically to any member of the human species. This usage is not in any way intended to be discriminatory or inconsiderate. Such would be inconsistent with the purpose of this book.

In some other related dictionary-type works, the authors have elaborated on the definition with some commentary. I have provided for this option here, but have generally noted such elaboration separately. When it occurs, it will appear on the line below the definition and follow the "*Comment:*" heading. Commentary should not, in my view, be included as part of the basic definition. The commentary is also used to provide positioning and measurement specifications in anthropometric work. While this could, in a strict sense, be considered part of the definition, it was done this way to provide easier reading due to the length of such definitions.

On occasion, mention may be made of some brand names for tests, software, equipment, etc. This does not endorse them in any way. The mention is made solely for the reader's information because these brands are commonly used within certain areas of the field. In addition, there is no intent to slight any names not mentioned. If this occurs, it is probably because I am either not aware of them or have an incomplete understanding at this time.

The references listed were those involved with more than a single term in the composition of this dictionary. Although many more books or periodical articles were used as sources, they are not referenced because they provided information on only a single term, and the inclusion of all such books and articles would make the length of the reference list unwieldy. Due to the complex integration of terms in this dictionary, I have not cited the references for each term individually.

Some "terms" are actually acronyms which resemble words. I have provided some of these acronyms within the text portion of the dictionary where I thought confusion could result. A list of the acronyms used in the book is provided in Appendix A, along with the term they represent. It might also be a good idea to check the acronym section before deciding a "term" is not in the dictionary.

Some organizations are referred to in the main body of the dictionary. Appendix B1 provides a list of all the national and international organizations and their acronyms I have accumulated which are involved in human factors work, or which occasionally publish information useful to human factors. This appendix should include all those organizations mentioned in the definition section of the book, plus many others not mentioned. Appendix B2 is the reverse. It provides a cross-reference of the organization acronym and its full name. Since apparently not all the organizations have acronyms (or I do not have an acronym for them), Appendix B2 is shorter. I have not provided addresses and phone numbers for these organizations because many of them change from year to year. Such information should be available in many local libraries. The reader should also be aware that not all of these organizations are still in existence, and some older ones are now known by different names. I hope I have included the newer name when appropriate.

INTERNAL ABBREVIATIONS

adj.	adjectival form
ant.	antonym
equiv.	equivalent
ger.	gerund
(n)	noun
occas.	occasional
opp.	opposite/counter term (not necessarily an antonym)
pl.	plural
(sl)	slang
syn.	synonym
(v)	verb

THE DICTIONARY FOR
HUMAN FACTORS/ERGONOMICS

NUMERICAL TERMS

0.7 convention
a recommended minimum correlation coefficient which should exist before accepting a relationship or making a projected anthropometric measurement between two sets of data for design purposes

1/3 octave band
see one-third octave band

10° observer
see CIE Supplementary Standard Observer

2° observer
see CIE Standard Observer

50th percentile
see median

8-hour time weighted average
see *time weighted average*

A scale
a sound weighting system which has response characteristics to reduce high- and low-frequency extremes and correlate well with the subjective judgement of the human auditory system for loudness

A shift
see *first shift*

A-weighted sound pressure level
(dB(A), L_A)
that sound pressure level measured using the A scale

Abbreviated Injury Scale (AIS)
a numerical rating system used in an attempt to quantify an automobile accident victim's severity of injuries

Rating	Severity
1	minor
2	moderate
3	serious
4	severe
5	critical
	(survival uncertain)
6	maximum
	(virtually unsurvivable)
9	unknown

abdomen
the lower anterior portion of the torso from the diaphragm to the pelvic region, including the overlying muscles; *adj.* abdominal

abdominal cavity
that cavity within the abdomen which contains such organs as the intestines, liver, and bladder

abdominal circumference
the surface distance measure of the lower torso at the level of the maximum anterior protrusion of the abdomen; *syn.* abdominal extension circumference
Comment: measured with the individual standing erect

abdominal depth, sitting
the transverse linear distance from the back to the front of the abdomen at the waist
Comment: measured with the individual sitting erect

abdominal depth, standing
the horizontal linear distance from the back to the front of the abdomen, at the level of the maximum anterior protrusion; *syn.* maximum abdominal depth, abdominal extension depth; *see also* **truncal depth, chest depth, waist depth**
Comment: measured with the individual standing erect; note the level at which the measurement is taken

abdominal extension circumference
see *abdominal circumference*

abdominal extension depth
see *abdominal depth*

abdominal extension height
the vertical distance from the floor or other reference surface to the level of the maximum anterior protrusion of the abdomen in the midsagittal plane; *syn.* abdominal extension level
Comment: measured with the individual standing erect

abdominal extension level
see *abdominal extension height*

abdominal extension to wall
the horizontal distance from the most laterally protruding point of the abdomen to a wall
Comment: measured with the individual standing erect against the

wall with minimal buttock compression

abdominal skinfold

the thickness of a horizontal skinfold centered at 3 cm lateral from and 1 cm inferior to the umbilicus

Comment: measured with the individual standing comfortably erect, the body weight equally distributed to both feet, and the abdominal muscles relaxed

abdominal wall

the covering of the abdominal cavity, consisting of fibrous and fatty tissue, muscles, and skin

abduct

move some body part away from the midline of the body or from some other body part; *n.* abduction; *opp.* adduct

abduction angle

that angle through which a joint is abducted from a specified reference position; *syn.* angle of abduction

abductor

any muscle which moves a body part away from the midline of a structure; *opp.* adductor

aberration

the failure of light rays to converge at a focal point in an optical system, resulting in blur; *see* **spherical aberration, chromatic aberration**

aberration

see **mental aberration, light aberration**

ability

having the physical and/or mental capacity to perform a given task effectively; *see also* **skill**

able-bodied

(sl) not having a physical handicap

ABLEDATA

a computerized database containing consumer product information on devices for disabled individuals

Comment: available from the National Clearinghouse on Technology and Aging

abnormal reading

see **abnormal time**

abnormal time

an observed elemental time value which is beyond typical statistical or policy limits; *syn.* abnormal reading

abort

terminate some ongoing process or activity prior to its scheduled or expected completion; *n.* abortion

above elbow (AE)

pertaining to an amputation at some level of the upper arm or a prosthesis which is fitted over the upper arm

above knee (AK)

pertaining to an amputation at some level of the thigh or a prosthesis which is fitted over the thigh

abrade

scrape away a region of a surface, skin and/or mucous membrane in the case of the body, by a mechanical process; *n.* abrasion

abrasive

any substance or material which contains fine particles of high hardness either intended for scraping the surface of an object or which may do so through accidental impact

abscissa

the horizontal or independent axis on a two-dimensional graph; *see also* **x axis**; *adj.* abscissal; *opp.* ordinate

absent

not present at some location when one is normally expected to be there; *n.* absence

absentee

an individual who is not present at his workplace when he is supposed to be there

absolute humidity

the mass of water vapor per unit volume of air-water vapor mixture

absolute maximum

the highest value anywhere on the total extent of a curve; *see also* **relative maximum**

absolute minimum

the lowest value anywhere on the total

extent of a curve; *see also relative minimum*

absolute pitch
a skill or ability of a person to identify the pitch of a pure tone without the use of any external reference

absolute temperature scale
see Kelvin

absolute threshold
that minimum stimulus intensity which represents the transition between a response and no response from an observer attending to a particular sensory/perceptual task under specified conditions; *syn.* lower threshold; *opp.* upper threshold; *see also* **threshold, threshold of audibility**

absorb
take up matter or an energy flux; *n.* absorption

absorbed dose
a measure of the radiation energy absorbed per unit mass

absorbent
any device, material, substance, or structure which absorbs an unwanted energy or substance

absorbent gas mask
any respirator which includes a container having some type of material to absorb toxic substances

absorber [1]
any material which is capable of taking up chemicals or radiation

absorber [2]
any device which attenuates vibration

absorption
see absorb

abtesla (abt)
see gauss
Comment: an older term

acceleration ($\vec{\alpha}$)
a vector representing the rate of change of velocity with time

$$\vec{\alpha} = \frac{d\vec{v}}{dt}$$

acceleration illusion
any perception of apparent motion or change in motion resulting from acceleration-induced stimulation of the vestibular apparatus, the visual system, or other mechanoreceptors

acceleration syndrome
any change in physiological and/or perceptual-motor-cognitive function due to the forces imposed on the body by changes in velocity; *syn.* g-force syndrome; *see also* **positive g, negative g, transverse g**
Comment: an older term

accelerometer
a force transducer used in measuring acceleration; *syn.* acceleration transducer, acceleration pickup

accelerometry
the quantitative measurement of accelerations of a structure or its components

accent lighting
any form of directional or other unique illumination emphasis as an attempt to bring attention to a segment of the field of view or some object within the environment

access time
that temporal interval required to gain an opening to or achieve a certain position within a given structure

accessibility
a measure of the ease with which a location may be reached, entered or viewed; *adj.* accessible

accessibility score
a rating based on the cross-sectional area of the access path available for an individual or body part to reach the desired point

accessory movement
see synkinesia

accident
any unplanned occurrence which results in damage to property or equipment, causes injury or illness to one or more persons, and/or adversely af-

fects an ongoing activity or function

accident frequency rate
the number of lost time accidents per 1,000,000 man-hours worked; *syn.* frequency rate

$$AFR = \frac{no.\ of\ lost\ time\ accidents}{1,000,000\ man-hours\ worked}$$

Comment: an older term

accident prevention
the design or application of countermeasures in an environment to reduce accidents or the accident potential

accident severity rate
the number of lost workdays per 1,000,000 man-hours worked; *syn.* severity rate

$$ASR = \frac{no.\ of\ lost\ workdays}{1,000,000\ man-hours\ worked}$$

Comment: an older term

accidental death
a death causally related to some accident

accidental impact
an undesired, other-than-functional impact; *syn.* nuisance impact

acclimatize
successfully adjust to a new environment, especially with regard to the thermal environment; *n.* acclimatization

accommodate [1]
adjust a sensory threshold or structure of a sense organ to best receive a signal of interest or to not perceive a continuous, constant stimulation; *n.* accommodation

accommodate [2]
change one's activities to minimize the possibility of revealing a weakness or use another method/mechanism to achieve the same or similar effect; *n.* accommodation

accommodation of workers
see *worker accommodation*

accumulative timing
a time study technique in which mul-

tiple timers are used with electrical or mechanical linkage to obtain task or work cycle times; *syn.* multiple watch timing

accuracy [1]
a measure of the closeness with which an object or end product is located, placed, or represented in some dimension, compared to a desired, known, or believed point

accuracy [2]
a measure of the extent to which a true or actual value of a measure is observed

acetylcholine
a bodily chemical used in neural and neuromuscular transmission between cells

achromatic
without hue (color, chroma); appearing white, black, or gray; *opp.* chromatic

achromatic color
a color which is without hue; *opp.* chromatic color

achromatic lens
a lens corrected to have the same focal length for two or more specified wavelengths

achromatic point
an equal energy white point on the CIE chromaticity diagram with coordinates of x = y = 0.33; *syn.* white point

achromatopsia
complete color blindness

acidosis
a condition of high concentrations of acidic metabolic by-products in tissue; *opp.* alkalosis
    Comment: may be caused by excessive muscular activity, inadequate respiration, or certain diseases

acne
any of several forms of disease or conditions which form pustules, nodules, or other such structures in the skin

acoustic
pertaining to sound

acoustic absorption coefficient (α_m)
the ratio of energy absorbed by a material to the energy incident to it

acoustic flanking
the structural transmission of vibrations to elements which re-radiate the sound in the acoustic range

acoustic intensity (I)
the rate of flow of acoustic energy per specified cross-sectional area

$$I = \frac{dW}{dA}$$

acoustic nerve
see **auditory nerve**

acoustic pressure (p)
see **sound pressure**

acoustic reflex
the contraction of the tensor tympani and stapedius muscles attached to the conducting middle ear bones to increase acoustic impedance in response to a high intensity sound

acoustic scattering
the irregular reflection, refraction, and/ or diffraction of sound in many directions

acoustic stimulus
any varying pressure from air or other fluid having sufficient intensity within the transducing frequency range of the object or organism; see **auditory stimulus**

acoustic trauma
a temporary or permanent hearing loss due to a loud noise in one or both ears or by a blow to the head which caused injury or damage to the ear

acoustics [1]
the study, measurement, or use of sound

acoustics [2]
the quality of a building or other structure which affects the character of sound perceived by the ear or other transducer

acromegaly
a condition in which an individual has an enlarged jaw and extremities, due to excessive secretion of growth hormone

acromial height, sitting
the vertical distance from the upper seat surface to acromiale
Comment: measured with the individual sitting erect and his arms hanging naturally at his sides

acromial height, standing
the vertical distance from the floor or other reference surface to acromiale
Comment: measured with the individual standing erect, his arms hanging naturally at his sides, and his weight equally distributed on both feet

acromiale
the most lateral/superior point of acromion

acromiale – biceps circumference-level length
the surface distance along the outer edge of the arm from acromiale to the level at which the relaxed biceps circumference measure is taken; *syn.* acromion – biceps circumference-level length
Comment: measured with the individual standing erect and the arms hanging naturally at the sides with the hands and fingers extended

acromiale – dactylion length
the vertical distance from acromiale to the tip of the middle finger; *syn.* acromion – dactylion length, shoulder-fingertip length
Comment: measured with the individual standing erect and the arms hanging naturally at the sides with the hands and fingers extended

acromiale – radiale length
the vertical distance from acromiale to radiale; *syn.* acromion – radiale length
Comment: measured with the individual standing erect and the arms hanging naturally at the sides

acromion
the flattened, expanded bony process

at the lateral end of the spine of the scapula; *adj.* acromial
> *Comment:* used as an anthropometric landmark

acrophase
the peak value in a biological rhythm cycle

acropodion
the most posterior fleshy point on the heel

act of God
an accident or disaster whose cause is generally considered beyond human control

actin
a globular protein involved in muscle contraction; *see also* **actomyosin**

actinic
pertaining to that range of ultraviolet wavelengths within the electromagnetic spectrum which is capable of causing chemical changes, generally below about 315 nm

actinic keratoconjunctivitis
an inflammatory condition of the corneal and/or conjunctival epithelium of the eye due to exposure of intense ultraviolet light; *see also* **welder's flashburn**

action limit (AL)
a NIOSH guideline for the maximum load which should be lifted manually by a healthy person under given conditions to maintain acceptable injury incidence and severity rates

action limit ratio (ALR)
the ratio of average lift weight to the calculated action limit

action potential
a rapid change in electrical potential via the exchange of ions across the cell membrane in nerve and muscle tissues due to an initial depolarization beyond the threshold potential, followed by a return to the resting potential; *syn.* nerve impulse, spike

action spectrum
the spectral sensitivity curve for a given

type of retinal photosensitive cell

activated charcoal
a form of charcoal which has been treated by very high temperature and steam to remove any absorbed impurities, thus making it possible to absorb impurities

active
resulting from internal causes and/or purposeful effort by an entity; *opp.* passive

active isolation
that energy attenuation or conversion to another form through the use of a system requiring its own energy to operate and acting near or within another system which is generating some undesired energy output; *opp.* passive isolation

active movement
the process of moving a limb or other body part by an individual under one's own control; *syn.* volitional movement; *opp.* passive movement

active restraint
a restraining device which has a positive locking feature and requires no action by an individual to be held in place; *opp.* passive restraint

active safety measure
any means of implementing safety precautions which requires an individual to take some action such as reading or comprehending; *opp.* passive safety measure

active window
that view on a display with which the user is currently interacting; *opp.* inactive window

activities of daily living (ADL)
those functions normally performed on a daily or near daily basis which are involved in sustenance of the individual; *see also* **daily living tasks**
> *Comment:* e.g., eating, grooming, urination/defecation, dressing, bathing

activity analysis
a study of the following set and any

interrelationships within the set: (a) involved individuals, (b) the environment, (c) the facilities or equipment present or available, and (d) the actions required to perform the particular activity under study

activity sampling
a sampling technique using many instantaneous observations of equipment or workers involved in an ongoing process to rate them as either functioning or non-functioning on each sample; *syn.* work sampling, snap reading technique, snap reading method, random observation method
    Comment: used for estimating the amount of time a machine or worker spends performing some function

actomyosin
a combination of actin and myosin which is involved in muscle contraction

actual coverage
see standard coverage

actual time
see observed time

actuation force
that force required to overcome static friction and begin the movement of a control or other mechanical device

actuation torque
that force directed at some distance from the center of rotation which is required to overcome static friction and begin rotation of a control or other mechanical device

acuity
the sharpness of a sense in resolving sensory stimuli; *see also visual acuity, vernier acuity, Snellen acuity, stereoscopic acuity, resolution acuity*

acute [1]
a dangerous condition or exposure which generally has a rapid onset and is of short duration, yet may be life threatening; *opp.* chronic

acute [2]
pertaining to a geometric angle less than 90°

acute oxygen toxicity
a central nervous system disorder due to breathing pure oxygen at higher than normal pressures for several minutes to a few hours, depending on the pressure, and characterized by a range of symptoms from muscle twitching to convulsions; *see also chronic oxygen toxicity*

acute radiation effects
any of one or more of types of illnesses or other bodily disorders that follow relatively high doses of ionizing radiation resulting in deaths of significant numbers of cells; *syn.* acute radiation syndrome

acute toxicity
those symptoms resulting from a relatively brief exposure period to high levels of a toxic substance

Adam's Apple
(sl); the anteriorly protruding portion of the thyroid cartilage
    Comment: larger in the male than the female

adaptation
a self-generated adjustment in a system in response to changes in the environment as an attempt to maintain functionality; *syn.* adaptive response; *see also sensory adaptation, perceptual adaptation*

adaptive control
a form of automated control equipped with a self-contained decision-making capability for modifying its own operation based on previous experience

adaptive equipment
any type of equipment which enables a disabled or other individual to operate a machine or system

adaptive response
see adaptation

adaptometer
an instrument designed to determine the degree of retinal adaptation or the time course over which adaptation oc-

curs by measuring changes in an observer's threshold for light detection

additional work allowance
see *excess work allowance*

additive color mixing
the addition of colored lights to an already illuminated surface or region, resulting in a change of apparent color; *syn.* light mixing; *opp.* subtractive color mixing

adduct
move toward the midline of the body or a body part; *n.* adduction; *opp.* abduct

adductor
any muscle which moves a bone toward the midline of a structure; *opp.* abductor

adenosine triphosphate (ATP)
a bodily chemical which is used for storing energy involving phosphate bonds

adhesive dirt
any form of dirt which tends to remain attached to a surface through an inherent stickiness

adipose tissue
that tissue composed primarily of fat cells with connective tissue for support

administration [1]
the process of carrying out more or less routine financial, personnel, and/or materials support types of operations in maintaining an organization; *adj.* administrative

administration [2]
a group of people performing administrative work

adolescent growth spurt
a phase in the maturing individual near puberty at which peak height and weight velocities occur, along with a change in body composition

adrenalin
see *epinephrine*

adsorb
take up and retain some material on the surface of a solid; *n.* adsorption

adult chorea
see *Huntington's chorea*

Advanced Dynamic Anthropomorphic Manikin (ADAM)
an anthropomorphic dummy developed by the Air Force to represent human anthropometry and vertical dynamic response for ejection seat testing

advisory
see *advisory signal*

advisory light
a visual indicator which provides information on the operation of essential equipment; see also *advisory signal*

advisory signal
any type of signal which indicates the condition of equipment or operations; *syn.* advisory; see also *advisory light*

aerobic
requiring or having adequate oxygen present in bodily tissue, especially muscle tissue; *opp.* anaerobic

aerobic capacity
see *maximal aerobic capacity*

aerobic endurance capacity
see *maximal aerobic capacity*

aerobic energy
that energy which can be derived from foodstuffs by aerobic metabolism; *opp.* anaerobic energy

aerobic metabolism
the normally complete physiological oxidation of glucose or other bodily fuels in the presence of adequate oxygen to water and carbon dioxide; *opp.* anaerobic metabolism

aerobic work capacity
see *maximal aerobic capacity*

aerosol
a gaseous suspension of very small solid particles or liquid droplets

aerospace
pertaining to equipment, vehicles, or activities in either or both a planetary atmosphere and space

aesthenic
see *asthenic*

aesthetics
having the sense of beauty, or the study of beauty

AFAMRL Anthropometric Data Bank
a computerized database of several anthropometric surveys, consisting of both American and foreign subjects

afferent
conducting a signal, information, or a substance toward a central point, usually referring to neural structures; *opp.* efferent

afferent nerve
a nerve which conducts sensory information from a receptor toward the spinal cord and/or the brain; *syn.* sensory nerve; *opp.* efferent nerve

affirmative action
the undertaking of special efforts for the recruitment, selection, and placement of women or members of minority groups in jobs to effect a balance in the workforce according to representation in the population

affricate
the type of sound produced on complete closure of the vocal tract followed by a constriction

afterimage
an aftersensation in the visual system; also after-image, after image; *see positive afterimage, negative afterimage*

aftersensation
a sensory impression in any modality which persists after cessation of the causing stimulus, but which may have different characteristics from the original stimulus; also after-sensation; *see afterimage*

age
see chronological age, mental age, developmental age

agent
the causative object, substance, entity, or location involved in an accident or disease

agent-specific disease
a disease known to be caused only by one factor

ageusia
an impairment in or loss of the sense of taste; also ageustia; *adj.* ageusic; *syn.* taste blindness

AGGIE
a general purpose, 3-D nonlinear finite element structures numerical modeling program

aggregate
the combination of the tool being manipulated or mass being lifted/carried with those primary body parts affected or used in the operation

agonist
see prime mover

agravic illusion
see oculoagravic illusion

air embolism
a form of decompression sickness in which an air bubble blocks blood flow in a blood vessel

Air Force pediscope
a system consisting of a pressure-transducing blanket and readout for measuring seat pressure on the ischial tuberosities

air monitoring
a form of environmental monitoring in which one or more quantities of environmental gases are taken and a determination of contents and proportions made

air pollution
the presence in the breatheable atmosphere of any one or more substances which might be harmful to health or negatively affect habitability
Comment: quantified levels for many substances

air purifier
a type of respirator which uses an air-purifying cannister to remove airborne contaminants from air via filtering, adsorption, and/or absorption prior to inhalation

air supplier
a type of respirator which provides

clean air or oxygen from a tank or external source for inhalation

Air Traffic Controller (ATC)
a certified individual responsible for regulating aircraft traffic within a specified region

air-blower noise
see *air-handler noise*

air-handler noise
that acoustic noise output from heating, ventilation, or air conditioning fans and ducts; *syn.* air-blower noise

air-purifying cannister
an air-tight module containing absorptive and/or adsorptive substances for use in an air purifier

airbag
a device which is pressurized to inflate on impact to protect the occupant in a vehicle

aircraft accident incidence rate (AIR)
a measure of the safety of flying, represented by the formula

$$AIR = \frac{no.\ aircraft\ accidents \times 100,000}{no.\ of\ flight\ hours}$$

airlock
a system comprised of two doors or hatches with an intermediate air space; also air lock

airport noise
that environmental noise in the vicinity of an airport due primarily to engine noise from approaching and departing aircraft

airspeed indicator
an aircraft display showing velocity relative to the surrounding air

airspeed/mach indicator (AMI)
see *airspeed indicator, mach indicator*

airway
the pathway through which air and other respiratory gases pass between the mouth or nostrils and the lung alveoli

airway resistance
that resistance which must be over-

come for air to flow through the airway

akimbo span
see *span akimbo*

akinesia
a movement disorder in which the person executes no voluntary movements or exhibits a pause prior to initiation of a movement

alarm
an indicator that some condition exists which may or will require human action to correct in order to prevent loss of life, property, or equipment

algorithm
a method or procedure for solving a problem or performing some task

alkalosis
an excessive amount of chemically basic substances in bodily tissues; *opp.* acidosis

all-or-none law
a rule that once an action potential is generated in a neuron or muscle cell, it will continue to the terminus of that cell unless damage is present; also all or none law

all-pass
pertaining to a condition or piece of equipment in which all frequencies of a signal are processed equivalently without attenuation; also all pass, allpass

allergy
an acquired alteration in reactivity to a substance which becomes apparent after some incubation period and re-exposure to that substance

allesthesia
the perception of a given peripheral tactile stimulus as occurring at a point different from the actual point of stimulation

allowable load
see *load limit*

allowance [1]
that specified minimal clearance between two parts which are to be assembled

allowance [2]

some time value or factor by which the normal time required to complete a task is increased to allow for such things as delays, policy, fatigue, or personal needs; *syn.* allowed time, time allowance

allowed time

see allowance

alpha (α)

the probability of making a Type I error; *see also* **beta**

alpha motor neuron

any large-diameter A-class motor neuron having the highest conduction velocity in the Erlanger-Gasser classification system which innervates muscle extrafusal fibers and is involved in muscle contraction

alpha particle

the nucleus of a helium-4 atom

alpha rhythm

a band of the EEG spectrum consisting of frequencies from about 8 Hz to 13 Hz

alpha testing

the preliminary testing phase of a new software product outside the facility or company in which it was developed; *see also* **beta testing**

alphanumeric

pertaining to a set of numbers, letters, or a combination of the two

alternating current (AC)

an electrical current flow which alternates in amplitude about a baseline; *opp.* direct current

alternative pointing device

an assistive device for a disabled individual's interaction with a computer; *syn.* alternative input device

alternative work schedule

any work schedule other than the standard work week

altitude sickness

a disorder resulting from hypoxia and reduced air pressure at higher altitudes, and which may consist of symptoms ranging from headache, vomit-

ing, and/or malaise to heart failure; *syn.* mountain sickness

alveolar gas exchange

that gaseous exchange through the thin walls of the alveoli and the capillaries, normally such that oxygen is absorbed by the blood and carbon dioxide is released into the alveolus

alveolar pressure

that combined air and water vapor pressure within an alveolus of the lung

alveolar ventilation

the replenishment of alveolar gases by atmospheric air

alveolus

a small air sac in the lung in which gaseous exchange takes place with the blood; *pl.* alveoli; *adj.* alveolar

ambidexterous

capable of functioning about equally well with either hand; also ambidextrous; *n.* ambidexterity

ambient

pertaining to the surrounding environment

ambulatory

able to move about; *n.* ambulation

Americans with Disabilities Act of 1990 (ADA)

a law passed by Congress with the intent of aiding those with physical and mental disabilities by preventing employment discrimination, providing for public access to public transportation, and providing for the use of other facilities and services used by the public at large

ametropia

a refractive defect in the eye lens or other eye structures with the resulting inability to focus images on the retina

amino acid

a class of organic compounds which contain an amine group and a carboxylic acid, and which provide the components of proteins

amp

see ampere

ampere (A)
a unit of electrical current; that amount of constant electrical current which, if maintained in two straight, infinitely long, parallel conductors having negligible cross-sectional area and separated by 1 meter in a vacuum, would produce a force of 2×10^{-7} newtons per meter of conductor; sl. amp

amplitude [1]
the instantaneous deviation or displacement from some baseline

amplitude [2]
the peak-to-peak difference, maximum value, or averaged value of a signal

amplitude modulation
the multiplication of an approximately constant higher frequency carrier signal by a second signal, usually of a much lower frequency

ampulla
the enlarged portion of a semicircular canal in the inner ear which contains the crista

amputate
remove a bodily structure or some part thereof; *n.* amputation

amputee
an individual with part or all of one or more limbs missing

anabolism
the synthesis of more complex living structures from simpler materials

anaerobic
pertaining to an oxygen-free environment, or an environment with an inadequate oxygen supply for normal processes; *opp.* aerobic

anaerobic energy
that energy derived from anaerobic metabolism; *opp.* aerobic energy

anaerobic metabolism
the partial physiological oxidation of glucose or other bodily fuels in tissues without adequate oxygen, forming lactic acid with the release of energy; *opp.* aerobic metabolism
Comment: can provide a brief re-

serve of energy under heavy physical workloads

anal
see anus

analog-to-digital conversion
the process of sampling the amplitude of a continuously varying signal at specified intervals and presenting a digital value to a resolution of some number of bits, typically carried out by an analog-to-digital converter; *opp.* digital-to-analog conversion

analog-to-digital converter (ADC)
any electrical and/or electromechanical device which performs an analog-to-digital conversion; *opp.* digital-to-analog converter
Comment: typically interfaced to a digital computer

analysis of covariance (ANCOVA)
a modified analysis of variance involving a compensation for covariates when random groups cannot be selected

analysis of variance (ANOVA)
any of a series of statistical tests in which variances are compared across two or more groups to make a determination as to whether the means of the groups are likely to be significantly different from one another

analytical estimating
a technique in work measurement in which element times are estimated from previous experience and knowledge of the concerned elements

analytical standard data
a set of time values represented in the form of or computed by a mathematical model; *see also standard data*

analytical workplace design
the process of using established human factors' concepts to design a workplace suitable for human interaction

anatomical position
a standard posture for defining certain aspects of the human body: the body is standing erect with the arms hanging at the sides and the wrists supinated

such that the palms face forward/anterior

anatomical reference point

see *landmark*

anatomy

the study of the geometrical and topographical features of all body structures and of the body as a whole; *adj.* anatomical

anechoic

the lack of significant reflected energy waves, usually with reference to sound; having no echo

anechoic chamber

a specially-outfitted enclosure in which the boundaries are covered with wedges of highly sound- or other energy-absorbing material to minimize reflections of incident energy; *syn.* anechoic room

anechoic room

see *anechoic chamber*

aneisekonia

a condition in which different image sizes are experienced in the two eyes

anemometer

any instrument which measures local air velocity

angle

a measure of the relative rotational difference in position between two intersecting linear or planar objects; *adj.* angular

angle diagram

a graphical plot of the angular relationship over time between two joints as the joints move in some specified way

angle of abduction

see *abduction angle*

angle of incidence

that angle from the perpendicular to the surface of an object at which a light ray or other entity strikes the surface of that object

angle of resolution

see *minimum resolution angle*

angle-torque curve

any graphical relationship in which the maximum isometric force exerted at a given angle is plotted against that angle for the range of motion

Angstrom (Å)

a unit of length; 10^{-10} meter

Comment: an older term

angular

see *angle*

angular acceleration (α)

the rate of change of angular velocity with time; *syn.* rotational acceleration

$$\alpha = \frac{d\omega}{dt}$$

angular deviation

that angle between the incident and transmitted light rays in a prism

angular displacement

a vector representing the change in angle by rotation about some origin

Comment: universal convention: counterclockwise is positive

angular frequency (ω)

the oscillation frequency in an oscillating system multiplied by 2π

angular momentum (L)

a vector representing the rotational momentum of an object about an axis

angular motion

the movement of a structure about its own local center of rotation

angular velocity (ω)

the rate of change of angular displacement about some axis of rotation with time; *syn.* rotational velocity

$$\omega = \frac{d\theta}{dt}$$

animal starch

(sl); see *glycogen*

anisomelia

inequality of length, as of a limb

anisotropic

having physical properties which vary in different spatial directions

ankle

the joint formed by the junction of the

distal ends of the fibula and tibia with
the talus, including all the surrounding soft tissues

ankle bone
see talus

ankle breadth
see bimalleolar breadth

ankle circumference
the minimum surface distance around
the lower leg just above the ankle
Comment: measured with the individual standing erect

ankle height
the vertical distance from the floor or
other reference surface to the level of
the ankle circumference measure
Comment: measured with the individual standing erect

annoyance
a condition or stimulus which causes
one to be disturbed, irritated, or
troubled

annual injury incidence
an OSHA formula used for determining the injury rate for comparison with
other companies or industries

$$AII = \frac{no.\ of\ OSHA\ form\ 200\ recordable\ injuries \times 200,000}{no.\ hours\ worked\ by\ company\ employees}$$

annulus
the fiber and cartilage structure surrounding the nucleus pulposus in an
intervertebral disk; also anulus; *syn.*
annulus fibrosus

anoxia
a lack of oxygen in the body or certain
tissues

antagonism
the competitive interaction or opposition of two or more agents to control or
lessen the effect of an agent's individual effect(s); *opp.* synergism

antagonist
an entity which opposes or competes

with the action of another entity
Comment: may be a person, group,
muscle, or drug

anterior
pertaining to the front portion of the
body or toward the front of the body

anterior neck length
the surface distance from suprasternale
to the junction of the posterior lower
jaw and the neck in the midsagittal
plane
Comment: measured with the individual standing erect and looking
straight ahead

anterior waist length
the surface distance from the most anterior point of the lower neck to the
waist
Comment: measured with the individual standing erect

anthracosilicosis
a lung disease caused by the inhalation
of dust containing silicon dioxide and
other components

anthro-
(prefix) like or pertaining to man

anthropology
the field comprising the overall study
of man, including physical and social
aspects

Anthropology Research Project
a DoD-sponsored project to provide
anthropometric surveys for USAF flying personnel

anthropometer
a device for measuring linear dimensions of the body

Anthropometric Data Bank
*see AFAMRL Anthropometric Data
Bank*

anthropometric measurement
any physical measurement derived
from the body or its various parts

anthropometrics
the study of anthropometric variables
and their relationships

anthropometrist
one who is qualified by education, train-

ing, and experience to practice anthropometry

anthropometry
that field which deals with the physical dimensions, proportions, and composition of the human body, as well as the study of the related variables which affect them; *adj.* anthropometric

anthropomorphic
having a form like a human or human parts

anthropostereometry
stereometry with direct measurement of the body

anti-exposure suit
any form of outer clothing to protect an individual from the elements, especially wind and cold temperatures

anti-*g* straining maneuver (AGSM)
any internally-generated technique for temporarily increasing blood pressure in an attempt to withstand high positive *g* stresses in high performance aerospacecraft; *see* **M-1 maneuver, L-1 maneuver**

anti-*g* suit
a special garment designed to apply counterpressure to the lower body during high positive *g* forces as an aid in preventing blackout of the wearer

anti-glare filter
a transparent device for reducing glare

anticipation error
an error produced due to an expectation of a change

antilogarithm
a number whose logarithm returns the original number; *opp.* logarithm

antinode
a point, line, or surface in a standing wave at which the wave has maximum amplitude

antipole
a point on the skull opposite to the point of impact in an incident

anus
the lower terminus of the digestive tract and rectum; *adj.* anal

anxiety
a state of apprehension regarding the perception of danger or a stressful situation

aorta
the large blood vessel which exits the heart carrying blood to the periphery, and from which all major peripheral arteries branch

aphakic
pertaining to an individual with the lens removed from one or more eyes

apocrine gland
a sweat gland whose ducts terminate in hair follicles; *see also* **eccrine gland**

aponeurosis
an expansion of a muscle tendon which serves to attach a muscle to bone at an origin or insertion, or to enclose a group of muscles

apophysis
an outgrowth, process, or projection from some structure

apostilb (asb)
a unit of luminance; equals $1/\pi$ or 0.3183 candelas per square meter
Comment: an older term

apparent motion
an illusion of motion, regardless of the cause, whether by certain patterns of non-moving stimuli, by certain conditions under which non-moving stimuli are observed, or by stimulation of sensory receptors or the nervous system

appetite
a desire, often for a particular type of food

application
a software package for performing a specific type of task other than direct system support or system utilities

applied
having a possible practical application; *opp.* theoretical

applied sciences
those disciplines involved in the use of information gathered by the basic sciences

aptitude

an innate ability for acquiring a particular skill or knowledge

aptitude test

any system or device for determining whether an individual is likely to be successful in an activity for which he has not yet been trained

aqueous humor

the clear fluid located in the front chamber of the eye between the cornea and the lens

arachnoid layer

a nonvascular membrane between the dura mater and pia mater surrounding the brain and spinal cord

arbitrary delay

an unscheduled interruption to work which is unrelated to the task or job being performed

arc

an anthropometric measurement following an open curved path, where the curve makes up the majority of the measurement value; *syn.* curvature; *see also* **circumference**

arc lamp

an illumination source which operates using the principles of discharge of low cathode voltages and high currents

arc welder's disease

see **siderosis**

arch

the curvature on the inferior surface of the foot

arch height

the maximum vertical distance from the floor or other reference surface on which a person stands to the bottom of the foot tissue between the anterior and posterior support points

Comment: measured with the individual standing erect, with the body weight equally divided between both feet

arithmetic mean

the sum of a set of values divided by the number of values in the set or the

duration of application; *see also* **mean**

$$\overline{X} = \frac{\Sigma X}{n}$$

arm

one of the pair of upper extremities, consisting of the humerus, radial, and ulnar bones and other associated soft tissues

arm circumference

see **forearm circumference, upper arm circumference, axillary arm circumference**

arm reach from wall

the horizontal distance from the wall to the tip of the longest finger

Comment: measured with the rear of both the individual's shoulders against the wall, and with both hands and arms extended forward parallel to the floor for symmetry

arm span

see **span**

arm work

that physical work which uses the arm(s), with essentially no or minimal trunk or leg involvement

arm-hand

involving both the arm and the hand, generally referring to internally-generated or motor activities; *see also* **hand-arm**

arm-hand steadiness

a measure of the ability to keep both the hand and arm steady, whether stationary or moving; *see also* **hand steadiness**

arm-hand-tool aggregate

the combination of the hand/arm and tool acting as a biomechanical unit

armrest

any structure extending approximately parallel to the sides of and higher than the seat pan of a seat for supporting the forearm while sitting

aromatic hydrocarbon

a hydrocarbon having one or more ringed structures derived from benzene or similar compounds

arousal
the degree of awareness of the environment

arrangement of workplace principles
see *workplace layout principles*

arrhythmia
see *cardiac arryhthmia*

arteriovenous oxygen difference
the difference in oxygen content between the blood entering and leaving the pulmonary capillaries

artery
a vessel which carries blood toward the periphery from the heart; *adj.* arterial; *pl.* arteries

arthritis
any of a set of diseases involving inflammation of one or more joint structures

arthroscope
an instrument having a small diameter tube for visualizing the interior of some body part

arthroscopy
the use of an arthroscope for examining interior bodily tissues

articular
pertaining to one or more joints

articulate
(v) produce speech sounds easily recognizable by another individual fluent in a given language; *n.* articulation; *adj.* articulate

articulated total body model (ATB)
a computerized model developed for examining the biodynamic effects of ejection from high-performance aircraft on the various body segments

articulateness
a measure of the number of joints and degrees of freedom possessed by a remote system

articulation
the linkage of two rigid structures by a movable joint

articulation index (AI)
see *speech articulation index*

articulator
any moveable structure used in the production of speech, such as the tongue and lips

artificial gravity
that relative downward acceleration experienced by an individual or object on the interior of a larger, rotating object as a result of centrifugal force

artificial horizon
a graphic or pictorial flight instrument display for providing the pilot with information about the orientation of the aircraft with respect to the ground

artificial pupil
a small aperture in a manufactured or cultured disk or diaphragm used to restrict the amount of light entering the eye

artificial radiation
the output from radioactive substances or from high energy electromagnetic wave production in instrumentation

artificial reality
see *virtual environment*

arytenoid
a skeletal muscle within the larynx which is involved with closing the posterior part of the glottis in speech

aryvocalis
a skeletal muscle of the larynx which is involved in controlling pitch by regulating the length of the vibrating segment of the vocal cord

as low as reasonably achievable (ALARA)
a principle that exposure of humans to radiation should be maintained as low as possible, given the risks involved

asbestos
a fibrous form of a hydrated calcium-magnesium silicate which acts as a good heat insulator, but is also hazardous in dust form

asbestosis
an increase in the amount of internal lung connective tissue due to the prolonged breathing of asbestos dust

ascending ramus
see *ramus*

aseptic bone necrosis
see *dysbaric osteonecrosis*

Asmussen dynamometer
a device using a piston drive for measuring strength, in either a push or pull mode

asphyxiation
a suffocation resulting from the deprivation of oxygen

assemble (A)
a physical work element involving the putting together of two or more components

assembly
the process of putting together individual pieces or components to make a whole or complete item

assembly line
a work arrangement in which the product being assembled is delivered to each person who then performs a somewhat specialized task or job at a specific worksite

assignable cause
any identifiable source of deviation from normal in some process or system

assistive device
any tool which either enables or enhances human machine interaction for an individual with a physical handicap

assumption of risk
the concept that an individual who is aware of the extent of some danger and still knowingly exposes himself to that danger assumes all consequences and cannot recover any damages, even if injured without fault to himself

asthenia
a physical weakness or lack of strength; *adj.* asthenic

asthenic
a Kretschmer somatotype characterized by a slender, feeble build; also aesthenic

asthma
a disease in which increased tracheal and bronchial sensitivity cause difficulty in breathing through a restriction of the airways

astigmatism
a nonuniform curvature in the optical surfaces of the eye or other optical system resulting in an unequal refraction of light in the different media and a blur on the retina or other intended focal point; see *stigmatism*

asymmetric lift
a manual lifting task in which the load is not equally shared by paired limbs

asymmetry
a lack of structural correspondence between two sides of a normally or expected symmetric structure, especially pertaining to paired members

asymptote
that value represented by approximately a horizontal straight line which a curve approaches as the axis approaches infinity; *adj.* asymptotic

at-will employee
an employee who works for a company under an at-will employment agreement

at-will employment
an agreement or understanding, either written or verbal, between an employer and employee that the employer may terminate an employee at any time, with or without good reason or notice, and with no legal liability of the employer; *syn.* employment-at-will

atelectasis
an incomplete state of expansion or a collapsed state of a portion or all of a lung

athetosis
a movement disorder characterized by almost continuous involuntary slow, sinuous movements

athletic
a Kretschmer somatotype having a stocky, muscular build with little body fat

atlanto-occipital joint
the junction of atlas with the occipital

bone of the skull

atlas

the first cervical vertebra

atmosphere

the sum of the gases and their proportions, partial pressures, density, etc. within a specified volume

atmospheric pressure

the force per unit area which air exerts on a given surface; *syn.* barometric pressure

> *Comment:* normal standard is 1.01325×10^5 N/m^2 or 760 mm Hg

atomic fission

see fission

atrophy

a reduction in the size of one or more body structures or tissues

attention

the general, but not highly directed, allocation of sensory-perceptual functions, possibly involving motor functions as well, to a subset of the possible inputs; *see selective attention, divided attention*

attention span

that length of time or number of items or tasks to which an individual can respond before performance deteriorates

attenuate

reduce in concentration, intensity, strength, force, or amplitude of some entity over time or space; *n.* attenuation; *occas. syn.* damp

attenuator

that which attenuates

attitude [1]

see posture

attitude [2]

the forward orientation of a vehicle which is capable of motion in all three spatial dimensions, especially an aircraft or spacecraft

attitude [3]

an individual's feeling or opinion about some issue or expected event which will shape his response

attractive dirt

any form of dirt which tends to remain attached to a surface through electrostatic forces

attributable risk

a measure of the occurrence of a specific disease or injury in those exposed to a particular situation or causal agent

attribute

some characteristic of an element or condition

audible

capable of being heard

audible frequency range

that band of sound frequencies which can be heard by the human ear, from about 16-20 Hz to 20 KHz; *occas. syn.* audible range, audio frequency range

audible range

see audible frequency range

audio frequency range

see audible frequency range

audiogenic

resulting from sound

audiogenic seizure

a reflex convulsion due to an intense or sudden noise

audiogram

a graph or table which indicates sound detection ability of pure tones at several frequencies, usually between 250 Hz and 8000 Hz

audiologist

a person trained and certified in the field of human hearing

audiology

the study of hearing

audiometer

an instrument for measuring auditory sensitivity as a function of frequency by presenting specified pure tones at different intensities above and below the absolute threshold

audiometric reference level

the sound pressure level and specification to which an audiometer has been calibrated

audiometric testing room
a specialized chamber insulated for sound and equipped for hearing acuity measurement

audiometry
the measurement of hearing ability

audition
the sense or process of hearing; *adj.* auditory

auditory absolute threshold
see threshold of audibility

auditory aftereffect
a phenomenon in which familiar sounds appear modulated for a period of time after listening to rapid, high-intensity impulses; also auditory aftereffect

auditory attention
the ability to focus on a single auditory source in the presence of distracting auditory stimuli

auditory canal
see Eustachian tube

auditory fatigue
a temporary increase in auditory threshold due to prolonged intense noise or a previous auditory stimulus

auditory lateralization
the determination by a person that apparent direction of a sound is either to the left or right of the midsagittal plane of the head when wearing earphones

auditory localization
the process of determining the apparent direction and/or distance of an external sound source

auditory nerve
that portion of the vestibulocochlear nerve which carries auditory information from the inner ear to the brain; *syn.* acoustic nerve

auditory ossicle
any of the three small bones in the middle ear used for hearing: the malleus (hammer), incus (anvil), and stapes (stirrup); *syn.* ossicle

auditory stimulus
any stimulus which excites the cochlea to convey signals indicating sound perception to the brain

auditory system
the combined structures of the external, middle, and inner ear which are involved in the function of hearing, and the acoustic nerve

auricle
the skin and cartilage comprising the visible flap of the external ear; *adj.* auricular; *syn.* pinna

auricular point
that location on the longitudinal axis of the external auditory canal at which it passes to the exterior

autokinesis
voluntary movement

autokinetic illusion
an effect in which a stationary point light source in a dark background or with no visual reference frame appears to move; *syn.* autokinetic phenomenon

autokinetic phenomenon
see autokinetic illusion

automated control
the use of feedback in a continuously-monitored, computerized system to self-correct any output deviations

automation
the increased use of mechanization and or computerization

automatism
a movement disorder in which non-reflex motor actions occur during abnormal states of consciousness

autonomic nervous system (ANS)
a generally efferent subdivision of the peripheral nervous system which is distributed to and directs the function of smooth muscle and glands throughout the body, normally at a subconscious level; *see parasympathetic, sympathetic*

autospectral density
see power spectral density

auxiliary process time
the time required for necessary processing operations in addition to those

incurred for basic manufacturing and which result in relatively minor physical appearance changes of a product such as polishing or deburring

availability
a measure of the likelihood of having a system in working order at any given time; *syn.* measure of availability

$$Availability = \frac{uptime}{uptime + maintenance\ time}$$

available machine time
that portion of the time during a task cycle in which a machine could be producing useful work but is not

Available Motions Inventory (AMI)
a series of tests using equipment developed at Wichita State University with the Cerebral Palsy Research Foundation of Kansas to make objective determinations of the physical capabilities of handicapped persons

available process time
that portion of the time during a processing cycle in which a worker or system could be performing useful work, but is not

avascular necrosis
see osteonecrosis

average
see arithmetic mean, mean, geometric mean

average acceleration ($\bar{a}$)
the result of the total change in velocity in a period of time divided by that time

average man concept
the idea that using the average measurement on a human dimension is adequate for describing a population
Comment: not generally valid, but can be useful as a guideline

average power
the total amount of physical work done, involving moving objects, divided by the period of time during which it is accomplished

average velocity
the total distance traveled in a period of time divided by that time value

aviator's breathing oxygen (ABO)
a grade of commercial oxygen for high-altitude flying which has no water content

aviator's vertigo
a disturbance in the pilot's orientation with respect to the earth caused by a conflict between gravitational and visual cues

avoidable accident
any accident which can be or could have been prevented by the implementation of appropriate controls/hardware, environmental conditions, or behaviors

avoidable delay (AD)
a work element involving a pause or interruption which is unnecessary, due to factors under worker control, and which is not calculated for in standard time figures

avulsion
the tearing away of a body part

axilla
the somewhat hollow region beneath the junction of the shoulder and trunk; *syn.* armpit; *adj.* axillary

axilla height
the vertical distance from the floor to the posterior axillary fold
Comment: measured with the individual standing erect and arms hanging naturally at the sides

axilla – waist length
the surface distance from the armpit to the waist
Comment: measured with the individual standing erect

axillary arm circumference
the surface distance around the arm at level of the axillary fold
Comment: measured with the individual standing erect and the arms hanging naturally at the sides

axillary fold
the junction of the torso skin and the arm skin beneath the shoulder at the axilla

axis [1]
the second cervical vertebra

axis [2]
a graphical or imaginary line representing one of the dimensions in a co-ordinate system, at which the value of all other dimensions is zero; *adj.* axial

axon
that generally relatively long tubular portion of a neuron which normally carries information from the cell body to a synapse or to a neuromuscular junction; *adj.* axonal

B

B

the unit of measure for a work performance measuring system developed by Bedaux in which 60 B units was the equivalent of one hour's output, taking into consideration allowances for fatigue and other necessities; *see Bedaux plan*

Comment: no longer used

B display

a radar display in which the data are presented on a rectangular coordinate system with range and azimuth comprising the axes; *syn.* range-bearing display

B scale

a sound weighting system which approximates the response characteristics of the human ear in the 40- to 70-phon equal loudness contour range

B shift

see second shift

B-weighted sound pressure level (dB(B))

that sound pressure level measured using the B scale

baby

a child under about one year of chronological age

back

the posterior aspects of the ribs, muscles, and all other tissues associated with the posterior trunk/torso from the thoracic vertebrae to the base of the spine

back curvature

the surface distance across the back as measured from the right midaxillary line at the posterior axillary fold level to the corresponding point on the left

Comment: measured with the individual standing erect and the arms hanging naturally at the sides

back injury

any injury involving the spine, the spinal cord, the nerves exiting from the spinal cord through the spine, a rib-vertebral junction, and/or the muscles of the back

backbone

(sl); *see spine*

background luminance

the luminous intensity of a region within which a target is to be viewed or detected

background noise

the total of all sources of interference in a system, apart from the signal; *syn.* noise, auditory ground

background processing

the data processing or transmission which is performed secondary to a primary operation or higher priority operations

background radiation

that electromagnetic or particulate radiation existing in the environment, exclusive of the instrumentation or radiation source being measured

backlash

a control system response in which the direction of movement is momentarily reversed when the movement of a control is stopped

backlight

the use of a lighting source behind an object to separate that object or region from the background; also back light

backrest

any structure which is capable of supporting the back

backrest reference plane
the plane established by a backrest

backrest-to-seat angle
see seatback angle

backward chaining
a reasoning or control strategy in which the beginning point is the final or desired state with the process extending backward to a known point; *syn.* goal-oriented problem solving; *opp.* forward chaining

backward masking
a type of masking in which the masking stimulus occurs following the test stimulus

bactericidal effectiveness
a measure of the ability of various regions of the ultraviolet spectrum to kill bacteria; *syn.* germicidal effectiveness

bactericidal lamp
a light source outputting a high level of ultraviolet-C radiation; *syn.* germicidal lamp

bactericide
any agent which kills bacteria

bacterium
any of a large variety of single-celled organisms having no nuclear membrane; *pl.* bacteria; *adj.* bacterial

baffle
a partition used to shield/absorb sound or light energy transmission

balance ¹
a condition in which working times, tasks, activities, and output are coordinated between the hands of an individual worker, between workers, or between groups so that an operation proceeds smoothly without building excessive inventory or wasting time

balance ²
a condition of stable posture in which muscle forces exactly counteract the gravitational or other forces imposed on the body

balance ³
a condition in which the output from all speakers in an audio system provide the same output intensity

balance delay
see balancing delay

balanced motion pattern
a sequence of movements using both the right and left hands/arms which enables the worker to establish and maintain coordination and an efficient rhythm

balancing delay (BD)
the waiting or non-productive time of one hand/arm of a single worker or of another worker or group in an operation due to a lack of balance; also balance delay

Baldrige Award
see Malcolm Baldrige National Quality Award

balk
that customer or user behavior of not entering a queue or line because of its length or waiting period

ball-of-foot circumference
the maximum surface distance measured around the distal ends of the protuberances of the metatarsal bones of the foot
    Comment: measured with the individual standing erect, with both feet on the floor, and the body weight equally distributed on both feet

ballast
an electrical transformer for producing the required current, voltage, and waveform to operate certain types of luminaires

ballism
a movement disorder characterized by flinging movements of the limbs; *syn.* ballismus

ballismus
see ballism

ballistic lift
a ballistic movement in which an object is being lifted, with the momentum resulting from the initial motion moving the object through much of the terminal portion of the trajectory

ballistic motion
see *ballistic movement*

ballistic movement
a rapid, gross, relatively smooth change in position of a bodily extremity which is initiated by one or more protagonist muscles which are active only during the initial phase of the motion; *syn.* ballistic motion, preprogrammed movement; *opp.* controlled movement

Ban-Lon
a fabric made from a combination of nylon, polyester, and other fiber blends

band pressure level
the sound pressure level within a specified frequency bandwidth

bandpass
pertaining to a limited range of frequencies which are transmitted or allowed beyond a certain point within a system; *opp.* bandstop

bandstop
pertaining to a limited range of frequencies which are not allowed to pass through a system, or which pass at a much lower intensity than the higher and lower frequencies; *syn.* band rejection; *opp.* bandpass

bandwidth [1] (BW)
that range of continuous frequencies capable of being processed or output by a system

bandwidth [2]
that maximum rate at which information can be transferred over a channel, typically with units in some multiple of bits per second

bang-bang control
a type of discrete system control using relays to control input and in which the operator moves a control from essentially maximum deflection in one direction to essentially maximum deflection in another direction

bank [1]
that excess amount or material or numbers of product which are allowed to accumulate at some point in a production line or operation without being currently worked in order to provide for reasonable fluctuations in flow; *syn.* float

bank [2]
the elevation of the outer margin relative to the inner margin of the curve on a roadway

bank [3]
roll an aircraft about its longitudinal axis

bar
a unit of pressure; equal to approximately one atmosphere, or 750 mm Hg

bar chart
see *bar graph, Gantt chart*

bar code
a set of parallel lines of differing widths which contain information

bar code reader
see *bar code scanner*

bar code scanner
a laser-based device which views a bar code and transmits the information to a computer; *syn.* bar code reader

bar graph
a graphical representation of the frequency of occurrence within a set of discrete groupings or values in which the length of the bar is proportional to frequency

Bárány chair
a rotating chair used for vestibular or nystagmus experimentation

barf bag
(sl) a plastic disposable bag for collecting vomitus during motion sickness

barn door
one of a set of adjustable light shields which may be used in conjunction with a luminaire to partially direct and control the luminance from that luminaire

barodontalgia
a form of decompression sickness resulting in tooth pain from the expansion of trapped air within a tooth or between a filling and the tooth material

barotalgia
that sensation of pressure or ear pain due to an inequality of air pressure between the middle ear and the environment; *syn.* ear squeeze

barotitis media
barotrauma to the middle ear

barotrauma
any injury resulting from expansion or contraction of gases in closed spaces within certain structures of the body due to pressure changes in the ambient environment

barrier
any object, individual, or structure which impedes progress toward a goal or which prevents entry to a region for safety reasons

barrier cream
a viscous substance which may be applied to the skin to prevent percutaneous absorption of toxic materials

barrier equivalent velocity (BEV)
the effective velocity at which a vehicle impacts a barrier in crash testing

barrier free
an ideal condition in which handicapped individuals have full and equal access to all facilities accessible to able-bodied individuals

barrier guard
any protective device designed to prevent access to hazardous areas, or to prevent inadvertant operation of controls

Barthel Index (BI)
a numerical score based on 10 items of a physically disabled individual's ability to care for himself by performing some of the activities of daily living

basal conditions
those conditions under which the basal metabolic rate measures are taken: (a) fasted for at least 12 hours, (b) following a night of restful sleep, (c) no strenuous exercise since sleep, (d) comfortable, relaxed conditions with the air temperature about 70–75°F, depending on clothing

basal metabolic rate (BMR)
that energy expenditure per unit time by the body while an individual is awake, but at rest under basal conditions

basal metabolism
that minimal metabolism required to maintain cellular function

basal temperature
the normal body temperature of a healthy individual following sleep in the morning

base period
that reference period of time (year, month, etc.) against which some current period is judged

base time
see normal time

base wage rate
that hourly monetary compensation paid to a normal operator working at a standard pace on a specified task

basic division of work
see therblig

basic element
see therblig

Basic Elements of Performance
a quantitiative technique for measuring the residual capabilities of disabled individuals

basic measurement scale
any of the four basic scales used for classifying data in statistical analyses; *see nominal scale, ordinal scale, equal-interval scale, ratio scale*

basic motion
any fundamental, complete motion using the primary physiological and/or biomechanical performance capabilities of the body or its member parts, as determined by motion analysis studies; *see also therblig*

Basic Motion Time Study (BMTS)
a predetermined motion time system

basic research
that fundamental research performed to acquire scientific knowledge, without concern for immediate practical

application; *syn.* pure research; *opp.* applied research

basic sciences

those disciplines involving the study of mathematics, chemistry, physics, biology, and psychology

Basic T

the arrangement of four basic flight instruments in a standard pattern, with the airspeed and attitude indicators and altimeter in a horizontal line across the top, and the heading indicator centered below

basic time

that time allowed or required for performing a work element at a standard rate

$$BT - \frac{observed\ time \times observed\ rating}{standard\ rating}$$

basilar membrane

that membrane in the cochlea to which the organ of Corti is attached

bass

those sound frequencies in the lower portion of the audio range, generally below about 250 Hz

batching

the process of scheduling work in small increments; *syn.* short-interval scheduling

bathtub curve

see life characteristic curve

batt

a section of insulating material

baud

a unit of serial transmission speed, usually equivalent to bits per second

    Comment: if each signal event is represented by one bit, baud is equivalent to bits per second

bauxite fume pneumoconiosis

a pulmonary fibrosis due to the inhalation of aluminum ore or processing dust fumes; *syn.* Shaver's disease

beam angle

a measure of light beam spread; the angle between diametrically opposed edges of a projected light beam at which the luminous intensity is some stated percentage of the maximum along the beam axis, with all measures taken in a wavefront equidistant from the source

beam axis

an imaginary straight line representing the direction along which a light beam is projected

beam element

a modeling structure which is capable of bending, torsion, and axial stiffness

beam spread

the lateral distribution of a projected light beam from the beam axis

beard

that long-term accumulation of hair growth on and around the chin and lateral portions of the face

beat

a periodic variation in the intensity of sound generated from the combination of two simple tones of slightly different frequencies having approximately the same orders of magnitude

beat frequency

the number of occasions per unit time at which a beat occurs

beat knee

a bursitis in the knee joint due to prolonged vibration, pressure, or repeated friction

Bedaux plan

a wage incentive plan in which performance above the rated standard 60 B units per hour would result in merit bonuses; *syn.* Bedaux system

bedrest

the maintaining of an individual in bed, either for therapy or experimental purposes

beginning spurt

a briefly higher than normal level of activity at the beginning of the work period or by an employee new to the job

behavior
any and all responses of an individual, group, or system

behavior-anchored rating scale (BARS)
any rating scale developed to evaluate individual behavior patterns

behavioral competence
having the ability to integrate those psycho- and sensorimotor patterns required to complete one or more specified tasks

behavioral dynamics
the behavioral operating characteristics of individuals and groups as they are conditioned by the external working environment and/or individual and group interactions

behavioral resistance
an opposition to carrying out an order, directive, or someone's expressed desire; *syn.* resistance

behavioral rigidity
an inability to effectively deal with new situations

behavioral toxicology
the study and assessment of neural impairment due to toxic chemical exposure through the use of psychological testing methods

Békésy audiometry
an auditory threshold determination procedure involving observer control in which an individual alternately, over several cycles, presses a switch to reduce signal level when the sound is heard and releases it when the sound becomes inaudible; *syn.* Békésy tracking procedure

Békésy tracking procedure
see Békésy audiometry

bel (B)
a dimensionless measure of the intensity of some energy, corresponding to the ratio of two intensity or power levels; *see decibel*

Belding-Hatch heat stress index
an estimate for body heat stress based on the ratio of actual evaporative heat loss to the maximum possible evaporative heat loss for the given environment, which may be determined by temperature, humidity, air velocity, workload, clothing, and their interactions; *syn.* heat stress index

belief
a concept or thought held as truth

belly [1]
the fleshy central portion of a muscle along its longitudinal axis

belly [2]
(sl) the stomach-abdominal region of the frontal portion of the body

belly button rule
a task design guideline that the hands should remain close to the abdominal region when lifting or handling items

below elbow (BE)
pertaining to a prosthesis for or an amputee/amputation for which some part of the forearm, and all of the wrist and hand are missing/taken

below knee (BK)
pertaining to a prosthesis for or an amputee/amputation for which some part of the lower leg, and all of the ankle and foot are missing/taken

benchmark
a thoroughly documented reference value or standard of measurement against which performance, response, or other characteristics may be compared with confidence

benchmark job
a job having enough common characteristics with one or more other jobs such that it may be used as a predictor for those jobs in such aspects as evaluations of worker output and time standards

bends
a form of decompression sickness in which pain occurs in joints, muscles, and/or bones

beneficial impact
a purposeful impact, as in performing some task; *syn.* functional impact; *opp.* accidental inpact

bent torso breadth
the horizontal linear distance across the shoulders with the individual in the bent torso position

bent torso height
the vertical linear distance from the floor or other reference surface to the highest point on the head in the bent torso position

bent torso position
a posture with the individual standing with his feet separated by 18 inches, leaning forward with his hands on his knees, and looking straight ahead

berylliosis
a pneumonia or pneumoconiosis caused by the inhalation of certain beryllium compounds

beta [1] (β)
a measure of a system's response bias in signal detection theory, represented by the ratio at the criterion level of the height of the signal + noise distribution to the height of the noise distribution alone

beta [2] (β)
the probability of making a Type II error; *see also* **alpha**

beta coefficient
the weighting factor preceding a variable in a regression equation

beta particle
a free electron or positron emitted from a radioactive element

beta rhythm
an EEG frequency band consisting of frequencies greater than 13 Hz

beta testing
the second release phase of software evaluation, just prior to release in the commercial market; *see also* **alpha testing**

bezel
a rim for holding a piece of transparent glass or plastic for a display on a meter or other indicator

Bezold spreading effect
see **color assimilation**; *syn.* assimila-

tion, similitude effects

Bezold-Brücke effect
see **Bezold-Brücke phenomenon**

Bezold-Brücke hue shift
see **Bezold-Brücke phenomenon**

Bezold-Brücke phenomenon
a change in the apparent color of a visual stimulus with a change in stimulus intensity or illumination; *syn.* Bezold-Brücke hue shift, Bezold-Brücke effect

bi-
(prefix) denoting a relationship to two symmetrical or approximately symmetrical parts

biacromial breadth
the horizontal linear distance across the shoulders from right to left acromion; *see* **shoulder breadth**
> *Comment:* measured with the individual standing erect, and the shoulders straight

bias [1]
an individual preference or prejudgment on an issue

bias [2]
a systematic error represented by the difference between the mean of repeated measurements and the true value; a tendency to over- or underestimate the true or actual value

bias [3]
a relatively constant voltage offset from zero

biauricular breadth
the horizontal linear distance from the most lateral point of the right ear to the same point of the left ear
> *Comment:* measured without auricular compression

bicanthic diameter
see **ectocanthic breadth**

biceps brachii
the large, two-headed muscle in the anterior upper arm

biceps circumference, flexed
the maximum surface distance around the biceps brachii

Comment: measured with both the shoulder and elbow flexed 90 degrees, such that the upper arm is horizontal, and the hand clenched into a fist

biceps circumference, relaxed
the maximum surface distance around the upper arm at the level of the biceps brachii belly
Comment: measured with the arm hanging relaxed at the individual's side

biceps femoris
a large, two-headed muscle in the posterior thigh; one of the hamstring muscles

biceps muscle
see biceps brachii, biceps femoris

biceps skinfold
the thickness of a vertical skinfold on the anterior midline of the upper arm over the belly of the biceps brachii muscle at the level of the upper arm circumference measure
Comment: measured with the individual standing erect and the arms hanging naturally at the sides

bicipital
pertaining to a muscle having two heads, often specifically to the biceps brachii and biceps femoris muscles

bicristale breadth
see biiliocristale breadth; also bicristal breadth

bicycle ergometer
a stationary cycle used to measure or work against a fixed or adjustable force
Comment: may only have one wheel in reality

bideltoid breadth
the horizontal linear distance across the maximum lateral protrusions of the right and left deltoid muscles; *see shoulder breadth*
Comment: measured with the individual standing erect and arms hanging naturally at the sides

bifocals
a pair of lenses in glasses having two correction portions, one for distance vision, the other for near vision

bigonial breadth
the horizontal linear distance across the gonial angles of the jaw
Comment: measured with the jaw muscles relaxed and the individual sitting or standing erect

biiliac breadth
see biiliocristale breadth

biiliocristale breadth
the horizontal linear distance across the torso measured between the superior points of the iliac crests; also biiliocristal breadth; *syn.* biiliac breadth, transverse pelvic breadth, pelvic breadth, bicristale breadth
Comment: measured with the individual standing erect, with his weight equally balanced on both feet

bilabial height
the vertical linear distance between the most superior point on the upper lip and the most inferior point on the lower lip

bilateral
pertaining to similar structures present on both sides of a symmetric or approximately symmetric body

bilateral teleoperator
a teleoperator system in which force and motion can be transmitted in both directions — from the operator to teleoperator and vice versa

bilingual
capable of speaking, writing, and understanding two languages

bimalleolar breadth
the horizontal linear distance across the protrusions of the medial and lateral ankle bones; *syn.* ankle breadth
Comment: measured with the individual standing erect and the body weight distributed evenly on both feet

bimanual
performed with both hands

bimodal [1]

pertaining to or affecting two sensory modalities simultaneously

bimodal [2]

a statistical distribution having two modal values

binary digit (bit)

see bit

binaural

having input to both ears simultaneously

binaural hearing

the perception of sound by both ears

binocular

pertaining to the use of or input to both eyes simultaneously

binocular accommodation

the process of both eyes accommodating simultaneously

binocular disparity

the difference in visual images on the right and left retinas resulting from the lateral separation of the eyes; *syn.* lateral retinal image disparity, binocular parallax

binocular fusion

the merging of images from the two eyes into a single perception; *syn.* fusion

binocular parallax

see binocular disparity

binocular portion of the visual field

see binocular visual field

binocular rivalry

a phenomenon in which an alternation of partial or entire images is perceived when the two eyes are stimulated simultaneously with different images; *syn.* retinal rivalry

binocular suppression

a loss of all or some portion of one eye's visual field resulting from conflicting information being presented to the fusional region of the other eye's visual field

binocular vision

that quality of vision existing by virtue of having two eyes in which the visual fields of the two eyes overlap; *syn.* stereopsis

binocular visual field

that portion of the visual field where the monocular visual fields of the two eyes overlap; *see monocular visual field*

binomial distribution

a theoretical discrete probability distribution for a binomial random variable

$$P = \binom{n}{r} p^n (1-p)^{n-r}$$

where:

n = total number of outcomes

r = number of successful outcomes

$\binom{n}{r}$ = the number of combinations of n outcomes, taken r at a time

Comment: approximates the normal distribution for large sample sizes

bio-

(prefix) pertaining to living systems or those components which may be or have been a part of a living system

bioacoustic

pertaining to the effect of sound on the body

bioacoustics

the study or use of the relationships between sound and living organisms

bioaerosol

any aerosol consisting primarily of biological entities such as microbes; the presence of biological entities in aerosol form

bioastronautics

the study of medicine, biology, and psychology in relation to space flight

biocontainment

any technique used to achieve bioisolation of one or more substances

biocular breadth

see ectocanthic breadth

biodynamicist

one who works in the field of biodynamics

biodynamics
that field concerned with the effects of external forces or dynamic conditions on biological systems; *syn.* impact biomechanics

bioelectrical impedance analysis (BIA)
a technique for estimating/measuring total body fat/lean body mass by observing the impedance of electricity passed through a part of the body

bioengineering
the integration and application of knowledge in the fields of human biology, medicine, and engineering

biofeedback
the use instrumentation to provide information to an organism which enables that organism to alter its behavior accordingly

biographical data
see biographical information

biographical information
any historical information about an individual, including physical characteristics, work experience, and involvement in other activities

biohazard
any biological agent arising from the environment which may present a threat to the health or well being of an individual or group

bioisolation
a condition in which biological systems are effectively separated from each other using any one or more of physical, chemical, or biological methods

biological clock
any hypothesized internal mechanism responsible for maintaining one or more biological rhythms; *syn.* body clock

biological half-life
the length of time required for a biological system to metabolize or eliminate half of a given administered or ingested substance via normal mechanisms; *syn.* half-life

biological monitoring
the sampling and quantitative analysis of air, biological fluids, or tissues for the presence of one or more hazardous biological agents or their metabolites

biological needs
the basic physiological needs for a living entity to function, including air/oxygen, water, and food

biological rhythm
a self-maintained, cyclic variation with a relatively fixed periodicity in a living organism; *see also **circadian rhythm, infradian rhythm, ultradian rhythm, circannual rhythm***

biomechanical profile
a combined set of biomechanical, electromyographic, motion, and other data recorded simultaneously during some activity

biomechanics
that field of study involving classical mechanical principles and their relationships as used by or applied to living organisms or biological tissues in motion; *adj.* biomechanical

biometry
the measurement of biological parameters and the use of simple descriptive statistics for the data obtained; *syn.* biometrics; *adj.* biometric; *see **biostatistics***

bionic
pertaining to any man-made device which has a function similar to or enhanced from a natural biological structure

bionics
the study of biological systems to derive technical knowledge for use in the design, modeling, development, and/or implementation of artificial systems

bioremediation
the use of biological organisms to reduce or eliminate pollution

biosensor
any sensor used in biotechnology which is composed of living tissue, biological materials, or fabricated from basic biological materials

biostatistics
the use of statistical methodology to describe biological data or draw inferences from those data

birth rate
the ratio of live births per some unit of existing population within a given period of time

birth-death process
a queuing system in which units to be served or worked on arrive and depart in a random fashion

biserial correlation
that correlation existing between two continuously distributed variables, but in which one of the variables has been scored as a dichotomous variable

bispinous breadth
the transverse distance between the centers of the anterior superior iliac spines
   Comment: measured with the individual standing erect and his weight evenly distributed on both feet

bit [1] (b)
a numerical value in the binary scale, either zero (0) or one (1)
   Comment: the basic unit in digital electronic systems; contraction of binary digit

bit [2] (b)
that amount of information obtained when one of two equally likely alternatives is given or specified; the basic unit of information; *see information theory*

bitragion breadth
the transverse width of the head as measured from right to left tragion

bitragion – coronal arc
the surface distance from right tragion, over the vertex, to left tragion
   Comment: measured with the individual sitting or standing erect with the scalp muscles relaxed

bitragion – crinion arc
the surface distance from right tragion, over the anterior hairline, to left tragion

Comment: measured with the individual sitting or standing erect with the scalp muscles relaxed

bitragion – inion arc
the surface distance from right tragion, over inion (including the hair), to left tragion
   Comment: measured with the individual sitting or standing erect with the scalp muscles relaxed

bitragion – menton arc
the surface distance from right tragion, under the anterior/inferior tip of the chin, to left tragion
   Comment: measured with the individual sitting or standing erect with the jaws closed and the facial muscles relaxed

bitragion – minimum frontal arc
the surface distance from right tragion, over the forehead just above the brow, to left tragion
   Comment: measured with the scalp and facial muscles relaxed

bitragion – posterior arc
the surface distance from right tragion, across the base of the skull, to left tragion
   Comment: measured with the individual sitting or standing erect with the scalp muscles relaxed

bitragion – submandibular arc
the surface distance from right tragion, under the gonial angles of the jaw, to left tragion
   Comment: measured with the individual sitting or standing erect with the jaws closed and the scalp and facial muscles relaxed

bitragion – subnasale arc
the surface distance from right tragion, across subnasale to left tragion
   Comment: measured with the individual sitting or standing erect and the scalp and facial muscles relaxed

bitrochanteric breadth
the horizontal linear distance between the most lateral projections of the right

and left greater trochanters; *syn.* bitrochanteric width

> *Comment:* measured with the flesh compressed and the individual standing erect with his weight distributed equally on both feet

bivariate regression

a special case of multiple regression in which the number of predictor variables is two

bizygomatic breadth

the transverse width of the face across the most lateral protrusions of the zygomatic arches; *syn.* face breadth

black

having the property of absorbing all or most of the incident visible light

black light

(sl) that form of electromagnetic radiation energy primarily having wavelengths from about 320 to 400 nm

black light lamp

a light source which emits black light

blackbody

a surface which, ideally, absorbs all incident visible light energy and emits radiant energy with a spectral distribution varying according to the absolute temperature of the surface; *syn.* ideal blackbody, blackbody source, Planckian radiator, blackbody radiator, full radiator, standard radiator, ideal radiator, complete radiator

blackbody locus

a set of points representing the chromaticities of a potential set of blackbodies with various color temperatures on a chromaticity diagram; *syn.* Planckian locus

blackbody radiator

see blackbody

blackbody source

see blackbody radiator

blackout

a temporary loss of vision, regardless of the cause; *see also grayout, gravity-induced loss of consciousness*

bland

not having a stimulating taste characteristic

bleed

drain or lose a fluid from a normally closed system, especially from the human circulatory system

blind [1]

not having certain information regarding ongoing activities in an experiment; *see also double blind*

blind [2]

having no visual capability, or having a Snellen visual acuity less than 20/200 even using corrective lenses

blind [3]

pertaining to a ship or other military vehicle which has lost its radar or other sensing capabilities

blind positioning

a movement which requires the placement of one or more objects at some orientation or point in space without visual cues; also blind-positioning

blind spot [1]

any region on or around a vehicle at which another object may not be readily seen due to lack of mirror coverage or inability to view directly

blind spot [2]

that region of the posterior eyeball where no photoreceptors are located due to the optic neural fibers exiting the eyeball

blinding glare

any extremely intense glare which interferes with vision for a significant period of time after removal of the glare source

blink [1]

a unit of time equal to 0.864 seconds or 10^{-5} day

blink [2]

turn quickly on and then off at approximately regular intervals

blink [3]

see eye blink

blink coding
the use of a blinking stimulus as a highlighting or attention-getting technique

blink rate [1]
that number of occasions which a light or segment of a display turns on and off within a specified interval

blink rate [2]
see *eye blink rate*

blip
a brief visual signal of higher intensity or different quality from the background, which may enable or enhance detection

block and tackle
a combination of a rope or other line material and an independent pulley
Comment: used to increase mechanical efficiency

blood
the viscous red bodily fluid consisting of plasma and the formed elements which carries nutrients, waste products, and body defensive mechanisms through the cardiovascular system

blood count
the number of erythrocytes or white blood cells in a cubic millimeter of blood

blood doping
the process of reinjecting an athlete's blood removed earlier to provide a greater hemoglobin content for an athletic event

blood forming organs (BFO)
the red bone marrow tissues and the spleen

blood pressure
that force exerted on the internal heart and vessel walls of the circulatory system by the blood; see *systolic blood pressure, diastolic blood pressure*

blue
a primary color, corresponding to that hue apparent to the normal eye when stimulated only with electromagnetic radiation approximately between wavelengths from 455 nm to 490 nm

blue blindness
see *tritanopia*

blue collar
pertaining to those workers typically doing production work, as opposed to management; *opp.* white collar

blur
a condition in which an image is not well focused; *adj.* blurred, blurry

bodily injury
any injury to an individual from mechanical or physical processes

body
the human frame, including all its organs, tissues, and other normal materials; *adj.* bodily

body breadth, maximum
the maximum linear horizontal distance across the body, including the arms
Comment: measured with the individual standing erect and the arms hanging naturally at the sides; specify the level at which the measure is taken

body clock
see *biological clock*

body composition
the proportions of tissue makeup in the body, generally classified by two primary categories as a function of body mass: lean body mass and body fat

body depth, maximum
the maximum horizontal distance between two vertical planes which represent the most anterior and posterior aspects of the torso
Comment: measured with the individual standing erect and arms hanging naturally at the sides; specify where the measurement is taken

body envelope
that volume which includes the body and any protective clothing or other items required during performance of a specific task

body fat
that portion of body composition which is composed of adipose tissue

body heat content (H_b)
the mathematical product of the body's heat capacity and the mean temperature of body tissues

body height
see stature

body mass
the mass of the body

body mass index (BMI)
a guideline for estimating the percentage of body fat and nutritional status of the body

$$BMI = \frac{weight}{(stature)^2}$$

body motion
the movement of one or more body parts which involves a mass redistribution within some coordinate system

body position
see posture

body proportionality
the distribution of an individual's circumferential measures of various body segments

body segment
any portion of the body located between two joints, or the terminal portion of a body part from a joint, which has a relatively constant geometry when moved

body surface
any part or all of the total surface area of the body

body surface area
the total surface area of the body

body temperature and pressure, saturated conditions (BTPS)
the air mixture saturated with water vapor at ambient body temperature, as found in the lung alveoli or exhaled air

body typology
any of various attempts to ascribe behaviors and personality characteristics

to the shape or composition of an individual's body; *see somatotype*

body versus machine rule
a task design guideline that the machine should not be capable of injuring the worker during any phase of a task

body volume
the total volume occupied by the body

body weight
the nude weight of an individual; *syn.* weight
> *Comment:* measured under standard conditions

body-load aggregate
the combined effect of the weight being manipulated and the weight of those parts of the body involved in a materials handling or lifting task

BOEMAN
a computerized human modeling package for aiding design and evaluation of reach capabilities in cockpits and other aircraft workstations

boilermaker's deafness
a form of hearing impairment in which an individual hears better under noisy conditions than in quiet
> *Comment:* caused by working for long periods around loud noises

bold [1]
a highlighting technique in which a wider stroke width or a greater number of dots/pixels are used to make text characters larger than normal

bold [2]
willing to take a great risk

bolus
a cohesive mass, either of food material for swallowing or of fecal material following defecation

bone
that rigid calcified connective tissue matrix which provides much of the structure for and supports body mass

bone conduction
the passage of sound waves to the inner ear via the skull bones

bone conduction test
a hearing test in which the audiometer oscillator or tuning fork is placed against the mastoid process of the temporal bone

bonus
any money paid by an employer to a worker in addition to a regular salary or wage; *syn.* bonus earnings

boot [1]
(n) a piece of footwear which is normally heavier or stronger than a shoe, has a higher heel, and extends above the ankle

boot [2]
(v) bring a computer system on line

borderline between comfort and discomfort
see comfort-discomfort boundary

boredom
a form of mental fatigue generally due to lack of stimulation, lack of interest in the ongoing activity, isolation, performance of a monotonous task, other similar situations, or some combination of these situations

Borg scale
see rating of perceived effort scale

Botsball
a small copper sphere, painted black and covered with a sized black mesh wetted fabric, which contains a thermometer for estimating heat stress; *see wet globe temperature*

bottom time
that length of time a diver has been at depth in an underwater dive or at maximum pressure in a hyperbaric chamber for treatment of decompression sickness

boundary representation
a technique used in solid computer modeling where the geometry is defined in terms of its edges and surfaces

bow
the deflection of a portion of a structure caused by a pressure differential on the two sides

brachium
the upper arm; *adj.* brachial

bradyarthria
see bradylalia

bradycardia
a lower than normal heart rate; *opp.* tachycardia

bradykinesia
any movement disorder in which body movements are slowed

bradylalia
a very slow articulation in speaking

bradylexia
an abnormal slowness in reading

Braille
a communication system for the blind which uses tactile characters

brain
that portion of the central nervous system located within the cranium

brain potential
any recordable electrical difference between two or more locations on the scalp or brain; *see electroencephalogram, evoked potential*

brain stem
that portion of the brain which is continuous with the spinal cord and lies beneath the cerebellum and cerebral hemispheres, containing neurons governing many of the body's vital functions

brain wave
the recorded or observed varying electrical potentials from the brain; *see also electroencephalogram, evoked potential*

brainstorm
propose and discuss ideas, ideally freely and without criticism, in an attempt to discover all possible approaches to a situation

Brayfield-Rothe Scale of Job Satisfaction
a commercially available standardized questionnaire for surveying job satisfaction among employees

breach
bypass, avoid, or dismantle a safety or security mechanism

breadth
width; a straight-line horizontal measurement having only lateral extent, from one side of the body or a body segment to the other

break time
see rest period

break-even analysis
any quantitative technique used to determine the sales necessary to achieve the break-even point; *syn.* break-even alternatives analysis

break-even chart
a graphical representation of the relationships between income and costs, usually based on different levels of volume for production and sales

break-even point
that economic level at which total operating costs equal total income, and the company neither makes a profit nor has a loss

breakdown [1]
a decomposition of some process or activity into its component parts

breakdown [2]
the ceasing of operation of a system, subsystem, or component due to some fault or failure

breakdown maintenance
see corrective maintenance

breaking strength
that stress level at which a material fails; *syn.* strength

breakpoint
that readily distinguishable point in time which represents a boundary between two task elements, at which one element is completed and the other is begun; also break point; *syn.* reading point, endpoint

breast [1]
the anterior thorax, especially in the vicinity of the nipple

breast [2]
the human female mammary gland

breastbone
see sternum

breathe
alternately inhale and exhale air from the lungs

bright
a highlighting technique in which one or more portions of a display appear brighter than the remainder

brightener
any colorless, fluorescent dye which causes washed clothing to appear brighter under certain lighting conditions by converting ultraviolet light into visible light, normally at the blue end of the spectrum; *syn.* whitener, whitening agent, optical brightener

brightness
a subjective judgment of the relative amount of light projected or reflected from a surface or object, ranging from brilliant to dark; *see also luminance*

brightness contrast
the subjective difference between the brightness of an object and the background against which that object is located; *see also luminance contrast*

brightness control
a potentiometer or other adjustment device for varying the luminance on a display; *syn.* brilliance control

brightness enhancement
the use of a flashing light within a certain flashing frequency range (about 2-20 Hz) to make a light appear brighter than if the same average light intensity were used from a steady light

bril
a subjective scale for judging brightness

brilliance
see brightness

brilliance control
see brightness control

British Thermal Unit (BTU)
the amount of heat required to raise 1 pound of water at 60°F to 61°F

broadband
containing many frequencies

broadcast
> a message sent to all stations connected to a computer network

broken shift
> see *split shift*

brow
> see *forehead, eyebrow*

brow ridges
> the bony ridges of the forehead that lie above the orbits of the eyes

bruise
> an injury characterized by capillary or venous hemorrhaging beneath an unbroken skin; see also *hematoma*

BRYNTRN
> a computer model for determining the effects of nucleons on target materials

BTPS conditions
> see *body temperature and pressure, saturated conditions*

bubble
> a trapped volume of air or other gas(es) within a more viscous fluid or solid

buffer ¹
> a region separating one area from another for safety, habitability, or other reasons

buffer ²
> a chemical or solution which has the capability to minimize wide fluctuations in pH

buffer ³
> a temporary computer storage location in which data may be kept while awaiting transfer to another, more permanent location

builder
> any chemical used in laundry which acts to soften water for improved detergent activity

building block
> one of a fixed group of elements or modules which may be joined to form a system or complete some activity

building code
> a set of regulations which provide standards to which structures must be built; see *code*

Built-in Test (BIT)
> a circuit or other equipment located within a system which automatically or on direction by an operator verifies system function

Built-in Test Equipment (BITE)
> that circuitry or other hardware incorporated into a system for monitoring that system's function and analyzing for faults when they occur

bulb
> the primary source of light in an electrically powered lamp

bulk arrival
> the arrival of several customers or users at a location at one time or as part of a single event

bulk materials
> any powdery, granular, or lumpy substance in loose form

burn-in test
> a period of time in which a completed system or set of subsystems are observed under expected operating or more extreme conditions to determine if any of the components will fail prematurely; *syn.* debug

burr
> a ragged edge or sharp point from processing

bursa
> a fluid-filled sac-like structure having a slippery surface and located at joints or other tissues to reduce friction in movement

bursitis
> an inflammation of a bursa, occasionally with calcium deposit development

burst
> a rapid decrease in pressure within a container of specified volume as it ruptures under pressure and the contents spread rapidly to the external environment

burst lung
> see *pulmonary hyperinflation syndrome*

bust depth
> the horizontal linear distance from the

most posterior protrusion at the bra tip level of an individual's back to the bustpoint

> *Comment:* measured with the individual standing erect; for females only

bustpoint

the most anterior external protrusion of the bra pocket

> *Comment:* for females only

bustpoint – bustpoint breadth

the horizontal distance between bustpoints

> *Comment:* measured with the individual standing erect; for females only

bustpoint height

the vertical distance from the floor or other reference surface to the bustpoint; *syn.* **chest height**

> *Comment:* measured with the individual standing erect and the weight distributed evenly on both feet; for females only

buttock

the mass of fleshy tissue posterior to the hip, consisting largely of the gluteus maximus and other muscles

buttock circumference

the surface distance around the body without tissue compression at the level of the maximum posterior protuberance of the buttocks; near *syn.* hip circumference, standing

> *Comment:* measured with the individual standing erect and the weight balanced evenly on both feet

buttock circumference, sitting

the surface distance around the buttocks and diagonally across the lap; *syn.* hip circumference, sitting

> *Comment:* measured with the individual sitting erect

buttock depth

the horizontal linear distance from the

maximum posterior protrusion of the buttocks to the most anterior portion of the torso at that level

> *Comment:* measured with the individual standing erect with the hip and thigh muscles relaxed

buttock – heel length

see buttock – leg length

buttock height

the vertical distance from the floor to the maximum posterior protrusion of the buttock

> *Comment:* measured with the individual standing erect and the weight balanced evenly on both feet

buttock – knee length

the horizontal distance from the rearmost point of the buttocks to the front of the kneecaps; *syn.* buttock – knee distance

> *Comment:* measured with the individual sitting erect, the knees flexed 90°, feet flat on the floor, and the upper leg parallel to the floor

buttock – leg length

the horizontal distance from the wall or the most posterior point of the buttocks to the underside of the heel; *syn.* buttock – heel length

> *Comment:* measured with the individual sitting erect on the floor or other flat surface (possibly against a wall but with no tissue compression), the knee fully extended, and the longitudinal axis of the foot perpendicular to the leg

buttock – popliteal length

the horizontal distance from the rearmost surface of the buttock to the back of the lower leg; *syn.* buttock – popliteal distance

> *Comment:* measured with the individual sitting erect, the knees flexed 90°, the feet flat on the floor, and the upper leg parallel to the floor

buttock protrusion

the point of maximum posterior protrusion of the buttock

byssinosis

a chronic lung disease due to prolonged inhalation of cotton and/or linen fibers

byte (B)

a group of bits which may be treated as a single unit in a digital computer

Comment: the number depends on the type of hardware, but is typically 8 bits

C

C scale

a sound weighting system having flat response characteristics for high sound pressure levels up to about 8 KHz

C shift

see third shift

C-weighted sound pressure level (dB(C))

that sound pressure level measured using the C scale

cabin

that occupied or occupiable portion of the interior of a passenger vehicle

cabin pressure

the atmospheric pressure within a cabin

cabin temperature

the dry-bulb temperature within a cabin

cabinet

an independent structure containing drawers and/or shelves

cadaver

the body of a deceased human

cafeteria benefit plan

see cafeteria plan

cafeteria plan

a means of handling fringe benefits in which the employer allocates a certain amount of money to each employee for such benefits, and the employee is able to select the distribution of those benefits to his own best advantage; *syn.* cafeteria benefit plan

caisson disease

see decompression sickness

CAL-3D crash victim simulator (CAL-3D CVS)

a computer modeling program for simulating the biomechanical responses of an individual in a vehicular crash

calcaneus

the heel bone; *pl.* calcanei; *adj.* calcaneal

calcification

the deposition of calcium-containing substances within the human body

Caldwell regimen

a procedure for static strength assessment, involving providing to the subject the detail of the experiment and the necessary instructions, noting the posture and muscles involved, and having the subject maintain a four-second hold on the measuring device

calender

a machine which passes some pliable material between rollers or plates to make a relatively smooth, continuous or long sheet

Comment: presents a nip point safety problem

calf

the fleshy part of the posterior lower leg, consisting largely of the gastrocnemius muscle

calf circumference

the surface distance around the lower leg in a horizontal plane at the vertical level which gives the greatest value

Comment: measured with the individual standing erect, his weight equally distributed on both feet

calf circumference, recumbent

the calf circumference of a reclining individual

Comment: measured with the individual supine, the knee and hip both flexed 90°, and the longitudinal axis of the foot perpendicular to that of the leg

calf depth

the linear horizontal distance from the posterior surface to the anterior surface on the lower leg at the level of the calf circumference

Comment: measured with the individual standing erect and his weight equally distributed on both feet

calf height

the vertical distance from the floor or other reference surface to the level of the maximum circumference of the lower leg

Comment: measured with the individual standing erect and his weight equally distributed on both feet

calf length

the linear distance parallel to the longitudinal axis of the lower leg between the knee joint level and the medial malleolus

calibration

a comparison or adjustment of a measurement device or system with unknown accuracy to a device or system of known or accepted accuracy

caliper

a device for obtaining accurate measurements of relatively short linear measures

calisthenics

a form of exercise to improve strength, endurance, and/or grace

call-out

a vocal method for presenting information to be heard by an individual; *see also* ***read-out***

calorie (cal)

that amount of heat required to raise the temperature of 1 gram of water 1°C; *syn.* small calorie; *see* ***Calorie***

Comment: the calorie used in nutrition and metabolism is the large calorie or Calorie

Calorie (Cal)

the unit for heat (energy) production in body nutrition and metabolism; equal to 1 Kcal; *syn.* kilocalorie, large calorie; *see* ***calorie***

CALSPAN

a computer modeling program for simulating crash victim dynamics

camera study

see ***memomotion study***

cancellation test

a clerical aptitude test for speed and accuracy in crossing letters or numbers in a sequence

cancellous bone

that interior portion of a some bones which contains a criss-crossed matrix of calcified bone tissue with the remaining volume filled with marrow; *syn.* spongy bone

cancer

any malignant tumor or growth; *adj.* cancerous

candela (cd)

an SI unit for luminous intensity; equal to the intensity of a 555 nm (5.40×10^{14} Hz) point source radiating 1.464×10^{-3} watts per steradian; *syn.* new candle

candle

1/60th of the intensity of a 1 cm^2 blackbody radiator at 2042 K; *syn.* international candle

Comment: an outdated term

candlepower (cp)

that measure of light intensity stated in candelas

canopy [1]

that large fabric part of a parachute which fills with air to slow the fall of an individual or object

canopy [2]

the transparent cover for the cockpit of an aircraft

canthus

the corner or angle formed by the junction of the eyelids; *pl.* canthi; *see* ***endocanthus, ectocanthus***

cap [1]

a soft type of headwear which is preformed and sized

cap [2]

a covering for a jar, bottle, or other rigid structure to contain the enclosed items or to prevent access by moisture, children, mold, or other entities

cap lamp
a lamp worn by miners or others working in dark areas which is attached to a safety cap or helmet

capacitance (C)
the value of the ratio the absolute charge of two equally but oppositely charged conductors to the potential difference between them

capacitive touchscreen
a display having a thin layer of material over its front which uses a change in capacitance to indicate a touch location

capacity [1]
see endurance

capacity [2]
the upper limit of an individual's ability to learn, understand, or perform through inherent ability, training, practice, and any other means

capitate bone
one bone of the distal group of bones in the wrist

capitulum
a smooth hemispherical protuberance at the anterior distal end of the humerus which forms part of the joint with the radius head

car sickness
that motion sickness due to travel in a road vehicle

carbohydrate
a class of organic substances consisting of carbon, hydrogen, and oxygen and which generally comprise the sugars, starches, and celluloses

carbohydrate loading
the intake of large amounts of carbohydrates prior to a long-duration, physiologically-fatiguing event in an attempt to generate additional glycogen reserves

carbon arc lamp
an arc lamp using carbon rods

carbon dioxide (CO_2)
a colorless and odorless gaseous product of bodily metabolism or combustion between organic material and oxygen

carbon dioxide production rate ($\dot{V}_{CO_2}$)
that rate at which carbon dioxide is exhaled from the lungs

carbon monoxide (CO)
a colorless and odorless toxic gaseous product of incomplete combustion of organic materials and oxygen

carboy
a large container surrounded by a wooden frame, which acts to protect the container

carcinogen
any agent or substance capable of causing cancer or accelerating the development of an existing cancer; *adj.* carcinogenic

carcinogenesis
the beginning or origin of a cancer

cardia
that region of the superior stomach which contains the esophageal sphincter; *adj.* cardiac

cardiac [1]
pertaining to the heart

cardiac [2]
see cardia

cardiac arrhythmia
an abnormality of the heart rhythm; *syn.* arrhythmia, heart arrhythmia

cardiac index
the cardiac output per square meter of body surface area

cardiac muscle
that branched, somewhat striated muscle comprising the wall of the heart which is involved in heart contractions

cardiac output (co)
the volume of blood pumped by the left ventricle of the heart in a given period of time
    Comment: usually expressed in liters per minute

cardiac pacemaker
the sinoatrial node; an electronic device which may be implanted in the body to provide regular stimulation to the heart

cardiac reserve
that ability of the heart to increase its output above normal to meet an increased workload

cardinal planes
the three standard planes used for describing the human body in the anatomical position — sagittal, coronal/frontal, and transverse/horizontal

cardiovascular
pertaining to the heart, blood, or blood-carrying vessels

cardiovascular shock
any condition exemplified by a sudden fall in blood pressure following an injury, operation, loss of blood, or administration of anesthesia; *syn.* shock

cardiovascular system
that bodily system consisting of the heart, blood, and blood-carrying vessels

carelessness
that behavior or mental functioning which does not exhibit adequate attention or concern for the task being performed

carpal tunnel
an internal passage in the wrist between the extensor retinaculum and the carpal bones through which the median nerve, finger flexor tendons, and blood vessels pass from the arm to the hand

carpal tunnel syndrome (CTS)
a painful disability in which the manipulative or gripping abilities of the hand and fingers are reduced due to median nerve compression injury within the carpal tunnel; *see repetitive motion injury*

Comment: generally caused by repetitive wrist motions, by unusual wrist motions, and/or by the use of awkward positions requiring main-

tained strength

carpus
a collective term for the wrist bones; *adj.* carpal

Cartesian coordinate system
see rectangular coordinate system

cartilage
a tough, fibrous, non-vascular connective tissue frequently found at the articulating ends of bones or as a forming material in tubular structures in the body; *adj.* cartilaginous; *syn.* gristle

cascade method
an experimental technique for determining visual stimuli relationships in which an observer sequentially adjusts the wavelength of one of a pair of visual stimului until a minimal difference exists

cascading failure
any secondary or other failure which results from the failure of another system or component

cassette loop analysis
the selection of some videotaped task or operation with cutting and splicing or copying to form a continuous loop for repeated viewing; *see also film loop analysis*

caster
a small, either fixed or swiveling, wheel attached to the base of an object for ease of movement across a surface

catabolism
that portion of the metabolic process involving the breakdown of complex compounds by the body to produce energy; *opp.* anabolism

catalepsy
a movement disorder in which the body experiences a loss of voluntary motion and a rigidity of passively-moved parts for prolonged periods of time

cataplexy
a movement disorder characterized by rapid onset of partial or complete loss of muscle tone as a result of extremely intense emotion

cataract
a loss of transparency in the eye lens or its capsule

catastrophe
an event resulting in injury, death, and damage or destruction of relatively great proportions
Comment: often also considered relative to the scope of activities, i.e., at an individual or system level

catch-up growth
a period of rapid growth following a growth retarding event, such as a severe illness or malnutrition

catecholamine
any of a group of chemical substances consisting of a benzene ring with adjacent hydroxyl groups and an amine group on a carbon chain which may serve as a neurotransmitter and/or a hormone; *see epinephrine, norepinephrine*

cathode ray tube (CRT)
an evacuated tube containing one or more electron guns, each of which directs an electron beam onto a surface coated with phosphors to create a light spot

cationic detergent
any of a group of detergents having a quaternary ammonium salt cation with a hydrocarbon chain

causal association
having a demonstrable connection between the occurrence of some factor and an incident, where the presence of that factor will increase the probability and the absence of that factor will decrease the probability of that incident

causal factor
any one of a combination of sequential or parallel circumstances which contribute either directly or indirectly to an incident

cause-effect diagram
a graphic display of the causes linked to an effect

caution
see caution signal

caution and warning (C&W)
a system or classification for providing information to the operator or crew of a vehicle that some life- or vehicle-threatening hazardous situation exists

caution signal
a signal provided for or presented to the operator or crew of a vehicle that some hazardous condition exists or will soon exist, and that action will be required to correct the situation; *syn.* caution

caveat emptor
let the buyer beware

cavitation
the behavior (including formation, activity, and decay) of bubbles in a liquid when the instantaneous local pressure within the liquid becomes lower than the vapor pressure of the liquid

ceiling [1]
the upper limit of performance measured by a test

ceiling [2]
the upper interior surface of a large enclosed volume, such as a room

ceiling area lighting
a form of general illumination in which the ceiling comprises essentially one large luminaire

ceiling limit
the maximum concentration of hazardous substances in the working atmospheric environment above which a worker should never be exposed

cell
the basic functional unit of a living organism, having the ability (at least at some point during its lifetime) of reproduction, growth, to respond to stimuli, and to function relatively independently given a suitable nutrient and respiration medium; *adj.* cellular; *see also tissue, organ*

cellular refractory period
that time following an action potential in a neuron or muscle cell during which the cell has reduced excitability or is

incapable of normal excitation; *syn.* refractory period

Celsius (C)

an empirical temperature scale having the 0° point at the freezing point of water under standard conditions and the boiling point at 100°; *syn.* Centigrade

> *Comment:* related to the absolute (Kelvin) scale by $T_C = T_K - 273.15$

Celsius degree (°C)

a division of the Celsius temperature scale which divides the range between the freezing and boiling points of water into 100 equal intervals; *syn.* Centigrade degree

center frequency

the geometric mean of a frequency band

center of gravity

a point representing a body or system at which the force due to a uniform gravitational attraction acts; *see also* *center of mass*

> *Comment:* center of mass and center of gravity can normally be assumed to be the same point for human factors work

center of mass

that point of an object or system which may be treated as if the entire mass of the object or system were concentrated at that point, and any external translational forces appear to act through that point; *see also* *center of gravity*

> *Comment:* center of mass and center of gravity can normally be assumed to be the same point for human factors work

center of rotation

that point about which a rotational movement occurs

centi-

(prefix) one one-hundredth or 10^{-2} of a base unit

Centigrade (C)

see *Celsius*

Centigrade degree

see *Celsius degree*

centile point

a point within a centile scale

centile rank

that position or score based on a centile scale; *see also* *percentile*

centile scale

a dispersion scale having a range of 100 in which each point represents one per cent of the population along some dimension

Centimeter-Gram-Second system (CGS)

that measurement subset of the metric system which uses the centimeter, gram, and second as its basic units; *syn.* CGS system

central blindness [1]

see *foveal blindness*

central blindness [2]

the lack of visual function due to optic nerve or visual cortex damage

central deafness

see *central hearing loss*

central hearing loss

a hearing impairment or deafness due to auditory nerve or auditory cortex damage; *syn.* central deafness

central nervous system (CNS)

that portion of the nervous system comprising the brain and spinal cord; *opp.* peripheral nervous system

central tendency

having a typical, average, or expected value within a frequency distribution

> *Comment:* a finer characterization of data beyond the distribution

central vision

see *foveal vision*

central visual field

that portion of the visual field which falls on the foveal or macula lutea portion of the retina; *opp.* peripheral visual field

central visual field blindness

see *foveal blindness*

centrifugal force

that outwardly directed radial force in a rotating reference frame; *opp.* centripetal force

centripetal force
that radial force directed toward the center of rotation of an object which keeps an object moving in a circular path; *opp.* centrifugal force

cerebellum
a lobed, heavily-fissured structure above and posterior to the brainstem which is concerned largely with muscular control, equilibrium, and coordination; *adj.* cerebellar

cerebral cortex
that superior and lateral portion of the brain generally overlying the thalamus, hypothalamus, and brainstem and consisting primarily of the frontal, parietal, occipital, and temporal lobes

cerebral palsy
any impairment of motor, perceptual, or behavioral functions dating from birth or infancy without worsening of symptoms

cerebrospinal fluid (CSF)
a clear, colorless fluid which shields the central nervous system from mechanical shock, distributes nutrients, and removes waste materials as it circulates

cerebrum
the telencephalon or most superior portion of the brain in the central nervous system, consisting primarily of the two large hemispherical bodies of neural tissue; *adj.* cerebral

Certified Safety Professional (CSP)
one certified by the Board of Certified Safety Professionals as being competent in his field and having professional integrity

cerumen
the waxy secretion from cells lining the external auditory canal

cervical
see cervix

cervical spine
that portion of the spinal column consisting of the seven cervical vertebrae in the neck

cervicale
the protruding tip of the 7th cervical vertebra at the base of the neck; *syn.* nuchale tubercle

cervicale height
the vertical linear distance from the floor to cervicale
Comment: measured with the individual standing erect

cervicale height, sitting
the vertical linear distance from the upper sitting surface to cervicale
Comment: measured with the individual sitting erect

cervix
a constricted structure in the body, typically referring to the neck or the inferior part of the uterus; *adj.* cervical

CGS system
see Centimeter-Gram-Second system

chafe
irritate the skin through friction

chair
a class of objects consisting normally of a seat, backrest, and having legs or other structure to raise the seat above a floor reference level

chair depth
see seat depth

chamfer
the bevel at the edge of an object

chance variable
see random variable

change direction (CD)
a basic physical work element characterized by a slight hesitation as the hand alters course while in motion

changeover
the process of modifying or replacing an existing workstation, workplace, or other facility, including the setup and teardown

changeover allowance
a special time allowance given a worker to compensate for the changeover time; *see also setup allowance, teardown allowance*

changeover time
that temporal period required to effect a changeover

channel capacity
the maximum rate at which information can be received, transmitted, or processed at a given point, for either the human or instrumentation

character height
the vertical distance assigned to or occupied by a character on a display

character width
the horizontal distance on a line of text from one point of one character to a corresponding point of the next character

characteristics of easy movement
see *motion efficiency principles*

chart
any form of graphical or tabular data which provides information about one or more variables or activities

chart recorder
see *oscillograph, kymograph, polygraph*

check
a mental skill involving the comparison of a finished product with what was planned to see if the goals were met or standards achieved

check study
a timing review of a job to evaluate the appropriateness of the standard time for that job

check time
the time period between the start time of a time study and the beginning of the first work element observed or between the completion of the last element and the stop time of the study

checkoff
the withholding of union dues from a worker's paycheck by agreement

cheek [1]
the tissues comprising the side of the face from the zygomatic bone to the mandible

cheek [2]
(sl) a buttock

cheekbone
see *zygomatic bone*

cheilion
the lateral corner of the mouth opening formed by the junction of the lips

chemical burn
the tissue damage or destruction directly as a result of chemical exposure

chemical element
the smallest substance into which some physical or chemical entity can be chemically divided and still retain its chemical properties; *syn.* element

Chemical Hygiene Plan (CHP)
part of a laboratory safety plan which must be established by labs handling hazardous chemicals due to a set of requirements dictated by OSHA

chemoreceptor
a portion of a large protein or other cellular molecule which has the three-dimensional capacity for accepting and/or binding to a specific chemical substance

chemotaxis
the response of an individual toward a chemical stimulus

chest
the thorax

chest breadth
the horizontal linear width of the torso without tissue compression at the nipple level (males) and at the level where the fourth rib meets the sternum (females)
Comment: measured with the individual standing erect with the arms hanging naturally at the sides, and breathing normally

chest breadth to bone
the horizontal linear width of the torso at the nipple level with tissue compression
Comment: measured with the individual standing erect, and breathing normally

chest circumference

the surface distance around the torso at nipple level; *syn.* chest/bust circumference

Comment: measured with the individual standing erect, breathing normally, and with the arms slightly abducted

chest circumference at scye

the surface distance around the torso at the level of the axillary folds

Comment: measured with the individual standing erect

chest circumference below bust

(females only) the surface distance around the chest just below the cups of the bra

Comment: measured with the individual standing erect and breathing normally

chest depth

the anterior-posterior horizontal linear depth of the torso measured at the nipple level (males) and above the breasts at the level where the fourth rib joins the sternum (females); *syn.* chest/bust depth

Comment: measured with the individual standing erect, the arms hanging naturally at the sides, and breathing normally

chest depth at scye

the anterior-to-posterior horizontal linear depth of the torso measured at the scye level

Comment: measured with the individual standing erect and breathing normally

chest depth below bust

the transverse depth of the chest at the level of the inferior margin of the xiphoid process

Comment: measured with the individual standing erect and breathing normally

chest height

the vertical distance from the floor to the center of the nipples (male) or point of the bra (female); *syn.* bustpoint height

(females)

Comment: measured with the individual standing erect and his weight balanced on both feet

chest skinfold

*see **pectoral skinfold***

chi square (χ^2)

a statistical test using differences in frequency data, especially for small samples, based on obtained vs. theoretical/expected frequency counts, to determine significance; also chi square test

$$\chi^2 = \sum_{i=1}^{n} \frac{(f_o - f_t)^2}{f_t}$$

where:

f_o = observed frequency
f_t = theoretical or expected frequency
n = sample size

chi square distribution

a mathematical or graphical function for chi square, having the probability distribution function

$$f(\chi^2) = G_v(\chi^2)^{\frac{(v-2)}{2}} e^{-\frac{\chi^2}{2}}$$

where:

v = degrees of freedom
G_v = a constant for a given v

Comment: shape of the distribution varies with degrees of freedom; approaches the normal distribution as degrees of freedom increases

chi square test

*see **chi square***

child

an individual younger than the age of puberty

chime

turn a cylindrical container on the edge of its base to assist in moving it from one location to another

chin

the anterior lower part of the jaw, including the anterior lower portion of the mandible and all surrounding tissues

chin prominence to wall

the horizontal distance from the wall to the most anterior protrusion of the chin

Comment: measured with the individual standing erect with his back and head against the wall, facing straight ahead

chinstrap

any thin flexible, strong material or device which is attached to headgear and can be passed underneath the chin for aiding in headgear retention

chloracne

an acne condition due to chlorinated hydrocarbon exposure

choice reaction time (CRT)

that temporal interval measured for an individual or group after the presentation onset of one or a group of stimuli to decide which of more than one possible responses is appropriate and initiate that response

Comment: generally represents an average time over several trials

chokes

a form of decompression sickness in which a choking sensation, difficult breathing, and/or substernal pain are experienced due to air bubbles in the lungs

cholesterol

a member of the sterol class of steroids produced naturally in the bodies of animals

chorea

a movement disorder in which a series of complex, involuntary writhing movements are made, generally involving distal extremities and/or the face, tongue, and swallowing muscles; *see Huntington's chorea, Sydenham's chorea*

choreologist

one who has been trained and is competent to record human movement in some system of notation

chroma

that apparent degree to which a color compares to a similarly illuminated white or achromatic reference; *see also Munsell chroma*

chromatic

having a hue; colored; pertaining to any color except white, black, or gray; *opp.* achromatic

chromatic aberration

an image containing colored fringes around the border, resulting from unequal refraction of light of different wavelengths causing focusing at different points in an optical lens system

chromatic adaptation

that modification of the color sensory properties of the visual system by observing colored stimuli

chromatic audition

see chromatism

chromatic color

any color except white, gray, or black

chromatic contrast

that apparent contrast due to the presence of differing adjacent hues or colors; *syn.* color contrast, hue contrast, simultaneous color contrast; *see also luminance contrast*

chromatic contrast threshold

that minimal difference in the combined aspects of luminance and chromaticity which is detectable for a given pair of adjacent stimuli; *syn.* color contrast threshold

chromatic vision

see color vision

chromaticity

a measure of the quality of colored light, defined either by its chromaticity coordinates or by its dominant wavelength and excitation purity

Comment: luminance or brightness is not involved

chromaticity coordinate

any of a set of numbers representing the proportions of two of the three normalized primary colors, usually x and y, required to produce a given color, with the brightness variable eliminated; *syn.* trichromatic coefficients, CIE chromaticity coordinates

$$x = \frac{X}{X+Y+Z} \quad y = \frac{Y}{X+Y+Z} \quad z = \frac{Z}{X+Y+Z}$$

Comment: represented as x, y, z in the CIE system

chromaticity diagram
a planar diagram based on the CIE color system and produced by using two of the chromaticity coordinates as axes in a rectangular coordinate system; *syn.* chromatic diagram

chromaticness
a visual attribute in which a perceived color appears more or less chromatic

chromatism
sensing an image of color when stimulated by a sensory modality other than vision; *syn.* chromatic audition; *see also* **synesthesia**

chrominance
the coloring power of a stimulus

chromostereopsis
see **color pseudo-stereopsis**

chronic
having a long duration; *opp.* acute

chronic oxygen poisoning
see **chronic oxygen toxicity**

chronic oxygen toxicity
a lung disorder due to breathing higher than normal oxygen partial pressures at normal barometric pressure for 24 hours or more and characterized by chest pain, pulmonary edema, and possibly damage to the alveoli and bronchi; *syn.* chronic oxygen poisoning; *see also* **acute oxygen toxicity**

chronobiology
the study of the effects of time on varying biological systems, including psychobiological rhythms

chronocyclegram
see **chronocyclegraph**; *also* chronocyclogram

chronocyclegraph
the single negative or photograph from

a chronocyclegraphic measurement; also chronocyclograph

chronocyclegraphy
the use of a motion tracking system comprised of (a) one or more small electric light bulbs which flash at known, regular intervals and are attached to the fingers or other body part and (b) a still camera, ideally using a stereoscopic camera to obtain three dimensional data, for recording motions on a single negative or print to determine velocities and accelerations of the body parts; also chronocyclography; *syn.* chrono-cyclegraph technique; *see also* **cyclegraphy**
Comment: typically the subject is in a darkened area; the exposure time is greater than or equal to one motion cycle

chronograph
a constant-speed recording device which marks a paper or tape at known intervals so that timing during an ongoing process can be determined; *syn.* marstochron, marstograph

chronological age (CA)
the age as of the previous birthday or the age as of the previous birthday plus 0.5 years; *syn.* age, physical age

chronological study
the observation and recording of events or data in the order in which they occur over time

CIE color rendering index (CRI)
a measure of the amount of color shift which an object appears to present when illuminated by one source compared to that of a reference source having a similar color temperature

CIE color system
a standard color reference system established in 1931 by the Commission Internationale de l'Eclairage based on the technique of flicker photometry and using a chromaticity diagram to specify color coordinates
Comment: generally the world standard

CIE Standard Observer
a table representing an observer having normal color vision which is developed from experimental data in color-matching using the primary colors with a 2° field of view; *syn.* standard observer, 2° observer; *see also CIE Supplementary Standard Observer*

CIE Standard Observer response curve
see spectral luminous efficiency function

CIE Supplementary Standard Observer
a variant of the CIE Standard Observer which accommodates a 10° field of view and better permits judgment of color matching in the shorter visible wavelengths (blue, violet); *syn.* 10° observer
Comment: adopted in 1964

ciliary muscle
an intrinsic smooth muscle of the eye, which is involved in lens accommodation

cinema verité
the use of only naturally available, not additional photographic, lighting for photography or videography

cinematography
motion picture photography

circadian
having a period of about 24 hours

circadian pacemaker
an internal timing mechanism which maintains circadian rhythms; *see also internal clock*

circadian rhythm
a biological rhythm with a period of about 24 hours; *see also biological rhythm*

circannual rhythm
a biological rhythm with a period of about one year

circulation ¹
the movement of blood through the circulatory system

circulation ²
the movement of people, information, supplies, equipment, or other items within a building or other structure

where work is being accomplished

circumduction
a basic type of joint motion occurring in those joints capable of three-dimensional movement in which the proximal end of a bone in its socket provides the apex of a cone and the distal end of that bone moves in a circular pattern, sweeping out a conical volume

circumference ¹
a curved, closed anthropometric measurement that follows a body contour; *see also arc*
Comment: need not be circular

circumference ²
the length comprising the perimeter of a circle

Civil Rights Act
an act designed to prohibit public accommodations and employment discrimination due to a person's color, race, religion, sex, or national origin

clarifying agent
any substance used to remove turbidity from drinks

clarifying lotion
a substance for removing oil and grease from the face

clash point
a point at which the human body or reach envelope, whether physically or in computer modeling, intersects some equipment, instrumentation, or workplace boundaries in a workplace

classical anthropometry
the measurement of various static body girths and lengths with measurement devices such as a simple tape measure, anthropometer, and calipers; *syn.* conventional anthropometry, traditional anthropometry

classical conditioning
a type of learning in which an initially neutral stimulus is paired with a natural stimulus and response such that after some number of trials the neutral stimulus will elicit the natural response

classification
a system for arranging entities into

groups based on one or more characteristics

clavicle
the bone which connects the sternum and the scapula; *syn.* collarbone

clean [1]
(v) remove dirt, impurities, or other undesired entities

clean [2]
(adj) pertaining to a condition in which specified or implied standards are met for cleanliness

clean room
an enclosed region in which extensive precautions are taken to minimize dust particles and other airborne contaminants

cleaning allowance
that paid time given an employee for personal hygiene required due to the working environment and for workspace and tool cleaning; *syn.* cleanup time

cleanup time
see **cleaning allowance**

clear [1]
(n) a function which removes the current selection from the display

clear [2]
(v) remove any turbidity from a fluid

cleavage line
any line of tension in the skin along which a tear will tend to occur from a penetrating object, producing a slit rather than a rounded opening; *syn.* Langer's line

Clerical Task Inventory (CTI)
a compilation of over 100 clerical or office-type tasks for job evaluation or wage determination purposes

click
press and release a button on an input device such as a mouse or trackball to provide a command or other input to a computer

clipboard
a temporary computer editing buffer which is independent of, but able to interface with, other system applications; *syn.* temporary editing buffer

clitoris
a touch-sensitive female erectile genital structure located anterior to the external urethral orifice, representing the female homologue of the male penis; *adj.* clitoral

clo
a unit for the thermal insulation provided by clothing, not counting the approximately 25% for heat loss via the respiratory system and passive diffusion through the skin; that amount of insulation needed in a sitting and resting average individual to be thermally comfortable in a normally ventilated room (approximately 10 cm/sec air velocity, 21°C temperature, and 50% relative humidity)

clockwise rotating shift
pertaining to a rotating shift work schedule in which the shift worked is periodically delayed by increments, i.e. from the first shift to the second, or from the second to the third; *opp.* counterclockwise rotating shift

Close View
a feature which permits enlargement of the display characters for easier reading by visually-impaired individuals

closed respiration system
a self-contained breathing gas system which provides a continuing and proper oxygen/nitrogen supply ratio and pressure for its personnel, with removal of carbon dioxide and excess water vapor; *syn.* closed respiratory gas system

closed shop
a workplace or organization permitting employment only to union members; *opp.* open shop

closed window
a display window not accessible to the user without taking some specific action to gain access; *opp.* open window

closed-loop system
any type of system in which the output

or some derivative of the output from the system is directed back into the system itself; also closed loop; *syn.* feedback control loop; *opp.* open-loop system

closure

the perception of a series of pattern elements as a single unit, rather than unrelated parts

clothes changing allowance

any work time for which an employee is paid due to a requirement for removing one clothing assembly and donning another; *syn.* clothes changing time

clothes changing time

see *clothes changing allowance*

clothing

any tailored or processed material or combination of materials which may be used to cover the body or its parts, for whatever purposes

clothing area factor (f_{cl})

that proportion of increased surface area over the nude body which is added by clothing

$$f_{cl} = \frac{clothed\ body\ surface\ area}{nude\ body\ surface\ area}$$

clothing assembly

see *clothing system*

clothing ensemble

see *clothing system*

clothing fastener

any device, mechanism, or system for attaching different articles of clothing or portions of a single piece of clothing together

clothing insulation value

see *thermal insulation value of clothing*

clothing system

the combination of garments and their arrangement being worn on the body at any one time; *syn.* clothing assembly, clothing ensemble

cluster workstation

a multi-person workstation built around a central core to provide some separation from co-workers

co-partnership incentive plan

an incentive plan in which workers have the opportunity to own a share of the business enterprise, thus obtaining some portion of the profits resulting from that ownership

coccyx

a triangular-shaped bone at the base of the spine formed by the fusion of the lowest five vertebrae

cochlea

that portion of the inner ear, resembling a snail's shell, which is involved in the transduction of mechanical sound energy to neural energy for hearing; *adj.* cochlear

cochlear duct

a tube-shaped structure within the cochlea which is filled with endolymph and contains the organ of Corti and the tectorial membrane

cockpit

the location within a vehicle from which control of the vehicle and observation of the external environment and events may occur

code [1]

(v) translate information or data from one form or symbol to another form or symbol which has meaning in its own context

code [2]

(n) a set of mandatory standards or regulations adopted by a local, national, or international governmental agency which have the force and effect of law; a set of recommended rules or guidelines within an industry

code [3]

(n) a sequence of steps in some process, such as a computer program or task

code [4]

(n) a system of symbols which can be used to organize and/or communicate information about conditions, processes, or entities

Code of Federal Regulations (CFR)

an organized presentation of rules which have been published in the *Federal Register* by the various agencies and executive departments of the U.S. government and which deal with areas regulated by the government

coefficient alpha

a measure of the internal consistency/reliability of a scale

coefficient of alienation (k)

a measure of the lack of relationship between two variables

$$k = \sqrt{1 - r^2}$$

where:

r = correlation coefficient

coefficient of concordance

see *Kendall's coefficient of concordance*

coefficient of correlation

see *correlation coefficient*

coefficient of determination (r^2, d)

the proportion of the variance accounted for by the Pearson product-moment correlation coefficient; equal to the Pearson product-moment correlation coefficient squared; *syn.* generality; *opp.* coefficient of non-determination

coefficient of evaporative heat transfer

see *evaporative heat transfer coefficient*

coefficient of friction (μ)

see *coefficient of rolling friction, coefficient of sliding friction, coefficient of kinetic friction, coefficient of static friction*; syn. friction coefficient

coefficient of kinetic friction

see *coefficient of rolling friction, coefficient of sliding friction*

coefficient of multiple correlation

see *multiple correlation coefficient*

coefficient of non-determination (k^2)

that proportion of the variance between two variables not accounted for by the coefficient of determination

coefficient of reflection

see *reflection coefficient*

coefficient of reliability

see *reliability coefficient*

coefficient of rolling friction

the ratio of the magnitude of the rolling force to the magnitude of the perpendicular force between two objects/surfaces at the point where their surfaces are parallel

coefficient of sliding friction (μ)

the ratio of the magnitude of the sliding force to the magnitude of the perpendicular force between two objects/surfaces

coefficient of static friction (μ)

the ratio of the magnitude of the static force to the magnitude of the perpendicular force between two objects/surfaces

coefficient of utilization (CU)

the value of the ratio of the luminous flux reaching the workplace to the total luminous flux emitted from a lighting source

$$CU = \frac{lumens\ reaching\ work\ surface}{lumens\ emitted\ by\ lamp}$$

coefficient of variation

a measure of the relative dispersion of the data compared to the mean, typically expressed as a percentage

$$CV = \frac{sd}{\overline{X}} \times 100$$

where:

sd = standard deviation of the sample

$\overline{X}$ = sample mean

coffee break

see *rest period*

cognition

those higher mental activities or intellectual function; *adj.* cognitive

cognitive disability

any disability involving literacy, mental capacity, learning, non-motor speech processes, or perceptual processes

cognitive dissonance

a discrepancy which exists between a

person's attitudes or statements and behaviors

cognitive reaction time
that temporal interval between the receipt of a stimulus and the initiation of a response in a task which requires some type of choice and which is presumed to involve cognitive processing; *syn.* decision time

cognitive restructuring
a mental exercise in which attempts are made by an individual to change certain personal beliefs

coherence (γ^2)
a measure of the correlation at each frequency, or within each frequency band, between two time series signals

$$\gamma_{xy}^2 = \frac{\left| S_{xy} \right|^2}{S_x S_y}$$

where:
S_{xy} = cross-spectral magnitude
S_x = power spectral density function for time series 1
S_y = power spectral density function for time series 2
Comment: ranges from 0.0 to 1.0

cohort
any two or more individuals having some characteristic(s) in common

coitus
the act of human sexual intercourse

cold flow
creep at room temperature

cold start fluorescent lamp
see *instant start fluorescent lamp*

cold stress
a form of environmental/thermal stress in which too much body heat is lost to a cold environment; *opp.* heat stress

collarbone
the clavicle

collective bargaining
a process in which labor and management establish operational rules and working conditions and define the compensation and benefits for employees

collimate
make parallel to a certain path; *n.* collimation

collimating optics
those optical components, such as lenses, used to produce parallel rays of light

collusion
that customer or user behavior in which one person waits in line for himself and one or more others

color
that aspect of visual perception due solely to stimulation of the retinal cones by different wavelengths of electromagnetic radiation within the visible spectrum, and neglecting such aspects of a stimulus such as structure, size, and pattern

color assimilation
a type of chromatic induction in which the difference or contrast between adjacent, differently-colored fields diminishes; *syn.* Bezold spreading effect

color blindness
having an inability to detect one or more of the primary colors; *adj.* color blind; *see monochromasia, protanopia, deuteranopia, tritanopia, monochromat, dichromat; see also color vision deficiency*

color coding
the use of multiple colors for easier, more rapid visual identification, access, and/or processing of groups of organized materials

color constancy
the phenomenon in which an object appears to have approximately the same color under different lighting conditions

color contrast
see *chromatic contrast*

color contrast threshold
see *chromatic contrast threshold*

color correction
an adjustment made for the presenta-

tion of a color image, usually to make perceived colors appear more natural

color deficiency
see *color vision deficiency*; *adj.* color deficient

color discrimination
the ability to perceive visual matches or note differences between hues, saturations, and brightnesses of two or more colored stimuli; *syn.* visual color discrimination

color formulation
the use of any one or combination of methods for making a desired color, including mathematical models, materials, colorants, particle sizes, absorption coefficients, and scattering coefficients

color grade
a measure of the color appearance of a product, which may be used to determine price or quality

Color Index
a publication by the American Association of Textile Chemists and Colorists which provides a large number of reference dyes and pigments

color match
(n) a condition in which two colored stimuli appear identical

color match
(v) make one variable color stimulus appear the same as a reference stimulus

color matching function
see *tristimulus value*

color mixing
the blending of colored lights or materials to alter an existing color; see *additive color mixing*, *subtractive color mixing*

color ordering system
any method for the unambiguous interpolation between closely-related colors within a large set of colors; *syn.* color system; see *CIE color system*, *Munsell color system*, *Federal Standard 595a*, *Coloroid color system*, *In-ter-Society Color Council – National Bureau of Standards color system*, *Natural Color System*

color pseudo-stereopsis
the visual perception of depth or of structure being out of the background plane from objects emitting or reflecting different dominant frequencies/wavelengths (especially blues and reds) within a dark background; *syn.* chromostereopsis

color rendering
that effect which a light source other than a standard illuminant has on the apparent color of an object

color rendering index
see *CIE color rendering index*

color saturation
a perceptual attribute pertaining to the strength or vividness of a particular hue; *syn.* saturation

color system
see *color ordering system*

color temperature (T_c)
the temperature (in Kelvin) of a radiating blackbody having the same chromaticity or spectral distribution as a given color light source; see *temperature color scale*

color temperature scale
a scale by which the color of light emitted from an incandescent source is related to temperature, normally corresponding to the Kelvin scale; *syn.* temperature color scale

color vision
see *photopic vision*

color vision deficiency
having some form of reduced color sensation ability; *adj.* color vision deficient; see *protanomaly*, *deuteranomaly*, *tritanomaly*; see *also* color blindness

color wheel
a disk consisting of multiple colored and appropriately interleaved radial segments, each segment being a single color, for providing a desired perceptual color mixture when spun rapidly

colorant
any substance added to a product to provide a different color; *syn.* coloring agent

Colorcurve®
a color ordering system based on the physical brightness of gray levels

colorfulness
that attribute of a visual sensation which appears to exhibit more or less of its hue

colorimeter
an instrument for measuring color, usually by finding the intensities of the three primary colors whose combination will give the color being analyzed

colorimetry
the study, measurement, specification, and use of color; *adj.* colorimetric
Comment: usually involving a numerical specification via a colorimeter or spectrophotometer

coloring agent
see **colorant**

colorless
having no chromatic color, achromatic

coloroid color system
a color ordering system which attempts to provide equal aesthetic spacing between colors and is based on specifications of hue (A), chromatic content (T), lightness (V)

column
a vertical arrangement of numbers, text, or other information in a matrix or table; *opp.* row

coma
a state of unconsciousness in which there is no objective response to stimulation

COMBIMAN
see **Computer Biomechanical Man model**

combination tone
an apparent secondary tone heard when two pure primary tones having widely separated frequencies are presented simultaneously

combined motions
two or more parallel elemental movements performed by a given body segment

combined work
a job or task involving any combination of two or more workers or workers and multiple machines

combustible
pertaining to any substance which is capable of catching fire and burning; *see also* **flammable**

combustible liquid
any liquid with a flash point greater than 100°F

comfort
a state of subjective well-being in relation to one's external environment; the absence of significant or excessive physical and/or mental stressors; *adj.* comfortable; *opp.* discomfort

comfort rating
an expressed measure of the level of satisfaction with one or more aspects of an individual's current environment

comfort rating scale
any of a number of ranking techniques for rating comfort

comfort-discomfort boundary
that threshold luminance in a glare condition at which visual discomfort becomes apparent; *syn.* borderline between comfort and discomfort

Comfort-Health Index (CHI)
a table based on a computed effective temperature, assuming a 50% relative humidity, for determining expected thermal sensations; *syn.* ASHRAE Comfort-Health Index

comfortable reach
that range through which an individual can reach without straining excessively against gravity or a restraint

command
any statement which may potentially be input to a computer system and which calls for one or more specific actions

command area
a region within a display in which user-input commands are presented for viewing

command error
an inappropriate or incorrect command entered into a computer

command input
the entering of a command to a system

command language
a clearly-defined specific set of terms for directing control of a computer

command line
a command area composed of a single line height on a text display which is reserved for user-entered commands

comminuted
broken into small pieces, as a type of bone fracture

commission
an incentive plan which represents an award to the employee of some specified portion of the selling price of some service or product

communication
the meaningful interchange using some form of language or other set of signals between individuals, groups, or instrumentation

compact bone
the dense outer tissue portion of a bone

comparison group
see *control group*

comparison stimulus
any variable stimulus which is presented in addition to a reference stimulus in certain experimental designs for determining difference thresholds

compass [1]
a sliding caliper

compass [2]
a magnetic sensing device used in navigation for determining one's heading relative to magnetic north

compatibility [1]
a measure of how well spatial movements of controls, display behavior, or conceptual relationships meet human expectations

compatibility [2]
that combination of characteristics which permit two or more individuals, groups, or systems to work together without significant interference or conflict

compensation [1]
a movement of a part of the body to restore or maintain equilibrium as another body part moves

compensation [2]
any behavior which attempts to minimize the effect of a weakness in one process by relying on another, stronger process or improving another process; *adj.* compensatory

compensation [3]
a form of payment for work performed or for a disability incurred

compensation policy
that rule or set of rules which an organization follows in setting payment rates for jobs or type of work done

compensatory
pertaining to the use of error information only in generating control inputs

compensatory damages
that monetary value awarded a victim by a court to pay for his injuries or losses; *see also* **punitive damages**

competition
that condition in which more than one person or group vie against each other for a limited number of prizes, positions, market share, or other reward

compilation
the development of higher order skills from lower level processes

complementary color
that perceptual color on the opposite side of the achromatic point in the chromaticity diagram from a given color, which, when mixed in proper proportions, will produce a gray or white; see also **complementary wavelength**

complementary wavelength (λ_c)
that wavelength designated on the spectrum locus of a chromaticity diagram by an extension of the line determining the dominant wavelength in the opposite direction from the achromatic point; *see also* **complementary color**

complete diffusion
a condition in which a diffusing medium so scatters the incident flux that no image can be formed from the transmitted flux

complete menu hierarchy
a menu hierarchy having the same number of menus along each branch from top to bottom

complex reaction time
the temporal interval required to react to a stimulus situation when a choice or discrimination needs to be made before responding

complex sound
any sound composed of a large number of multiple sinusoidal components and their harmonics/overtones

complex spectrum
those coefficients resulting from a Fourier or other transform of a time series signal which contain both real and complex values

complex tone
an auditory signal composed of multiple simple sinusoidal components with different frequencies

complexion
the color and overall appearance of facial skin

compliance [1]
the degree of match between the manipulatory requirement of a teleoperator task and movement capabilities of the teleoperator

compliance [2]
acting in accordance with some rule or regulation

compliance [3] (C_m)
a measure of the softness of a system or structure, represented by the reciprocal of the stiffness

composite maintenance
the integration or simultaneous use of several types of maintenance

compound
any chemical substance formed by the combination of two or more different chemical elements

compound fracture
a broken bone in which at least one of the ends protrudes through the skin surface; *syn.* open fracture

compress
reduce the volume of a substance or material, or the duration of some event; *n.* compression; *adj.* compressive

compressed seat height
the height of a cushioned chair seat pan from the floor or other reference surface when an individual is seated in it

compressed spectral array (CSA)
a three-dimensional display or hardcopy of a sequential series of spectra as a function of time, with time being the depth axis

compressed workweek (CWW)
a work schedule in which employees provide approximately 40 hours of work in less than five days

compressibility
that property of a tissue or other soft, loose material to be locally depressed or of a gas to be reduced in volume when external pressure is applied; *adj.* compressible

computer anxiety
a state of apprehension or fear when required to interact with a computer, which is out of proportion to any reasonable danger posed by the computer

computer anxiety scale
a survey consisting of 10 test items dealing with feelings about computers, on which an individual judges a rank for each item according to a Likert scale; *syn.* Raub scale

Computer Assessment of Reach (CAR)
a crew station modeling program which attempts to determine what percentage of aircrew will be able to function in a given design

computer assisted tomography (CAT)
see computerized axial tomography

Computer Biomechanical Man (COMBIMAN)
a three-dimensional, interactive computer graphics modeling software package capable of enfleshment which can be used in the physical evaluation of pilots and other aircrew members for crew station design, including sizing, reach, strength, and visual field

computer graphics
the input, processing, or output of any pictorial or graphical data displayed on a computer monitor or hardcopy

computer input device
any type of hardware tool which can be used by an individual to get text, graphics, commands, or data into a computer

computer model
any numerical or graphical representation of objects, systems, or processes using a computer

computer vision
the integration of one or more video cameras with appropriate software into a computer processing system for any purpose, such as electronic scene comparison, to simulate human vision for mobile robots, or other uses

computer-aided instruction (CAI)
the use of computers and displays for presenting information to be learned; *syn.* computer-assisted instruction

computer-aided manufacturing
see computer-integrated manufacturing

computer-assisted instruction
see computer-aided instruction

computer-human interface (CHI)
see human-computer interface

computer-integrated manufacturing (CIM)
the use of computers in the actual manufacturing process; *syn.* computer-aided manufacturing

Computerized Accommodated Percentage Evaluation (CAPE)
a modeling tool for determining what percentage of the aircrew population could function satisfactorily in a given crew station design
Comment: an old model, no longer used

computerized axial tomography (CAT)
the use of computers for control, acquisition, storage, processing, and display of a series of single planes of X-ray images along the longitudinal axis of the body or other X-ray transparent objects; *syn.* computer assisted tomography

Computerized Relationship Layout Planning (CORELAP)
a computer model for developing a plant layout based on relationships when large numbers of groups are involved

Computerized Relative Allocation of Facilities Technique (CRAFT)
a computer model for improving a plant layout, with the priority of minimizing transportation costs

concave function
a mathematical relationship or graph which has a negative second derivative during the interval of interest, resulting in an inverted U-shaped curve; *opp.* convex function

concentric action
a dynamic muscle action which involves active muscle shortening against a resistance; *syn.* concentric contraction, concentric muscle contraction; *opp.* eccentric action

concentric contraction
see concentric action

concentric muscle contraction
see concentric action

concept

an abstract idea or notion which enables an individual to generalize from known specific examples of some classification to examples not previously encountered; *adj.* conceptual

concept hierarchy

an organization in which the most general aspects of a concept are located at the top, with subsidiary aspects branching beneath

concept hierarchy analysis

the examination of a concept hierarchy to determine if a better arrangement can be made or to compare with related structures

concept trainer

a training aid used when the principles to be learned are too complex to be easily understood from verbal descriptions or when simulation with actual physical objects appears to be the optimum method

concordance coefficient

see coefficient of concordance

concurrent loading

a test or working condition in which an individual is required to perform both a fatiguing exercise and a criterion task simultaneously

concurrent validity

having a high correlation between job incumbent test scores and performance on the job

concussion

a clinical condition caused by a sudden, strong mechanical force applied to the head and characterized by temporary impairment of neural function such as an alteration in consciousness or disturbances of vision, equilibrium, and/or reflexes

condiment

any flavoring added to food to improve taste or increase stimulation of the taste buds (such as spice, salt, etc.), or an item having such effect (such as gum or a mint)

conditional cues

any information displayed which provides the user with a brief indicator of the current operating rules or conditions

conditioned reflex

a learned response to a stimulus which did not originally cause that response

conditioning

any physical/mental activity or training which prepares an individual for a given task

conduction deafness

see conductive hearing loss

conductive deafness

see conductive hearing loss

conductive hearing loss

a hearing impairment due to an inability to mechanically transmit sound waves through the tympanic membrane and/or middle ear; *syn.* conduction deafness, conductive deafness, conduction hearing loss

conductive heat loss

that amount of heat eliminated from the body via heat conduction, indicated by an equation of the form

$$H = kA \frac{\Delta T}{\Delta x}$$

where:

H   =   heat loss
k   =   thermal conductivity coefficient
A   =   body surface area in contact with another object
$\frac{\Delta T}{\Delta x}$ = temperature gradient

condyle

a rounded projection on a bone surface, often associated with a joint; *adj.* condylar

cone

an approximately conical sensory structure concentrated in the foveal region of the retina which provides for fine visual acuity and the detection of color under high luminance levels

cone monochromatism

a condition in which an individual has

only a single type of retinal cone, thus seeing only one color, while having normal color brightness discrimination

confidence interval
a certain range of values within which there is a specified probability that a given value will lie

confidence limits
the boundaries of a specified confidence interval for a given mean and standard deviation

configuration control
a design or procedure for the controlled development, operation, and maintenance of a system

confined space
a region which has a limited number and size of openings, poor natural ventilation, or was not designed for continuous human occupancy

conflict
a state resulting from an individual having incompatible desires or two or more individuals or groups having different goals or means to a goal

confounding variable
a variable which is uncontrolled and which has, or is likely to have, some effect in an experiment

congenital abnormality
any defect in the structure or function of an individual existing before or at birth

congestive hypoxia
a form of hypokinetic hypoxia in which venous blood flow is reduced

conjunctiva
a thin membrane covering the interior eyelid and anterior portion of the eyeball

conjunctivitis
an inflammation of the conjunctiva, from any cause; *syn.* pinkeye
Comment: may be due to bacterial infection, air pollution, excessive glare, ultraviolet light, wind

connected word recognition
a capability in which a phrase or a sequence of a few meaningfully-connected words may be understood by an artificial system

connective tissue
any bodily tissue having the primary function of supporting and restraining other tissues or organs

consciousness
an awareness of one's external environment

consensual standard
see consensus standard

consensus standard
an agreement by various interested and/or affected organizations or individuals on a requirement for certain specifications to be met in the design of equipment when those specifications have not been empirically determined to be appropriate; *syn.* consensual standard

consequation
any aspect of the environment which changes the behavior of an individual encountering it

conservation of angular momentum
the principle that the angular momentum of an object will remain unchanged unless the object is acted on by a net torque

conservation of linear momentum
the principle that the linear momentum of an object will remain unchanged unless the object is acted on by a net force

conservation of momentum
see conservation of angular momentum, conservation of linear momentum

consistency
a level of performance which repeatedly falls within certain specified limits

constant
a fixed numerical value, or a symbol representing such a value

constant element
a job or task element in which a worker

exhibits consistency of performance time, even if minor changes in processing or product dimensions are made

constant error
the difference between the point of subjective equality and the known standard value in psychophysical testing

constraint
any compelling force or limiting structure which restricts freedom of action to within certain bounds

constrictor
a muscle which contracts to close or reduce the cross-section of an opening

construct
(n) a postulated attribute of an individual assumed to be reflected in observable behaviors

construct validity
the extent of the relationship between what a test measures and how test scores are reflected in behavior or performance

constructive solid geometry (CSG)
a technique in solid modeling where primitive solids are generated and combined to produce more complex forms

consultant
an individual or group who is uniquely qualified or claims expertise in a particular field and may be called upon to perform some specialized technical function on a one-time or an occasional basis

consumer
one who purchases goods or services for final use, not having the intent to reprocess or repackage for resale

consumer product
a product intended for final use primarily by the general public, as opposed to industrial use

contact dermatitis
an inflammation from either a direct skin irritant or allergic reaction from some material to which the skin has become sensitized coming into contact with the skin

contact grasp
see finger touch

containerize
place a material within a container; *n.* containerization

containment
a process, structure, or system within a specified area or volume for preventing an entity from spreading and/or interacting with other materials or another environment

containment level
that degree of independence or separation in containment provided by a specified system

contaminant
any unwanted material, substance, or entity

contaminate
place one or more contaminants in a location where they may degrade the environment; *n.* contamination

content validity
the extent to which a test samples a domain of important job behaviors

contingency
a possible situation or event, usually referring to an undesirable or abnormal situation or event

contingency allowance
a small time allowance included within the standard time to cover for legitimate, expected additional work and delays
> *Comment:* usually not measured precisely because of its infrequent occurrence

contingency analysis
an analysis performed to identify what abnormal situations, errors, or malfunctions a system may develop or encounter to improve system performance or establish what special human responses may be required under those circumstances

continuous forms
having each individual form or sheet attached to the next, usually with

guides for a printer, and which necessitates separation after printing

continuous function
any mathematical function which has no breaks or gaps in its extent; *opp.* discontinuous function

continuous noise
that noise which is persistent over long periods of time; *opp.* impulse noise

continuous passive motion machine
a device which repeatedly cycles automatically to passively flex and extend one or more joints through their ranges of motion

continuous reading method
see cumulative timing

continuous spectrum
a range of frequencies within which all frequencies are present

continuous speech recognition
see speech recognition

continuous timing
see cumulative timing

continuous timing method
see cumulative timing

continuous variable
a variable which may take any value within a specified range of values; *opp.* discrete variable

continuous work
a sustained workload without any rest pauses

contra-
(prefix) pertaining to the other or opposite side with reference to a structure or point

contractile tissue
any tissue which is capable of shortening in response to stimulation

contraction
a shortening or reduction in some dimension of a structure

contralateral
located on or pertaining to the opposite side of the body

contrast
see chromatic contrast, luminance contrast

contrast attenuation
a decrease in the amount of contrast over space or time

contrast detection
a basic visual task in which the visual system perceives a difference in luminance, creating an object and a background

contrast ratio
a mathematical relationship involving some form of a ratio between figure luminance or reflectance and background luminance or reflectance; *see also luminance contrast*

contrast sensitivity
a measure of the ability to perceive a visual contrast between two regions; the reciprocal of the contrast threshold

contrast threshold
the smallest difference between two visual stimuli which is perceptible to the human eye under specified conditions of adaptation, luminance, and visual angle on a certain proportion of a set of trials; *syn.* liminal contrast, liminal contrast threshold, threshold contrast

control [1]
any device which enables a user to direct the action or operation of some equipment or system

control [2]
a condition or individual which serves as a reference to limit or test probable sources of error in an experiment

control [3]
a psychomotor skill involving fine positioning

control [4]
a state of being in charge, of having command or authority, and the ability to manage a situation

control arrangement
see control layout

Control Assessment Protocol (CAP)
a systematic procedure for clinicians to follow in the evaluation of disabled individuals for assistive devices

control coding
the use of any of a variety of coding methods for labeling a control; *see color coding, shape coding, size coding, label coding, location coding*

control force
that amount of force required to operate a control; *see also control torque, actuation force*

control group
a group of individuals or items selected from what is believed to be the same population as an experimental group, but which are not exposed to the experimental treatment(s) under consideration; *syn.* comparison group

control layout
the grouping of manual controls within a location at a workplace; *syn.* control arrangement; *see also control location, display-control layout*

control limit
that boundary value which a measurement on some aspect or dimension of a product or system must not exceed

control location
the general placement of controls for use by an operator; *see also control layout*

control placement
see control location

control precision
a psychomotor ability involving the positioning of larger muscle groups to make rapid, repeated adjustments to one or more controls

control sensitivity
the ratio between the amount of movement or change on a display and the control movement; *opp.* control-display ratio

control spacing
that distance between the human-operated mechanism for two or more control devices

control stick
the primary control device on many types of aircraft, generally consisting of a rod-shaped structure extending

from the floor in front of the pilot's seat with aircraft handling and other controls; *syn.* stick

control system
a system whose primary function is the monitoring of outputs from a given set of functions and using that data/information to regulate that set in some specified manner or propose new regulations

control torque
that amount of torque required to operate a rotary control; *see also control force, actuation torque*

control-display layout
see display-control layout

control-display ratio
the ratio of movement of a control to the movement or change of an indicator on a display; also control/display ratio; *syn.* control-response ratio; *opp.* control sensitivity

control-response ratio
see control-display ratio

controlled experiment
an experimental investigation in which the relevant independent variables are directly and systematically manipulated and/or controlled and the effects of such manipulation are measured; *syn.* control experiment, controlled study

controlled motion
see controlled movement

controlled movement
any controlled bodily movement in which prime mover and antagonist muscles are integrated using muscle contraction throughout the range of the motion to generate a desired force and/or velocity; *syn.* nonballistic movement, tension movement; *opp.* ballistic movement

controlled time
that elemental time which is governed solely by some external process

controller
any device used for operating and/or regulating a system

contusion
an injury from a direct mechanical impact to the surface of some body structure which results in bruising but does not break the surface

convective heat loss
that amount of heat eliminated from the body via convection, indicated by an equation of the form

$$H = h_c A(T_s - T_a)$$

where:
H = convective heat loss
h_c = convective heat transfer coefficient
A = body surface area
T_s = weighted mean skin temperature
T_a = air temperature

convective heat transfer coefficient (h_c)
a number which includes factors for clothing thermal characteristics and environmental conditions

conventional anthropometry
see classical anthropometry

convergence
a coordinated inward rotation of the eyes about their vertical axis to fixate on a point near the observer to obtain fusion

convergence angle
that angle formed by the intersection of the line of sight of each eye when both eyes are fixated at a single point

convergence point
that location on a curve at which a worker's learning curve achieves standard performance

convergent phoria
a tendency for an observer to fixate in front of a stationary target

convex function
a mathematical relationship or graph having positive second derivative over a specified interval of interest, resulting in a U-shaped curve; *opp.* concave function

cool color
a blue or green color, or a color which

appears less bright than another for a given intensity; *opp.* warm color

Cooper Scale
a rating scale with a range of 1 (excellent) through 10 (fatal) which was developed in an attempt for pilots to provide more objective evaluations of aircraft handling qualities; *syn.* Cooper Rating Scale
Comment: no longer used

Cooper-Harper scale
an ordinal rating procedure using a decision tree on a scale of 1 (excellent) through 10 (major deficiencies) for task difficulty; *syn.* Cooper-Harper aircraft handling characteristics scale
Comment: designed originally for use by test pilots for evaluating aircraft handling, but has been used in other physical workload situations as well

Cooper-Harper Scale, modified
an ordinal rating procedure using a decision tree on a scale of task difficulty ranging from 1 (very easy) through 10 (impossible) for mental workload determinations

coordinate ¹
(n) a position in space, time, amplitude, or some other dimension

coordinate ²
(v) cause separate entities to act together harmoniously toward a final goal; *n.* coordination

coordinate system
a spatial reference system with a defined origin and rules for defining locations within that system; *see also rectangular coordinate system*

coordinate transformation
any mathematical or graphical process for modifying or shifting a coordinate system

copy
a computer operating system function which duplicates a file or segment in another location while leaving the original file or segment intact

core temperature

a temperature recorded from deep enough in the body so as not to be normally affected by external conditions

core-shell model

a simple thermodynamic concept in which the human is treated as having a heat-producing core and a surrounding shell, with heat exchange occurring through the shell to the environment

Coriolis acceleration

that acceleration generated by the simultaneous exposure to rotational motion about two axes in an inertial reference frame

Coriolis effect

the misperception of body orientation, commonly accompanied by nausea and vertigo on exposure to Coriolis acceleration

cornea

a transparent structure forming the anterior portion of the eyeball, and through which light passes en route to the retina for vision; *adj.* corneal

corneo-retinal potential (CRP)

the bioelectric potential between the anterior and posterior eyeball

coronal plane

see frontal plane

Comment: often used instead of frontal plane in conjunction with the brain

coronary

pertaining to blood vessels or nerves which encircle an organ or other structure, especially the heart

coronary occlusion

the blockage of an artery supplying blood to the muscle tissue of the heart

coronoid process

a projection from the proximal end of the ulna which fits into the coronoid fossa on flexion of the elbow

corrected effective temperature (CET)

a measure of environmental heat stress which includes average radiant temperature and globe temperature effects

corrective maintenance

a form of maintenance which is intended to return a system or piece of equipment to proper operating status after it has failed; *syn.* breakdown maintenance, unscheduled maintenance, remedial maintenance

correlated color temperature

that temperature of a Planckian radiator whose perceived color most closely resembles that of a given stimulus source when viewed at the same brightness and under specified viewing conditions

correlated work crew

a group of workers who interact with each other or work together on a task, such that each individual's work is not independent

correlation

a measure of the degree of relationship between two or more variables

correlation coefficient

a number between 1.0 and −1.0 which represents the degree and direction of correlation between two variables

correlative kinesiology

see electromyographic kinesiology

cortex

the outer portion of an organ or structure, usually referring to the brain, adrenal gland, or bone; *adj.* cortical

cortical bone

the compact bone tissue next to the surface of a bone

cosine

a trigonometric function; the value of the ratio of the adjacent side of an acute angle to the hypotenuse in a right triangle

cosine law of illumination

a rule that the illumination on any surface changes according to the cosine of the incident light angle from perpendicular to the surface

$$E = \frac{I\cos\theta}{d^2}$$

where:

E = illumination level

I = intensity of light source

θ = the angle of incidence of the light from perpendicular

d = the distance from the light source

cosmic radiation

the combined flux of electrons and atomic nuclei which are generated in deep space

cosmic ray

see **cosmic radiation**

cost

those expenses incurred in producing a product, delivering goods, or providing a service, whether financial, human, or metabolic

cost – benefit analysis

the determination or estimation and evaluation of the weighted relative financial, social, and/or other costs to the same or other categories of rewards or compensation; also cost/benefit analysis

Comment: should be performed prior to undertaking the endeavor being considered

cost of accidents per employee (CAE)

the cost of the accidents incurred per year spread across the average number of employees

$$CAE = \frac{total\ accident\ costs}{average\ number\ of\ employees}$$

cost-effectiveness

the relative financial or other benefits obtained compared to the cost of alternatives; adj. cost-effective

costal cartilage

that segment of cartilage which attaches a rib to the sternum or, in some cases, to adjacent ribs

coulomb (C)

the amount of electricity passing a given point in one second at a current of one ampere; syn. absolute coulomb

coulomb friction

that friction from movement between dry surfaces

counter [1]

the top surface of a workspace

counter [2]

any device or system which keeps track of incrementing or decrementing numbers of objects or events

counterclockwise rotating shift

pertaining to a rotating shift work schedule in which the shift worked is periodically regressed by shift increments, i.e., from the first shift to the third, or from the second to the first; opp. clockwise rotating shift

coupling

any of a variety of possible interfaces between the hand or robotic grapple fixture and another object for purposes of gripping or touching

courseware

that application or system software and the programmed/coded information base which are used to provide the information and interactions in a computer-aided instruction system

covariate

a variable which is related to and varies as the predictor and outcome variables do

coverage

the number of jobs or the number of personnel whose jobs have been assigned standards during a particular period

covert behavior

any behavior consisting of actions not directly viewable by an external observer; opp. overt behavior

covert lifting task

an operation in which body parts are moved, thus involving biomechanical aspects of the body, but which doesn't involve the handling of a load other than the body parts themselves; opp. overt lifting task

coxal bone
> a bone consisting of the fused ilium, pubis, and ischium making up part of the pelvic girdle; *syn.* hip bone, pelvic bone, innominate bone

cramp
> any painful, involuntary contraction of a muscle, particularly of a skeletal muscle or the intestine

crane
> a mechanical device or machine for lifting and/or moving loads

cranial length
> the linear distance from glabella to opisthocranion

cranial nerve
> any of 12 pairs of nerves or fiber tracts generally associated with sensory and/or motor functions of the head and neck

cranial suture
> a suture between two bones enclosing the brain

craniosacral
> pertaining to the cranium and sacral portions of the spine or spinal cord, especially referring to the parasympathetic division of the autonomic nervous system

craniostat
> a device for measuring the facial angle

cranium
> that portion of the skull without the mandible, generally that which surrounds the brain; *adj.* cranial

Crash Injury Research project (CIR)
> a U.S. government-sponsored project intended to determine the causes of aircraft accidents and record the injuries sustained in each accident
>> *Comment:* an older program; now Aviation Safety Engineering and Research

crash safety
> a measure of a vehicle's ability for the occupant(s) to survive an impact and evacuate the vehicle following the impact

Crash Survival Design Guide (CSDG)
> a multi-volume document providing information on various aspects of aircraft design criteria which enhance crew and passenger survival during and following a crash

crashworthiness
> a measure of the capability of a vehicle to act as a protective container and energy absorber during impact

crawl
> a type of locomotion which involves moving in approximately a prone position, using the hands/elbows and knees for support and movement

crawlspace
> a region of low height, generally under a large structure of some type, through which a worker may access certain utilities connections or other equipment; also crawl space

creativity
> the ability to generate ideas for novel approaches, devices, or artistic works through imagination, thinking, or considering a situation from a different perspective; *adj.* creative

creep
> a change in method within a task by a worker over time

crest factor
> the ratio of the peak value of a vibratory motion to the root mean square value of that motion over a specified time interval

CREW CHIEF
> a computerized, three-dimensional human modeling program for simulating an aircraft maintenance person with respect to accessibility of components for maintenance and ultimately to the incorporation of such data into aircraft design

crew load
> the number of personnel used to perform work on a certain product or component

crew station
> any workstation or worksite within a

vehicle intended for use during vehicular operation by one or more members of the crew of that vehicle

crew-induced load
the reaction forces exerted by an individual on a structure as a result of that individual exerting effort with or reacting to external forces caused by another object on another portion of the body

cricoarytenoid
see *posterior cricoarytenoid, lateral cricoarytenoid*

cricoid cartilage
a ring-shaped piece of cartilaginous tissue encircling the airway passage in the larynx

cricothyroid
a skeletal muscle in the larynx involved in producing tension and elongation of the vocal cords

crinion
the point in the midsagittal plane where the hairline meets the forehead
Comment: in a balding or hairless individual, estimate where the hair growth line would be if he had normal hair

crista
the sensory structure within the ampulla of a semicircular canal, which detects motion of the head; composed primarily of the cupula and sensory hair cells

cristale
see *iliac crest*

criterion variable
the variable consisting of the observed result in a correlation or regression study; *opp.* predictor variable
Comment: analogous to the dependent variable in experimental work

criterion-related validity
the usefulness of some test as a predictor in job performance

critical
pertaining to an aspect of such importance that an operation can't proceed without it or a situation may be come

life-threatening

critical damping
the minimum viscous damping that will allow a displaced system to return to its initial position without oscillation

critical flicker frequency (cff)
see *flicker fusion frequency*

critical function
an activity or operation which can have a major impact on system performance or can endanger workers or the project if it fails

critical fusion frequency (cff)
see *flicker fusion frequency*

critical incident method
a performance appraisal technique for either a system or employee; *syn.* critical incident technique
for a system: the process of gathering data by asking the users of that system to describe significant incidents, according to some established criteria;
for an employee: the maintenance of a log documenting both favorable and unfavorable behaviors exhibited during an evaluation period

critical organ
that organ or tissue for which radiation injury will be of greatest detriment to health; the limiting organ for a particular radiation circumstance

critical path analysis
see *Critical Path Method*

Critical Path Method (CPM)
the development and use of a networked model containing the times required for different phases of a job, from which the critical path is determined and a decision made as to how the job will be carried out; *syn.* critical path analysis

critical ratio (CR)
the value of the ratio of a deviation from a mean to the standard deviation for that distribution

critical score
that score which appears to separate those most likely to be successful from

those most likely to fail

critical value

that value which lies on a boundary for rejection or acceptance of a hypothesis

criticality

a scale or ranking of the possible types of failures in a system as to the importance of continued functioning of that system; *syn.* criticality index

cross light

provide equivalent illumination on a subject using a pair of luminaires arranged at equal angles from the plane generated by the subject and the viewing axis of the viewer/camera

cross-boundary interaction analysis

a study of the work-related interactions between workers on different tasks to determine the interdependence between tasks

cross-coupling

a situation in which an event occurring in one aspect affects or causes an event to occur in another aspect

cross-modality matching

see cross-sensory matching

cross-sectional area

that exposed area when an object or image is cut perpendicular its longitudinal axis and viewed along the longitudinal axis

cross-sectional design

a research methodology in which all samples are taken at approximately the same point in time

cross-sectional study

a study using a cross-sectional design

cross-sensory matching

a research technique in which the intensity of stimulation on one modality is compared or matched to the intensity of stimulation in another modality; *syn.* cross-modality matching

cross-sequential design

a research methodology in which independent groups of individuals from the same birth cohort are measured at different times/ages

cross-training

a technique in which a worker may be trained on the job of one or more co-workers, usually with his co-workers being likewise trained

crossover analysis

an evaluation for costing purposes of what alternative work methods should be used for different production levels

crosstalk

a signal which is communicated to another channel in a system where it is not desired

crotch

a location between two structures which emanate from adjacent points and are interconnected by some tissue or other material; *see pubic crotch, thumb crotch, interdigital crotch*

crotch height

see pubic crotch height

crotch length

see pubic crotch length, thumb crotch length

crown

see vertex

crown – rump length, reclining

the linear horizontal distance from the top of the head to the bottom of the buttocks

> *Comment:* measured with the individual supine on a recumbent-length table, the hips flexed 90°, and the head oriented so the Frankfort plane is perpendicular to the board surface

crush injury

any injury in which bodily tissues are severely compressed and possibly torn due to mechanical forces

cuboid bone

one of the foot tarsus bones, lying between the calcaneus and the lateral two metatarsals; *syn.* os cuboideum

cue

a stimulus which is a signal to respond

culture

the totality of behaviors, values, attitudes, and customs shared by a group

cumulative distribution function (CDF)

see cumulative probability distribution

cumulative error

an error whose sum doesn't converge to zero as the number of samples increases

cumulative exposure

a weighted sum intended to represent an individual's effective exposure to some environmental condition over a period of time when the levels or intensity of that condition vary throughout the period of interest

$$E_c = \sum_{i=1}^{n} L_i T_i$$

where:

L_i = level of exposure (intensity, concentration, etc.)

T_i = length of time at exposure level i

n = number of exposure intervals used

cumulative frequency distribution

a graphical or tabular representation of an ever-increasing curve corresponding to the summation of all scores of a data set, such that for each point on the distribution, the ordinate value represents the sum of all scores less than the corresponding point on the abscissa; *see also cumulative probability distribution*

cumulative pathogenesis

the development of some type of trauma through continued stress on one or more parts of the body

cumulative probability distribution

a graphical, mathematical, or tabular representation of the integration or summation of some probability distribution function, yielding the cumulative probability of all events occurring in that set; *see also cumulative frequency distribution*

cumulative sum chart

a statistical quality control chart where the sum of product deviations is plotted against time; *syn.* cusum chart

cumulative timing

a work timing technique in which the timing device is permitted to run continuously across all elements of the task being measured; *syn.* continuous timing, continuous timing method, continuous reading method, cycle timing

cumulative trauma disorder (CTD)

see repetitive motion injury

cupula

a gelatinous mass enclosing the sensory hair cells of a crista for detecting motion within the semicircular ducts

curb cut

a section of curb at which a ramp has been laid, usually at an intersection, from the sidewalk to the street for the passage of wheeled vehicles or handicapped individuals

curie

a unit for radioactive decay; equals 3.7 $\times 10^{10}$ disintegrations per second

current

the flow of electrons or electrical charge; *syn.* electrical current

cursor

a movable symbol, icon, or other element on a display to indicate position or pointing; *see pointing cursor, placeholding cursor*

curvature

see arc

curve

a mathematical function or graphical representation of the relationship between two or more variables

curve fitting

the process of determining which particular curve/line or function best fits the known data points

curvilinear

pertaining to one or more lines which are not straight

curvilinear correlation

see non-linear correlation

curvilinear regression

see non-linear regression

cushion

any form of soft material which acts to increase body tactile comfort

cusum chart

(sl); *see* **cumulative sum chart**

cut [1]

the removal of a selected block of text, data, or graphics from the display for storage in a temporary buffer, for possible recall and placement in another location; *opp.* paste

cut [2]

a tissue injury of varying depth but with much greater length than width

cutaneous

pertaining to the skin, its sensory receptors, or to the sensations produced by those receptors

cutaneous lip

the area between the upper lip and the nose

cuticle

see **eponychium**

cutoff frequency

that frequency at which an electrical filter begins to attenuate a signal

Comment: the direction of the attenuation depends on the type of filter

cutting plane

an imaginary surface along which a computer model is "sliced" to yield a cross-section

cyanosis

a condition in which the skin and certain membranes have a bluish-purple tint due to a lack of oxygen in the tissues

cybernation

the use of computers in automating industry

cybernetics

the study of communication and automated feedback control functions between living organisms and machined systems with an emphasis on gaining an understanding of living organisms by using machine analogies

cyberspace

an abstract version of a virtual environment which extends beyond three dimensions

Cybex dynamometer

a commercial dynamometer which can measure static or dynamic isokinetic strength

cycle

a complete sequence of elements or events making up a unit process or activity in a repetitive, periodic operation; *adj.* cyclic

cycle life

the number of times a material can be stressed at a given level before it fails or is expected to fail

cycle per second (cps)

see **Hertz**; *pl.* **cycles per second**
    Comment: an outdated term

cycle time

that time required or used, whether by man or machine, to perform all the elements in a complete work cycle

cycle timing

see **cumulative timing**

cyclegram

see **cyclegraph**

cyclegraph

a photographic record of the motion obtained in cyclegraphy

cyclegraph technique

see **cyclegraphy**

cyclegraphy

the process of making a single photograph using one or more small light bulbs which are on at all times during the process for tracking the body or its parts with an exposure time on the same negative of at least one cycle of a repetitive motion; *syn.* cyclegraph technique; *see also* **chronocyclegraphy**
    Comment: typically the subject is in a darkened area

cyclic

pertaining to a periodic event

cyclic element

an element of some operation or

process which occurs at least once in every period of that operation or process

cyclic timing
see cycle timing

cyclograph
see cyclegraph

cylindrical grip
a type of grip in which the flexed fingers and the palm are used as if to hold an object of constant diameter with an extended length, where the degree of flexion of each finger joint is similar for each finger

D

d′
a statistical index of an individual's sensitivity in estimating the distance between the mean of a noise distribution alone to the mean of the signal plus noise distribution; also d prime; *see signal detection theory*
Comment: units are in standard deviations

dactylion
the most distal point of the fleshy part of the middle finger, excluding the nail

dactylion height
the vertical distance from the floor to the tip of the middle finger
Comment: measured with the individual standing erect and the arm, hand, and fingers extended downward at the side

daily living tasks
those necessary tasks for normal housekeeping, cleanliness around the home; *see also activities of daily living, instrumental activities of daily living*

damage
any loss of material value or usefulness

damages
that compensation within the legal system for an individual or group which has suffered some loss, damage, or injury due to the attributed fault of another individual or group; *see also compensatory damages, punitive damages*

damping
the dissipation of the energy within a dynamic system over time for whatever reason

damping factor
the ratio of actual system damping to critical damping for a system

danger
a situation in which a hazard and the potential risk exist for damage, injury, illness, or pain; *adj.* dangerous

danger zone
a physical location in which some type of hazard exists

dark ¹
having little or less reflected light

dark ²
a severely reduced light level in the visual environment

dark adaptation
the process of undergoing neurochemical changes in the eye after being placed in darkness or low light levels, during which the visual system becomes more sensitive to light; *syn.* scotopic adaptation; *opp.* light adaptation

dashpot
a symbol for a viscous damper for mechanical modeling, representing a vane placed within a viscous fluid

data
a formalized representation of numbers or characters which have meaning for communication, interpretation, or processing purposes

data display code
a graphical symbol representing a data point on some type of graphic output

data entry
the process of inputting data into a computer using a pre-established format, regardless of the technique used

data inquiry
the process of requesting and retrieving information from a computer and viewing it on a display or on a hardcopy

Data Store

a database on human reliability

databank

any location, but typically in a computer system, where large amounts of a specific type of data are stored for retrieval by users; also data bank

database (db)

a collection of related data; also data base

day shift

see **first shift**

day-night average sound level (DNL, L_{dn})

a single value which represents an average measure of noise exposure based on sound pressure levels over a 24-hour period, with the assumption of an active period from 7 A.M. to 10 P.M. and a sleep or quiet period from 10 P.M. to 7 A.M.

$$L_{dn} = 10 \log_{10} \left[\frac{1}{24} \left(\int_{7AM}^{10PM} 10^{SPL_A(t)/10} dt \right. \right.$$
$$\left. \left. + \int_{10PM}^{7AM} 10^{\{SPL_A(t) + 10\}/10} dt \right) \right]$$

where:

SPL_A = A-weighted sound pressure level measurements

Comment: a 10 dB factor is added to the night integral due to the increased annoyance from being being disturbed during sleep

daylight

that light present during the daytime hours from the sun, or the corresponding artificial illumination in terms of spectrum and intensity

daylight availability

that amount of sunlight received from the sun with reference to certain conditions, such as location, intervening substances, date, and time

daylight lamp

any artificial light source with an output having a spectrum similar to that of a certain type of daylight; see also **standard illuminant**

daywork

that work for which compensation is based on time present, not output

dazzle

experiencing a condition of extreme brightness due to reflected and scattered light from particles in the atmosphere, resulting in viewing difficulties

de minis violation

a violation of some standard not having any immediate or direct threat to the safety or health of a worker

dead band

see **dead zone**

dead hand

see **Raynaud's phenomenon**

dead man control

a device requiring a constant force of a minimum magnitude applied to the device for operation a piece of equipment, and having a default mode which turns off or stops the equipment if that force is not applied

dead zone

that region, usually around the neutral position of a knob, handcontroller, or lever, where there is no output from a device, even though an input may be provided

deaf

unable to hear any airborne or bone-conducted sounds due to some defect or damage in the auditory system or the brain; *n.* deafness

Comment: many terms use "deafness" but are in reality only hearing reductions

deafened

having a loss of hearing ability after normal speech and hearing patterns had been established

deafness

see **deaf**

death

the cessation of biological life without the current possibility of resuscitation

death rate

see **mortality rate**

debug

check a system for errors or operational failures and correct any flaws discovered

deceleration

acceleration in the direction opposite to that of the velocity vector to effect a slowing of motion; *syn.* negative acceleration

deci-

(prefix) one-tenth of a base unit

decibel (dB)

one tenth of a bel; a non-dimensional logarithmic ratio of the measured quantity and a reference quantity for expressing power, pressure, or amplitude

$$P = 10 \log \left(\frac{I}{I_0}\right)$$

Comment: the most common unit of sound or other signal intensity

decision delay

see cognitive reaction time

decision making

the process of evaluating information which results in the selection of a course of action

decision time

see cognitive reaction time; syn. decision delay

decompression sickness

any of a range of conditions in which inert gas bubbles form in the tissues with possible tissue damage due to too rapid a decrease in the body's external environmental pressure; *syn.* dysbarism, caisson disease; *see also bends, chokes, paresthesia, baro-dontalgia, dysbaric osteonecrosis, air embolism, staggers*

decontaminate

remove some harmful substance from a medium or surface; *n.* decontamination

decrement [1]

a deterioration in some performance measure; *see performance decrement*

decrement [2]

a decrease of a counter value in computing

deductive reasoning

the ability to apply general rules to specific problems and arrive at a logical conclusion

default

a value, condition, or state which is automatically selected by a computer or other system unless overridden by an operator or program

defecate

eliminate solid waste from the digestive system via the anus; *n.* defecation

defect

any physical condition of a material or system, whether recognized or not, which is outside of specifications or standards and affects either function or appearance; *adj.* defective

defensive response (DR)

see startle response

deformable element

any structure, whether physical or modeled, which is not rigid

degree [1] (°)

a unit of angular displacement; $1° = 1/360$ of circle

degree [2] (°)

a unit of temperature, either Fahrenheit, Celsius, or Kelvin/Absolute

degree of freedom [1] (df)

the minimum number of independent generalized coordinates required to completely define the positions of all parts of a system at any given time; also degree-of-freedom; *pl.* degrees of freedom

degree of freedom [2] (df)

the number of values which are free to vary within a sample, given specified sampling constraints and experimental design; also degree-of-freedom; *pl.* degrees of freedom

delay

one of a set of basic work elements which involves some pause or interruption in an ongoing process or activity; *syn.* interruption, stoppage

delay allowance [1]

a credit of time or money given the operator to compensate for incentive on a specific delay incident not covered by the piece rate or standard

delay allowance [2]

a period of time which is added to the normal time to compensate for contingencies and minor delays beyond the control of the operator; *syn.* unavoidable delay allowance

delay time

any temporal interval during which a worker is idle due to any cause beyond the worker's control, such as an equipment breakdown, a lack of tools or parts, or a shortage of materials; *syn.* waiting time, lost time, inherent delay; *see also* ***idle time***

DeLorme boot

a special boot used to exercise the quadriceps muscles

DeLorme exercises

see ***progressive resistance exercises***

Delphi method

see ***Delphi technique***

Delphi technique

a process designed to obtain a consensus of experts by successive iterations of questioning interspersed with feedback on others' opinions and supporting reasons; *syn.* Delphi method

delta ray

the track of electrons recoiling from ionization or atomic reactions in tissue

delta rhythm

an EEG frequency band consisting of frequencies less than 4 Hz

deltoid arc

the surface distance from acromiale to the point where the deltoid muscle disappears from view

deltoid muscle

the large skeletal muscle extending over the superior and lateral part of the shoulder

demand variability

a change in the desire to purchase a product over time

demanded motions inventory

the motions which are required to perform a given task

demographics

the gathering, analysis, and/or use of information such as occupation, income, education, family size, and ethnic background from those populating a certain region

demography

the sociological/statistical study of populations to understand their vital statistics and distribution; *adj.* demographic

denitrogenation

remove nitrogen from a system, in particular the body by breathing a nitrogen-free gas mixture

densitometer

an instrument for determining the optical density of translucent materials

densitometry [1]

the study and/or measurement of optical densities

densitometry [2]

the hydrostatic determination of total body fat

Department of Defense Standard (DOD-STD)

a U.S. Department of Defense Standard which uses metric values; *see also* ***Military Standard***

dependent variable

a response variable whose value is determined, wholly or in part, by one or more independent variables within an experimental situation; *opp.* independent variable

deployment

the distribution of workers to specific work sites

depolarize

reduce the amount of electrical charge across some structure, usually with reference to a neuronal or muscle cell membrane

depreciate
spread the cost of a system, piece of equipment, structure, or facility over time, usually for tax or accounting purposes, to allow for its reduction in value; *n.* depreciation

depressor [1]
any muscle producing a downward movement

depressor [2]
any device effecting a downward movement of some structure

depth
a straight-line measurement with anterior to posterior extent in any sagittal plane and perpendicular to the frontal plane of the body

depth cueing
the process of making a complicated image more readily understandable by distinguishing between elements in the foreground and background

depth perception
the ability to distinguish relative distances of two or more objects or the distance of a single object from the observer

derived requirement
a requirement not imposed by some original or high-level document or management, but which is imposed by secondary documents or lower levels of management

dermal
pertaining to the skin

dermatitis
an irritation or inflammation of the skin

dermatome
that region of the skin innervated with sensory fibers from a single spinal nerve

dermatome chart
a graphic or visual display of the regions of the body surface innervated by specific spinal nerves

dermatosis
any skin disease; *pl.* dermatoses

dermographic pencil
an instrument for marking landmarks/pointmarks on skin for taking anthropometric measurements

describing function
any mathematical model or representation of a time-varying system involving humans, generally consisting of some transfer function plus remnants

descriptive statistics
the collection of data and use of statistical methods to present certain information describing a group or population; *opp.* inferential statistics

design
the process of developing the requirements, structure, dimensions, tolerances, and materials to be used for an entity

design driver
a requirement which causes a system to be designed in a specific way

design eye point
a fixed point providing a line of sight within which all controls and displays at a workstation should be located; *syn.* design eye position; *see design eye volume*
   Comment: only recommended for use when head position is severely constrained in the task/job

design eye position
see design eye point

design eye volume
that region within which an operator's head is free to move and provide appropriate lines of sight while at a workstation

design for maintainance
a design priority concept which emphasizes the future maintenance aspects of a product's structure; *syn.* design for maintainability

design for manufacturing
a design priority concept which emphasizes the assembly aspects of manufacturing in structural design; *syn.* design for manufacturability, design for producibility

design for producibility
  see design for manufacturing

design for reliability
  a design priority which emphasizes minimizing the chances of failure and/or maximizing the mean time between failures

design for use
  a design priority which emphasizes the ease of use of a product

design solution
  an engineering design which meets or exceeds a set of requirements

design speed
  that rate at which a mechanically-driven operation is intended to occur or at which a piece of equipment is intended to move or rotate

desquamation
  a shedding, peeling, or casting off, especially of skin cells

destructive test
  a procedure in testing product quality in which the material or product being tested is destroyed; *opp.* non-destructive test

desynchronize ¹
  change the electroencephalogram from a low frequency, high amplitude rhythm to a higher frequency, lower amplitude rhythm; *n.* desynchronization

desynchronize ²
  change a biological rhythm from a normal or typical phase relationship to another; *n.* desynchronization

desynchronosis
  see jet lag

detectability
  one or more qualities of a signal, display, or other stimulus which affect its probability of being perceived, either in isolation or against a background

detection threshold
  see threshold

detector tube
  a transparent tube which contains specific chemical materials adsorbed onto

inert granules designed to undergo a color change if certain chemicals or certain amounts of chemicals are drawn through the tube

detent
  a part which stops or releases a movement

deterministic
  pertaining to those data which can be explained or predicted with reasonable accuracy via some explicit solvable mathematical relationship; *opp.* non-deterministic

deuteranomaly
  a color vision deficiency involving a reduced ability to discriminate green in colors

deuteranope
  one having deuteranopia or deuteranomaly

deuteranopia
  a form of color blindness involving an inability to discriminate the green content of colors; *syn.* green blindness; *adj.* deuteranopic

development
  the growth, improvement, or maturation of one or more characteristics of some product or individual

development time
  the temporal period required to design, engineer, and prepare the manufacturing documentation for some device

developmental age (DA)
  an index of growth using an age equivalent determined by standardized observations; *see also* **chronological age, mental age**
    Comment: may include body measures, mental, emotional, social, and mental observations

developmental anthropometry
  the study of growth in size and/or proportions of the human body

developmental quotient (DQ)
  the value of the ratio of the developmental age to chronological age

device-independent
an operation or procedure which has similar functions may be executed or performed on a variety of pieces of equipment which may differ in structure, method of operation, and/or appearance

dew point
the temperature at which condensation begins to appear as an air-water vapor mixture is slowly cooled at constant pressure

dexterity
the degree of manipulative ability via perceptual-motor coordination; *adj.* dexterous

dextrality
prefering the right-hand

Diagnostic Rhyme Test
a forced-choice test in which an individual is required to select the word he believes was spoken from two rhyming options

diagnostic study
a preliminary investigation or study of some operation, process, individual or group in an attempt to learn the causes of problems; *syn.* diagnostic survey

dial caliper
a caliper which uses a rotary dial to indicate the distance between two points

dialog
see dialogue

dialogue
the content of a structured sequence of steps in an interaction between a user and a computer; also dialog

dialogue box
a pop-up display window which requests user input regarding some computer system function; also dialog box

diameter
a straight line in a plane or section connecting two opposing points on a circle, approximately circular, or elliptical structure and passing through the center of that structure

diaphysis
the shaft or central portion of a long bone

diastolic blood pressure
the minimum arterial blood pressure occurring during that portion of the cardiac cycle when the heart relaxes and a ventricle fills with blood; *opp.* systolic blood pressure

dichoptic
pertaining to viewing conditions in which the visual display to each of the two eyes differs with respect to some property of the stimulus

dichotic
pertaining to listening conditions in which differential stimulation of the two ears occurs according to some definable physical property of the stimulus such as duration, frequency, phase, intensity, or bandwidth

dichromat
one who has dichromatopsia

dichromatopsia
a form of color blindness involving an inability to discriminate only one of the three primary colors: red, green, or blue; *see also deuteranopia, protanopia, tritanopia*; *adj.* dichromatopsic
    Comment: therefore capable of seeing two of the primaries

difference limen (DL)
see difference threshold

difference spectrum
that spectrum which is obtained by the subtraction of one spectrum from another

difference threshold
the degree or intensity by which two suprathreshold stimuli must differ if a difference is to be noted on a specified percentage of the trials; *syn.* differential threshold, difference limen; *see also just noticeable difference*

Differential Aptitude Test (DAT)
a commonly-used test for determining verbal, abstract, and mechanical reasoning ability, spatial relations,

clerical speed and accuracy, and grammatical/spelling skills

differential equation

any equation containing one or more derivatives of a mathematical function

differential piecework

that form of compensation in which the piece rate is variable, and based on the total number of pieces produced during a specified period

differential threshold

see difference threshold

differential timing

the use of subtraction or simultaneous equations for obtaining the time value of an extremely short duration work element in a time study by combining the time values of elements preceding and following the element in successive cycles

differentiate [1]

distinguish between one or more conditions

differentiate [2]

mathematically determine the ratio of a small change in a dependent variable as a function of change in the independent variable; *n.* differentiation; *opp.* integrate

diffraction

the change in direction of a wave front by some obstacle or non-homogeneity within a medium by a means other than reflection or refraction

diffuse

(v) re-direct or scatter energy transmission in multiple directions over a region

diffuse lighting

that light which is not incident from any particular direction and is of approximately the same intensity within the volume of consideration; also diffused lighting

diffuse reflectance

the value of the ratio of diffused flux leaving a surface to the incident flux; *opp.* specular reflectance

diffuse reflection

the distribution of an incident energy flux in many directions from a surface on which it is incident; *opp.* specular reflection

diffuse sound

a region in which the temporal average of the mean-square sound pressure is the same everywhere, and the flow of energy is equally probable in any direction; also diffuse sound field

diffuse transmission

the passage of an incident energy flux through a material with a wide distribution either internally or on emergence from that material; *opp.* specular transmission

diffuse transmittance

the value of the ratio of the flux passed as diffused to the incident flux; *opp.* specular transmittance

diffuser

any covering for the light source(s) within a luminaire which scatters the light on leaving the luminaire to distribute it more evenly; *syn.* diffusing medium

diffusing medium

see diffuser

dig

a bubble defect which lies on the surface of a transparent material or window

digit [1]

any of the fingers or toes; *adj.* digital

Comment: a convention is to number the digits with Roman numerals, beginning with the thumb and big toe — for example, the thumb is digit I, index finger is digit II, etc.

digit [2]

a numerical symbol, having a potential range of the integers from 0 through 9; *adj.* digital

Comment: the actual range may vary with the number base being used

digital dermatoglyph

see fingerprint

digital display
　　see numerical display

digital-to-analog conversion
　　the process of changing numerical data
　　from a sequence of bits to a continuous
　　graphical curve; *opp.* analog-to-digital
　　conversion
　　　　Comment: typically done with time
　　　　series data

digital-to-analog converter
　　that electrical or electromechanical
　　equipment used for digital-to-analog
　　conversion; *opp.* analog-to-digital con-
　　verter

dihedral angle
　　that angle between two planes

diluent
　　a substance added to another to reduce
　　the concentration of the latter

diluent gas narcosis
　　*see inert gas narcosis, nitrogen narco-
　　sis*

dim
　　reduce the light intensity from a source

dimension [1]
　　any orthogonal spatial axis, typically
　　representing length, width, or depth;
　　adj. dimensional

dimension [2]
　　an aspect of a picture, concept, or other
　　entity for consideration

dimmability
　　having the capability of reducing light
　　intensity without turning off one or
　　more sources; *adj.* dimmable

dimmer
　　any device used for dimming

DIN color system
　　a color ordering system based on the
　　relative importance of hue (T), satura-
　　tion (S), and darkness (D) using a stan-
　　dard daylight (D_{65}) and CIE tristimulus
　　values

diopter
　　the reciprocal of the focal length of a
　　lens; *adj.* dioptric
　　　　Comment: uses the meter as the unit
　　　　of measure

diotic
　　pertaining to listening conditions in
　　which both ears are stimulated by iden-
　　tical stimuli

diplacusis binauralis
　　a condition in which a single tone, pre-
　　sented to both ears, is perceived as
　　having a different pitch in each ear

diploe
　　the spongy bone between the inner
　　and outer layers of the flat skull bones

dipthong
　　a vowel sound which involves some
　　articulator movement

direct anthropometric measurement
　　the measurement of some anthro-
　　pometric dimension using one or more
　　tools in physical or near physical con-
　　tact with the body; *opp.* indirect
　　anthropometric measurement

direct component
　　that portion of an energy flux which
　　arrives at a given location on a path
　　from the source, without reflection; *opp.*
　　indirect component

direct cost
　　a cost due to or for supporting direct
　　labor; *opp.* indirect cost

direct current (DC)
　　a non-oscillating current flow, travel-
　　ing only in one direction; *opp.* alternat-
　　ing current

direct glare
　　a type of glare experienced when a
　　bright light source is within an
　　individual's field of view

direct labor [1]
　　that effort expended on a product or
　　service which advances that product
　　or service toward its specifications or
　　completion; *syn.* productive labor

direct labor [2]
　　any effort which is readily identified
　　with and chargeable to a specific prod-
　　uct or project; *opp.* indirect labor

direct lighting
　　that illuminated environment in
　　which approximately 90% or more of

the luminous flux is directed onto a work or other surface; *opp.* indirect lighting

direct manipulation

a user-computer interface in which the entity being worked is continuously displayed, the communication involves button clicks and movements instead of text-like commands, and changes are quickly represented and reversible

direct manipulation control

having command of an object or cursor via the use of a direct manipulation device

direct manipulation device

any device intended for use in controlling a cursor or other responding object on a display; *syn.* cursor control device, control device

direct manipulation dialogue

the manipulation of symbols in the display via a cursor

direct radiation effect

any of those cellular effects in which radiation damage is caused by ionization of the DNA molecules without an intermediate step; *opp.* indirect radiation effect

direct ratio

the value of the ratio of the luminous flux actually reaching a given surface to the luminous flux emitted from a luminaire

direct viewing

having an object, especially one being manipulated, within sight of the unaided eye; *syn.* direct vision; *opp.* indirect viewing

direct worker

an employee involved in direct labor; *opp.* indirect worker

directional lighting

that lighting exposing an object or a surface primarily from a given direction

directional microphone

a microphone whose response/sensitivity varies significantly by design with the direction of incident sound

directional response

a description, usually graphical, of a transducer response as a function of the direction of the emitted/incident energy in a specified plane and/or at a specified frequency; *syn.* directivity pattern

dirt

any material or substance which causes an unclean condition; *see adhesive dirt, attracted dirt, inert dirt*

dirt depreciation

a reduction in light transmission or reflection due to dirt accumulation; *see luminaire dirt depreciation, room surface dirt depreciation*

disability

any reduction in normal capacity due to mental or physical factors which prevents an individual from experiencing or performing a full complement of activities; *adj.* disabled

disability glare

a viewing condition in which glare interferes with visual clarity, thus reducing visual performance

disabled

see disability

disabling injury

any injury which prevents an individual from performing his normal job function(s) for one or more days beyond the date of the injury

disassemble (DA)

a therblig involving the separation of parts or components

discometry

the study or process of measuring the pressure in the nucleus pulposus of an intervertebral disk

discomfort

a state other than well-being due to the presence of one or more undesirable environmental stressors; *opp.* comfort

discomfort glare

a viewing condition in which glare from one or more high-intensity sources within the field of view causes an observer to experience visual pain or annoyance

discomfort index
a method for estimating effective temperature as a heat stress measure

discomfort threshold
that stimulus intensity at which, in a specified proportion of the trials and/or in a specified proportion of individuals, will sufficiently activate a sensory system to cause a reported change from a typical sensation for a given modality to a sensation of being uncomfortable; *syn*. threshold of discomfort

discontinuous timing
see repetitive timing

discrete
having separate, clearly distinguishable components

discrete spectrum
a presentation of the amount of energy in a complex waveform at each frequency present in the waveform; *opp*. continuous spectrum

discrete variable
a variable which can assume only a specified, finite number of values; *opp*. continuous variable

discrete word recognition
see word recognition

discretion
having the freedom to make decision(s)

discriminate [1]
distinguish reliably between conditions, stimuli, or divisions on a measurement scale; *n*. discrimination

discriminate [2]
treat differently based on some attribute of a person; *n*. discrimination

discrimination reaction time
the temporal interval required to discriminate between two or more stimuli and decide if a response is appropriate

disfigurement
any defect which degrades the appearance of the body or other material structure

disinfect
destroy most or all disease-causing microorganisms, except viruses

disinfectant
a chemical or other agent which disinfects

disk [1]
a round, flat magnetic or optical medium for storage of digital data

disk [2]
see intervertebral disk

dislocated shoulder
an injury in which the head of the humerus has been forced out of the glenoid cavity of the scapula

dismantling allowance
see teardown allowance

disorientation
confusion as to one's position in space and/or time; *adj*. disoriented

dispersion [1]
the spread of scores or other quantitative results in a given sample or frequency distribution; *see also measure of dispersion, variability*

dispersion [2]
an indication of the rate of change of the refraction index on the various wavelengths of energy passing through a transparent medium

displacement [1]
the amount of fluid discharged when an object is placed wholly within or in buoyant equilibrium with that fluid

displacement [2]
a vector quantity specifying the position or change in position of a point with respect to some reference coordinate

displacement joystick
see isotonic joystick

display
the presentation of data and/or graphics from a system or device in a format designed for human perception through one or more of the senses

display density
the proportion of the total screen area

which is used to present information or data

display format

that arrangement of the data, command areas, messages, and other features on a display

display layout

the grouping of displays at a workplace; *see also* ***display-control layout***

display-control layout

an aspect of workstation design involving both the location and grouping of an integrated layout involving both controls and displays for the human operator; *syn.* control-display layout

disposable

an item which is intended for use only once

distal

remote; a point or region which is farther from the trunk or point of attachment than some reference point; *opp.* proximal

distance

the length of the path between two entities

distilled water

water which has been heated to its boiling point or above to form steam, then condensed with cooling, the process intending to remove minerals and other materials

distort

cause a (usually undesirable) change in the natural shape or form of a physical entity, image, information, or energy waveform; *n.* distortion

distract

divert attention from, prevent concentration on, or inhibit a timely or correct response on some task; *n.* distraction

distractor

any environmental feature which distracts

distributed control

having controlling mechanisms or subsystems at other than a central location

distributed practice

a training or experimental procedure in which practice periods are separated by rest periods or periods of different activity; *syn.* spaced practice; *opp.* massed practice

distribution

see ***frequency distribution***

distribution temperature (T_d)

that temperature of a blackbody radiator whose relative spectral power distribution is essentially the same as that of the radiation source being considered

disturbance input

an undesired input affecting the value of an output signal for which control is being attempted

disuse osteoporosis

an osteoporotic condition induced by lack of use rather than a metabolic dysfunction

diurnal

pertaining to the daytime, or active during the day

diurnal rhythm

see ***circadian rhythm***

    Comment: an older term

divergence

an outward rotation of both eyes to focus on a point further away from the observer

divergent lateral retinal disparity

see ***divergent disparity***

divergent phoria

a tendency for an observer to fixate behind a stationary target

divided attention

that form of attention in which an individual must perform two or more separate tasks concurrently, all of which require attention; *see* ***timeshare***

division of labor

the separation of a job into smaller tasks; *syn.* division of work

do (DO)

a physical basic work element in which a worker performs some operation which results in a change in the form, physical condition, or chemical composition of a product

dock

move a vehicle adjacent to another compatible vehicle or a compatible facility and join the two

dominant eye

the preference for the use of one eye over the other when given a choice

Comment: may be subconscious

dominant wavelength (λ_d)

that visual wavelength represented on a chromaticity diagram by the point of intersection with the spectrum locus of an extended straight line from a sample chromaticity through the achromatic point

Donaldson scale

a scoring system based on a large number of variables for judging how well an individual is capable of performing the activities of daily living, of caring for himself, and of mobility

door

a structure commonly having a thickness much less than its length and width, and which is attached on one side by hinges for use in closing off one volume from another

doorway

a short passageway surrounded by a frame, in which a door may be mounted

Doppler effect

an observed change in pitch or frequency due to a difference in relative velocity between an energy source and a receiver

dorsal

pertaining to the back of the hand or the upper portion of the foot; pertaining to the back or that direction in quadruped animals

Comment: occasionally used synonymously with posterior in the human

dorsal flexor

see **dorsiflexor**

dorsal hand skinfold

the thickness of a skinfold at the middle of the back of the hand and parallel to the long axis of the hand

dorsiflexion

a motion involving raising the toes and upper part of the foot; *opp.* plantar flexion

dorsiflexor

any muscle which raises the toes and upper foot about the ankle joint; *opp.* plantar flexor

dose [1]

a measure of the ionizing radiation absorbed by the body

dose [2]

the amount of a drug presented to the body

dose limit

see **maximum permissible dose**

dose rate

the dose delivered per unit time

dosimeter

an instrument or device which measures exposure to ionizing radiation

double click

press a button on a computer input device two times within a specified brief time period to command two operations at once, such as specify and open a file; *see also click*

double shift

the working of two shifts during a 24-hour period

double underline

a highlighting technique in which two horizontal lines are drawn below a line of text

double vision

see **diplopia**

double-blind

an experimental condition in which neither the adminstrator nor the subject knows the true experimental treatment on a given trial; also double blind

downgrade [1]

a dilution or reduction of the skill level required for a task or job

downgrade [2]

the lowering of a particular job in such aspects as responsibility, scope, degree of difficulty, or wage category

downtime

the time during which an operation cannot proceed or a piece of equipment or instrumentation cannot be used productively due to maintenance, breakdown, lack of materials, or other causes; *also down time*

Draeger tube

see detector tube

drag

move a computer input device such as a mouse such that a screen element or cursor moves across a display; a direct manipulation operation

drawer

a structure which is usually open on one side and closed on all other sides and bottom and which is designed to slide into and out of a cabinet, rack, or other housing

drive [1]

maneuver or control a vehicle designed for essentially 2-dimensional travel, as on the ground or a relatively hard, fixed surface

drive [2]

(sl) motivation

drop delivery

the simple release of an object after being transported to some location where it is to be transported further, stored, disposed of, or processed

drug tolerance

the progessive decrease in susceptibility of the body to a drug's effects resulting from repeated administrations or addiction; *syn.* tolerance

dry bulb temperature

the temperature of a gas or gas mixture as indicated by a thermometric device shielded from heat radiation or conduction; *syn.* air temperature

dual shift

an operating mode in which workers

are working two shifts, usually with the employees divided into two teams

duct velocity

the air velocity within a duct carrying that air to some location

dumb terminal

a CRT and keyboard having no local processing capability other than simple input/output

dummy

see mannikin; see also mannequin

dummy variable

a discrete variable in regression analysis which is not continuously distributed and has at least two distinct values

dura mater

the tough, outermost membrane which covers the surface of the brain and spinal cord; also dura

dust

a collection of small airborne particles either naturally occurring or generated by processing of various types of materials

duty

a set of operationally related tasks

Dvorak keyboard (DSK)

a keyboard with a letter distribution pattern of AOEUIDHTNS on the home row; *syn.* Dvorak simplified keyboard; *see also QWERTY keyboard*

dwell time [1]

that length of time for which the eye is fixated on a given point or within a specified region

dwell time [2]

that period of time which an aircraft or other vehicle is capable of staying at or over its destination/target before having to return

dynamic

involving motion; *opp.* static

dynamic action

any muscle contraction or elongation; *see isotonic action, isoinertial action, isokinetic action, eccentric action, concentric action*

dynamic anthropometry
the study and/or measurement of the changes in body dimensions during motion; *syn.* functional anthropometry; *opp.* static anthropometry

dynamic display
any display containing one or more screen structures which are updated at or near real time; *opp.* static display

dynamic equilibrium
the ability to maintain and control body position while in motion through the integrated involvement of the cristae in the semicircular ducts, vision, and the cerebellum and muscle activity; *see* **static equilibrium**

dynamic flexibility
the ability to perform extent flexibility rapidly and repetitively

dynamic modulus
the ratio of stress to strain under vibroacoustic conditions

dynamic moment
see **angular acceleration**

dynamic muscle work
see **dynamic work**

dynamic response index model (DRI)
a model representing the human torso as single degree-of-freedom system for predicting probability of spinal injury for a given $+g_z$ acceleration time history, assuming a restrained seated crewman in an ejection seat

dynamic strength
a measure of the ability to apply force through a range of motion

dynamic vision
the ability to interpret moving visual stimuli

dynamic visual acuity
a measure of the ability to resolve detail in a changing or moving stimulus; *see also* **visual acuity**

dynamic work
the work performed when one or more muscle lengths change, producing external motion; *syn.* dynamic muscle work; *opp.* static work

dynamics
the study of the body in motion, whether due to internal generation or external forces; *opp.* **statics**

dynamograph
see **chart recorder**

dynamometer
a device for measuring external force or torque, especially that generated by human muscular contraction; *see* **Asmussen dynamometer, Cybex dynamometer**

dyne
that amount of force which will accelerate one gram at one cm/sec^2; *see* **newton**

dysbaric osteonecrosis
a form of decompression sickness resulting in bone lesions, especially near joints; *syn.* aseptic bone necrosis
 Comment: probably due to air embolism

dysbarism
see **decompression sickness**; *adj.* dysbaric

dysfunction
any impaired function of some body part or of the body as a whole

dyskinesia
any of a variety of abnormal involuntary movements, generally due to some pathology in the extrapyramidal system; *adj.* dyskinetic; *see also* **tremor, athetosis, chorea, ballism, movement disorder**

dyslalia
any speech impairment due some defect in the speech-generating structures, especially the tongue

dyslexia
a difficulty in reading, or an inability to learn how to read; *adj., n.* dyslexic

dysmetria
an inability to perform accurate control of a range of voluntary movement, especially of the hand

dysplastic
   a body type which cannot be readily classified as any of Kretschmer's standard athletic, asthenic, or pyknic somatotypes; misshapen

dyspnea
   difficult, labored, or uncomfortable breathing

dystonia
   a movement disorder involving lack of normal muscle tone

E

ear

a structure within and external to the side of the head consisting of three major aspects, which is used for hearing and equilibrium; *see external ear, middle ear, inner ear*

ear breadth

the horizontal linear distance from the most anterior point to the most posterior point of the external ear

> *Comment:* measured with the head level and the scalp and facial muscles relaxed

ear clearing

the process of equalizing pressure between the middle ear and the external environment

ear length

the vertical distance between the highest point of the upper rim and the most inferior point of the ear lobe of the external ear

> *Comment:* measured with the head level and the scalp and facial muscles relaxed

ear length above tragion

the vertical distance along the long axis of the auricle from tragion to the level of the upper rim

ear protrusion

the horizontal distance from the bony eminence directly behind the auricle to the most lateral protrusion of the auricle

> *Comment:* measured with the head level and the scalp and facial muscles relaxed

ear squeeze

see barotalgia

earblock

the failure of the middle ear to equalize pressure with the external environment due to blockage of the eustachian tube

earcon

the auditory counterpart of the visual icon

earcup

the cavity on the lateral interior structure of a helmet, headphone, or other headgear into which the pinna is expected to fit when the headgear is worn

eardrum

see tympanic membrane

earflap

any piece of cloth, fur, or other soft material designed into headwear for protecting the auricle from cold, sun, or other environmental stressors

earlobe

the fleshy tissue at the base of the auricle

earned time

the standard time, in a specified time unit (usually hours), which is credited to one or a group of personnel on completion of one or more jobs

earnings profile (I_t)

an individual's anticipated future annual income from employment

earplug

any device which fits into the external auditory canal for the purpose of reducing the acoustic intensity reaching the eardrum

earring

a piece of jewelry worn on the earlobe

eccentric action

a dynamic muscle action which involves muscle lengthening with an increase in muscle tension; *syn.* eccentric contraction, eccentric muscle contraction; *opp.* concentric action

eccentric contraction
see eccentric action
> Comment: an older term

eccentric muscle contraction
see eccentric action

eccrine gland
a sweat gland whose ducts terminate on the free skin surface; *see also apocrine gland*

echo
(v) display on a computer screen the character or other symbol typed on a keyboard; *n.* echo

echo
(n) an acoustic or electromagnetic reflected energy signal which has sufficient magnitude and delay to be distinguishable from the original emitted signal; *pl.* echoes

echography
see sonography

echoic memory
a sensory memory associated with the auditory system

ecological stress vector
see environmental stressor

economic life
that period of time which either minimizes an asset's total equivalent annual cost or maximizes an asset's equivalent annual net income; *syn.* minimum cost life, optimum replacement interval

economy of scale factor
the ratio of the change in investment cost to the change in capacity

ectocanthic breadth
the horizontal linear distance from ectocanthus of the right eye to ectocanthus of the left eye; *syn.* biocular breadth, bicanthic diameter; *opp.* endocanthic breadth
> Comment: measured with the individual sitting or standing erect, and the facial musculature relaxed

ectocanthus
the junction of the most lateral parts of the upper and lower eyelids, with the eyelids open normally; *syn.* external canthus, lateral canthus; *pl.* ectocanthi; *adj.* ectocanthic; *opp.* endocanthus

ectocanthus to back of head
the horizontal linear distance from ectocanthus to the back of the head
> Comment: measured with the individual standing or sitting erect and looking straight ahead, and the facial musculature relaxed; equivalent to ectocanthus to wall

ectocanthus to otobasion
the horizontal linear distance from ectocanthus to otobasion superior
> Comment: measured with individual sitting or standing erect, with the facial musculature relaxed

ectocanthus to top of head
the vertical linear distance from ectocanthus to the vertex level of the head
> Comment: measured with the individual standing or sitting erect, with the facial musculature relaxed

ectocanthus to wall
the horizontal distance from ectocanthus to a reference wall
> Comment: measured with the individual standing erect with his back and head against the wall, looking straight ahead, and the facial musculature relaxed; equivalent to ectocanthus to back of head

ectomorph
a Sheldon somatotype having characteristics of a thin, frail-appearing body build with little fat or muscle, small bones, and thin chest; *adj.* ectomorphic

eczema
a non-contagious irritation of the skin

edema
an excessive accumulation of fluid in body interstitial spaces which produces a swelling

edit
manually change the data or information in a file, document, or other form of textual or graphic material

effective sound pressure

the root mean square value of the pressure exerted at a given location by an acoustical waveform over a complete cycle; *syn.* root mean square sound pressure, sound pressure

effective temperature [1] (ET)

that temperature in a 50% relative humidity environment resulting in the same total skin heat loss as the actual environment, *syn.* heat stress index

effective temperature [2]

a single number representing a subjective temperature which combines the effects of dry bulb temperature, humidity, and air velocity, with any surrounding structures at the same dry bulb temperature; *syn.* heat stress index

Comment: an older definition

effective thermal insulation value of clothing

see total thermal insulation value of clothing

efferent

conveying information away from a central point, pertaining especially to neural signals; *opp.* afferent

efferent nerve

a collection of one or more axons which conducts signals primarily from the central nervous system to the periphery; *opp.* afferent nerve

efficiency

the effectiveness of some process, usually measured with respect to the amount of output compared to energy, cost, or other measure input

effort [1]

that point of force application on a lever

effort [2]

the expenditure of physical and/or mental energy in the performance of some task

effort arm

that portion of a lever arm from the fulcrum to the point at which an effort is applied; *syn.* force arm

effort rating

see performance rating

effort time

that part of the cycle time during which an employee is required to use his skill and effort

effort-controlled cycle

see self-paced work

egress

exit from a region or space; *opp.* ingress

Eiband tolerance curve

a graph developed from both human and animal data illustrating the likelihood and severity of injuries based on uniform accelerations of short duration

Comment: an older concept

ejection seat

a seat structure which uses rockets or explosive devices to propel a crewmember from a high performance aircraft in a life-threatening, emergency situation

elapsed time

the temporal interval from the beginning point of some activity to a specified or current point of that activity

elastic limit

the level of physical deformation beyond which damage to a structure occurs and/or the structure will not return to its original condition

elbow [1]

the junction of the humerus with the radius and ulna and all surrounding tissues

elbow [2]

that joint in a robotic arm capable of planar motion and corresponding by analogy to the human elbow in function

elbow breadth

the horizontal linear distance between the medial and lateral epicondyles of the humerus; *syn.* humeral breadth

Comment: measured with the flesh compressed, the individual standing erect, and the arms hanging naturally at the sides in the anatomical position

elbow circumference, flexed
the surface distance around the flexed elbow over the olecranon prominence and through the elbow crease
Comment: measured with the elbow flexed 90°, the shoulder flexed 90° laterally such that the upper arm is horizontal, and the hand clenched into a fist

elbow circumference, fully bent
the surface distance around the olecranon prominence and the crease of the elbow
Comment: measured with the elbow maximally flexed and the fingers extended touching the shoulder

elbow circumference, relaxed
the surface distance around the extended elbow
Comment: measured with the individual standing erect and the arm hanging naturally at the side

elbow – elbow breadth
the horizontal distance across the body from the lateral surface of the left elbow to the lateral surface of the right elbow; also elbow-to-elbow breadth
Comment: measured with the individual sitting erect, the elbows flexed 90° and resting lightly against the body

elbow – fingertip length
see forearm – hand length

elbow – grip length
the horizontal distance from the posterior tip of the elbow to the center of the clenched fist
Comment: measured with the elbow flexed 90°

elbow height
the vertical distance from the floor or other reference surface to the height of radiale; *syn.* radiale height
Comment: measured with the individual standing erect and the arms hanging naturally at the sides

elbow rest height, sitting
the vertical distance from the sitting surface to the bottom tip of the elbow; *syn.* elbow rest height
Comment: measured with the individual sitting erect, the upper arm resting vertically at his side, and the elbow flexed 90°

elbow – wrist length
the horizontal linear distance from the posterior tip of the elbow flexed 90° to the tip of the styloid process of the radius
Comment: measured with the individual sitting or standing erect, the upper arm vertical, and the palm facing medially

electric discharge lamp
a source of radiant electromagnetic energy within or near the visible spectrum resulting from the passage of electrical current through one or more materials in the gaseous state

electrical ground
an electrical reference point or return path for current flow; *syn.* ground

electrical impedance (Z)
the total opposition to an alternating current in an electrical circuit; *syn.* impedance

electrical muscle stimulation (EMS)
the stimulation of muscles or muscle tissue with electrical current/voltage

electrical resistance (R)
a measure of the opposition to electric current flow; *syn.* resistance; *see also impedance*

electrical shock
the passage of electrical current/voltage through the body, resulting in the abnormal stimulation of muscles and nerves; *syn.* shock

electrical skin resistance (ESR)
see skin resistance response

electrical stimulation
any form of artificial activation of nerves, muscles, or other materials by the application of electrical current/voltage

electro-silence

the absence of measurable electrical potentials in biological tissues

electrocardiogram (ECG)

a graphical record or other visual display of the electrical activity of the heart as recorded from various points on the body surface, usually consisting of a P wave, a QRS wave complex, and a T wave, depending on the recording locations; *syn.* Electrokardiogram [EKG]

electrocardiograph

the instrumentation used to obtain a graphical recording of heart electrical activity

electrocardiography

the study, measurement, recording, analysis, and/or interpretation of the electrical activity of the heart

electrode

any electrically conductive device used for sensing or applying electrical current/voltage

electroencephalography (EEG)

the study, measurement, recording, analysis, and/or interpretation of electrical activity from the brain

electroencephalograph

the instrumentation used to obtain a graphical recording or the graphical recording itself of brain electrical activity

electroencephalogram (EEG)

a graphical recording or other visual display of the electrical potentials generated by the brain and measured by electrodes attached to the scalp or implanted within the brain itself

electrogoniogram (EGG)

the electronic display or hardcopy record of changes in a joint angle using a potentiometer-equipped or other type electrical goniometer

electrogoniography (EGG)

the measurement, study, or analysis of changes in joint angles using potentiometer-equipped or other type of electrical goniometers

electrogoniometer

an electromechanical goniometer, normally using changes in electrical resistance across a potentiometer to indicate the joint angle

electrokardiogram (EKG)

see electrocardiogram
    Comment: an older term

electroluminescence

the emission of light due to the application of an electromagnetic field to certain materials, and which is not due to heating effects alone; *adj.* electroluminescent

electromagnetic field (EMF)

any combination of an electric field and a magnetic field which occur as a result of natural or artificially-generated electromagnetic radiation

electromagnetic interference (EMI)

a disturbance of some system due to the presence of electromagnetic fields

electromagnetic radiation

electromagnetic waveforms or energy which is propagated through any region; *syn.* radiation

electromagnetic spectrum

the entire, theoretically infinite, range of electromagnetic radiation, typically subdivided into regions according to wavelengths or frequency for practical purposes

electromyogram (EMG)

a graphical recording or other visual display of the electrical potentials generated by a muscle, muscle group, or a large segment of muscle tissue and measured by electrodes placed in or over the tissues involved

electromyographic kinesiology

the use of electromyography in the analysis of human motion; *syn.* correlative kinesiology

electromyography (EMG)

the study, measurement, recording, analysis, and/or interpretation of the electrical activity of muscles; *adj.* electromyographic; *syn.* myography

electron
a small subatomic particle which possesses a negative charge and normally orbits the atomic nucleus

electronystagmogram (ENG)
a graphical recording or other visual display of the electrooculogram during nystagmus

electrooculogram (EOG)
a graphical display or recording of eye movements as detected by surface electrodes positioned on the skin around the eye socket, which is due to the relative orientations between the eyeball (corneo-retinal potential) and the electrodes; also electro-oculogram

electrooculography
the study, measurement, recording, analysis, and/or interpretation of the electrical activity associated with eye movements; also electro-oculography

electrophysiological kinesiology
the use of electrophysiological techniques in biomechanical and kinesiological research and training

electrophysiology
the study of any form of electrical activity of the body, either associated with natural processes or due to external stimulation

electroretinography (ERG)
the study, measurement, recording, analysis, and/or interpretation of the electrical potentials from the retina

electrostatic precipitator
a device which causes deposition of airborne particles through the use of electrical potentials

electrotherapy
the use of various aspects of non-ionizing electromagnetic radiation or conduction in an attempt to heal, reduce pain, or create other beneficial effects

element [1]
a basic division of work, whether for man or machine, consisting of one or more basic, describable, and quantifiable motions or processes

element [2]
see chemical element

element breakdown
a descriptive listing of work elements, with or without certain parameters for each

element time
that period of time required or allowed to perform a specified work element or other portion of a process or task

elemental motion
see therblig

elimination [1]
defecation or urination

elimination [2]
the reduction in the use or importance of an impaired process as proficiency in an alternate process is developed

embedded measure
a hidden process, operation, or test which an individual completes as a subset of a regular job or task, and which is intended to provide another individual or group with information about that person's performance

embolism
a blockage of a blood vessel by some substance

emergency button
a type of emergency stop consisting of a pushbutton installed on or near a piece of equipment which is capable of quickly shutting off electricity to that equipment

emergency lighting
that lighting intended to illuminate an area on failure of the routine supply for the protection of life or property

emergency mover
a skeletal muscle which may be used to assist a prime mover when a very high force level is required

emergency stop [1]
a pushbutton, switch, or other control device installed in or on a piece of equipment which is capable of quickly cutting power to that equipment in an emergency

emergency stop [2]
a rapid cessation of the forward motion of a vehicle to avoid undesirable consequences

emergency switch
a type of emergency stop consisting of a switch located in some readily-accessible position for quickly shutting down a system in an emergency

emmetrope
one who has normal refractive vision

emmetropia
a condition of normal optical vision in which parallel light rays are brought to an accurate focus on the retina without the need for accommodation; *adj.* emmetropic

empirical distribution
a distribution of sampled events or data; *opp.* theoretical distribution

empirical workplace design
the evolutionary design of the working environment based on a combination of human factors engineering and experience

employee
an individual who has an agreement to work for an employer and is compensated by that employer for his time and/or effort

Employee Aptitude Survey (EAS)
a commonly-used test for determining symbolic, verbal, and numeric reasoning abilities, word fluency and comprehension, spatial visualization, visual pursuit, speed and accuracy abilities, and manual speed and accuracy

employee hours
the total number of hours worked by all employees in a facility or company; *syn.* exposure hours

employee participation team
see **Quality Circles**

Employee Retirement Income Security Act (ERISA)
a government regulation with the intent of guaranteeing employees

pensions if they leave a company before retirement age and that sufficient funds will exist to pay pensions when due

empower
give an individual the challenge or opportunity to show creativity, demonstrate personal responsibility, and provide quality work; *n.* empowerment

empty field myopia
the condition of eye accommodation for near, as opposed to far, vision when viewing a homogeneous field; *see* **space myopia**; also empty-field myopia

encoder
any device for coding one or more values for use by another device or computer

end effector
a remote mechanical latching device for gripping, holding, and/or performing work

end item
the final manufactured product, typically built to certain requirements or specifications

end plate [1]
a specialized region of muscle cell membrane in which an axon terminates with extensive branching; also end-plate; *syn.* motor end plate

end plate [2]
a layer of cartilage at the top and bottom of each intervertebral disk

end-plate potential (EPP)
a prolonged potential change from the resting potential across the membrane of a muscle cell which may or may not result in a muscle action potential

endocanthic breadth
the horizontal linear distance between the right and left endocanthi; *syn.* interocular breadth; *opp.* ectocanthic breadth

endocanthus
the junction of the most medial parts of the upper and lower eyelids, with the eyelids open normally; *syn.*

internal canthus, medial canthus; *pl.* endocanthi; *adj.* endocanthic; *opp.* ectocanthus

endolymph
the fluid within the semicircular ducts, the utricle, saccule, and cochlear duct of the inner ear

endomorph
a Sheldon somatotype characterized generally by a soft, rounded body, with greater amounts of fatty tissue, little muscle, and an abdominal protrusion; *adj.* endomorphic

endpoint
see breakpoint

endurance
a measure of the ability to maintain some specified level of effort, usually specified in units of time; *syn.* capacity

energy
the capacity for work, or the amount of work done

energy expenditure
see metabolic rate

energy management
the allocation or use of energy

enfleshment
the use of volumes surrounding body segments or links in human computer modeling to simulate the presence of body tissues

engineer
an individual qualified by education, training, and/or experience to practice in one or more fields of engineering

engineered performance standard
see standard time

engineering
a discipline in which knowledge of the mathematical and natural sciences, gained by some combination of education, training, and practical experience, is integrated with various natural materials and forces to shape the environment

engineering anthropometry
the application of anthropometric data for designing products to be used by humans; *see human factors engineering*

engineering model
a full-size structural model which is functionally identical to and dimensionally corresponds with the intended or actual final production item

engineering psychology
see human factors engineering

engineering tolerance
the maximum degree of variation permitted or allowed on a given specification, drawing, or part; *syn.* tolerance, tolerance specification, tolerance limits

English system
the measurement system predominating in the English-speaking countries such as the United States and Great Britain which uses units of feet, inches, pounds, gallons, etc.; *opp.* metric system

engram
a postulated neural pathway representing the trace of a memory in the brain

engulfment
the process of loose granular materials collapsing to bury an individual

enhancement coding
any technique for increasing the chances that a particular item will stand out against a background
    Comment: e.g., color coding, blinking, bolding

enter
a user operation which signifies the end of a sequence of keystrokes or other operations and directs the computer to take action based on the content of that sequence

entity [1]
one of the more basic graphical elements, such as a line, arc, or circle

entity [2]
an individual, organism, or other object having existence

entraining agent
any event, signal, or cue which is a

driver for maintaining periodicity in biological rhythms; *syn.* Zeitgeber, synchronizer

entrainment
the process of an entraining agent driving a biological rhythm

envelope
a specified volume as determined by some methodology or required function

environment
the combination of all influences impinging on an entity; *adj.* environmental

environmental anthropometry
the study or measurement of changes in an individual's anthropometry due to his physical environment

environmental control
the regulation or alteration of the environment to maintain certain conditions

environmental inputs
the economic, social, psychological, managerial, mechanical, and climatic variables which cause an individual to respond, either physiologically or behaviorally

environmental monitoring
the process of taking of samples from a specified environment, usually air, water, and/or ground, and analyzing them to determine if any hazards are present

environmental sampling
the taking of samples from the environment for analysis; *syn.* sampling; *see also environmental monitoring*

environmental stressor
any condition in the environment which produces stress in an organism, whether climatological, biological, chemical, mechanical, or particulate; *syn.* ecological stress vector

enzyme
a protein which serves to reduce the activation energy required in physiological or chemical reactions, and is chemically unchanged at the conclusion of the reaction

epicondyle
a bony protrusion at the distal end of bones such as the humerus, radius, and femur

epicondylitis
a general inflammation or infection in the area of an epicondyle

epidemiology
the study of the control and distribution of disease or other condition in the community or population, especially pertaining to worker occupational diseases; *adj.* epidemiological

epiglottis
a large piece of cartilage at the top of the larynx which closes the tracheal entrance when swallowing to prevent food from entering

epinephrine
a catecholamine which may act as a neurotransmitter or hormone, depending on the location and source; *syn.* adrenalin

epiphysis
the region at the end of a long bone having an expanded cross-section

eponychium
the thin layer of tissue which overlaps the lunula at the base of a nail; *syn.* cuticle

Equal Employment Opportunity (EEO)
a series of government regulations intended to prevent discrimination in hiring, firing, and promotion of minorities and women

equal-energy white point
see achromatic point

equal-interval scale
a measurement scale which meets the criteria for an ordinal scale and in which items can be classified by value on a linear magnitude measure, with equal distances between measures, but providing no information as to the absoluteness of the magnitudes; *syn.* interval scale

equilibrium
a state in which the body maintains a desired posture or retains control in

body movement through continuous sensory monitoring and the balancing of muscle tensions; *see static equilibrium, dynamic equilibrium*

equinus

a deformity where the foot is continuously plantarflexed

equipment-type flow process chart

a flow process chart which provides a plan or usage record for equipment

equivalent form

any of two or more forms of some test which are very similar in content and difficulty and which are expected to yield similar means and variability for a given group

equivalent groups method

see matched groups design

equivalent mean luminance

the transformed luminance output by a flickering light compared to an equivalent steady light

equivalent sound level (L_{eq})

a calculated value intended to represent a time-varying background noise for comparison to the effects of a constant noise energy level

erect

pertaining to a standing posture in which the individual's shoulders are back and the neck is fully extended

erg

a CGS unit of work; equals 1 dyne acting through a distance of 1 cm

ergograph (Kelso-Hellebrandt)

a device for measuring muscle work output in a series of repetitive movements

ergometer

any device which permits some determination of the work performed by an individual over a period of time

ergonometrics

see work measurement

ergonomic analysis

see human factors analysis

ergonomic design of jobs

see job design

ergonomic job analysis

see human factors analysis

ergonomic lifting calculator

a sliding rule device distributed by the National Safety Council for determining whether or not a lifting task is acceptable

ergonomics

see human factors

ergonomist

a human factors specialist or engineer

Erlanger-Gasser classification

a method for classifying motor neurons, based on conduction velocity, into three primary groups: A, B, and C, with the A group being further divided into four subgroups: α, β, γ, and δ

error [1]

an inappropriate response by a system, whether of commission, omission, inadequacy, or timing

error [2]

any discrepancy between an observed or calculated value and the expected value or a value known to be correct

error rate

the number of errors per division, in which the division may be time, number of products output, motions, or other quantifiable variable

erythema

a reddening of the skin, usually due to heat or ultraviolet radiation exposure; *adj.* erythemal

erythemal threshold

that level at which erythema becomes apparent; *syn.* minimal perceptible erythema

erythrocyte

the discrete red-colored element which contains hemoglobin and carries oxygen and carbon dioxide in the blood; *syn.* red blood cell, red blood corpuscle

esophagus

that portion of the digestive system composed of the passageway extending from the lower part of the pharynx to the stomach; *adj.* esophageal

esophoria
a condition in which the eyes tend to turn inward, preventing binocular vision

esthesiometer
an instrument for measuring touch sensitivity

esthetic
pertaining to the senses, especially when pleasing to the senses

ethics
that moral code practiced by an individual or group, typically referring to a moral code involving honesty, integrity, and other qualities generally judged to be good; *adj.* ethical

ethmoid bone
a relatively complex, irregular-shaped bone within the anterior medial region of the skull behind the nose

ethnic group
a group of people which either maintains affiliation due to strong racial and/or cultural ties or is descended from a certain race or culture

etiology
the study of the causes of disease

eustachian tube
a hollow tubular structure connecting the middle ear with the nasal/oral cavity

evaporative heat loss
the dissipation of body heat through perspiration, indicated by an equation of the form:

$$H = kA(P_s - P_a)$$

where:

H = evaporative heat loss
k = evaporative coefficient
A = body surface area
P_s = saturated vapor pressure of water at skin temperature
P_a = ambient water vapor pressure

evaporative heat transfer coefficient
the value of the ratio of the permeability index to the total thermal insulation value of clothing; *syn.* coefficient

of evaporative heat transfer, evaporative transmissibility

evening person
(sl) an individual who generally likes to go to sleep late at night, likes to sleep late, and has trouble waking early in the morning

evening shift
see second shift

event
a collection of one or more sample points

eversion
a turning of the bottom of the foot outward such that the more sagittal portions are also elevated slightly; *opp.* inversion

evertor
any muscle which is involved in eversion of the foot; *opp.* invertor

evoked potential (EP)
an electrophysiological response recorded from the brain or scalp which is time-linked to peripheral sensory stimulation; *syn.* evoked response

evoked response
see evoked potential

examine (E)
a mental basic work element involving examining a part or product

excess work allowance
a special time allowance given a worker for additional work required beyond that specified in his normal task or job or due to some alteration from usual working conditions; *syn.* additional work allowance

exchange rate
a tradeoff for an increased sound pressure level above recommended limits for a proportionately reduced period of time

excitation
the application of an external input/force to a system which causes that system to respond in some way

excitation purity (p_e)
the distance between a color sample

and neutral white in the 1931 CIE chromaticity diagram relative to the distance between neutral white and the spectrum locus or the purple boundary in the same direction; *syn.* purity

exercise
the use of muscular exertion to maintain conditioning, train for an athletic event, or in an attempt to maintain health

exercise physiology
the study of the metabolic activities and changes ongoing during exercise, including the aerobic and anaerobic mechanisms, and respiratory, neuromuscular, and cardiovascular mechanisms

exhaust air
that air rejected to the outside from a ventilation system

exhaust ventilation
the removal of exhaust air from any space, usually by mechanical means, with replacement air filling the space to dilute the remaining substances

exhausting work
that level of work activity which has a gross metabolic cost of over 380 Calories per square meter of skin surface per hour in young men

exophoria
a condition in which the eyes tend to turn outward, preventing binocular vision; *opp.* esophoria

expectation
a mental set in which an individual anticipates a certain outcome in a given situation

expected attainment
see fair day's work

expected work pace
the rate of work output required to achieve a certain level of earnings or production standards

experience
the verifiable, objective history of one's work performance

experience curve
a graphical plot of a worker's perfor-

mance over time, especially in the learning phase of a job

experience rating
see merit rating

experimental variable
see independent variable, dependent variable

experimenter (E)
one who designs, supervises, and/or conducts research

experimenter error
any error resulting from an experimenter's inappropriate action or inaction, regardless of its nature

expert
an individual who (a) possesses certain knowledge, wisdom, and/or skills in a particular subject not likely to be possessed by ordinary persons, (b) acquired such knowledge, wisdom, and/or skills by study, investigation, and/or experience, (c) is capable of reasoning, inference, and drawing conclusions based on hypothetical facts relating to that subject, and (d) can offer reasonable opinions regarding one or more situations dealing with that particular subject

expert evidence
any testimony given by an expert witness based on objective data or information, or information derived directly from such objective data or information

expert opinion
a statement of belief by an expert witness, based on a given situation

expert system
a decision-making job aid, generally developed in consultation with experts in a given field and which typically contains a computer-based model and database generated from that human expertise

expert testimony
one or more statements made by an expert witness containing opinions and/or evidence

expert witness

an expert who presents testimony or evidence in a court of law

expiratory reserve volume (ERV)

the maximum volume which can be exhaled from the lung after a normal inhalation; *syn.* lung expiratory reserve volume

explosimeter

any instrument which detects and/or measures the presence of flammable gases or vapors in an environment

explosion

a rapid increase in pressure due to physical and/or chemical changes occurring within a confined volume, followed by a sudden expansion into the surroundings; v. explode; *opp.* implosion

explosive

a substance or combination of substances which is capable of exploding

explosive decompression

a rapid and significant decrease in barometric pressure

explosive limit

see flammable limit

explosive range

see flammable range

explosive strength

that force expended in a very short burst of intense muscular activity

exponent

a number conventionally placed to the right and above a base number, representing the power to which the base number is raised for evaluation

exponential distribution

a distribution having the probability distribution function of

$$f(x) = ae^{-ax}$$

where:

a   = 1/mean, and a > 0 for x > 0
f(x) = 0 for x ≤ 0

exposure

a measure representing some combi-

nation of the amount of time an individual or object has been located in some environment and the severity of that environment

exposure hours

see employee hours

exposure limit

the maximum vibrational acceleration as a function of frequency and duration

Comment. an older term

extend

move adjacent body segments connected by a common joint such that the angle between the segments increases in the direction opposite to that of maximum flexion; *n.* extension; *opp.* flex

extended duty hours

see extended work hours

extended functional reach

see thumb-tip reach, extended

extended hours

see extended work hours

extended source

any energy source whose dimensions are significant relative to the distance between the source and the point of observation; *opp.* point source

Comment: significant usually refers to greater than about 10′ of arc for visual work

extended work hours

that working time beyond the normal workday hours; *syn.* extended duty hours, extended hours

extensor

any muscle whose contraction normally causes joint extension; *opp.* flexor

extensor retinaculum

a membranous band of fibers in the posterior hand/wrist which forms the carpal tunnel through which the finger extensor tendons pass; *syn.* transverse dorsal ligament

extent flexibility

the ability to twist, stretch, bend, or reach out with one or more parts of the body on a one-time basis

external
> beyond the outer or surface portion of the body or a body segment; *opp.* internal

external auditory canal
> the tubular structure leading from the external environment to the tympanic membrane; *syn.* external auditory meatus

external auditory meatus
> *see external auditory canal*

external canthus
> *see ectocanthus*

external ear
> the visible, most lateral aspects of the ear, including the auricle, external auditory canal, and the tympanic membrane; *syn.* outer ear

external element
> any work element in a process or operation which is performed by the operator outside the machine- or process-controlled time; *see external work*

external mechanical environment
> the man-made physical environment, consisting of tools, equipment, etc.; *opp.* internal mechanical environment

external naris
> the entrance from the exterior to the air passageway of the nose; *syn.* nostril; *opp.* internal naris

external occipital protuberance
> *see inion*

external pacing
> pertaining to externally-paced work; *opp.* self pacing

external time
> that amount of time required to perform manual work elements when a machine is not in operation

external viewing
> having the capability for seeing outside a vehicle, either to view the vehicle itself or the surrounding environment

external work
> any work element or combination of work elements in a process or opera-

tion which is performed by the operator outside the machine- or process-controlled time; *syn.* outside work; *see external element*

externally-paced element
> a work element whose completion is beyond a worker's control; *syn.* restricted element

externally-paced work
> any manual or human-machine work in which the work pace and/or output is at least in part beyond a worker's control; *syn.* restricted work; *opp.* self-paced work

exteroceptor
> any sensory receptor at the body surface which receives information about the external environment

extinguishing agent
> any substance capable of performing a fire extinguishing function

extorsion
> a rotation of one or both eyes about their vertical axes away from the midline; *opp.* intorsion

extra allowance
> that additional time allowed for the completion of work which is not specified in the standard allowance

extracanthic diameter
> the horizontal linear distance between endocanthus and ectocanthus of one eye

extracellular water (ECW)
> that bodily water external to the cells; *opp.* intracellular water; *see also total body water*

extrafusal fiber
> the contractile fiber of muscle tissue which is capable of generating motion or tension; *see also intrafusal fiber*

extraocular muscle
> any of the six voluntary muscles which are capable of positioning the eyeball within the orbit; *see inferior/superior oblique, inferior/superior rectus, medial/lateral rectus*

extrapolate
> estimate a value beyond current knowl-

edge by using known current values and a predictor; *n.* extrapolation; *opp.* interpolate

extrapyramidal system
a collection of subcortical neural structures involved in skeletal muscle activities which generally have more central integration, are slower than and supportive of pyramidal system motor function, and have involvement with postural motions; *opp.* pyramidal system

extrasystole
a premature heartbeat

extravehicular activity (EVA)
that activity outside a support or transportation vehicle, especially referring to space flight which requires a space suit; also extra-vehicular activity

extravehicular mobility unit (EMU)
an enclosed, self-contained clothing set for protecting the occupant outside a protective vehicle in a hazardous environment

extremely high frequency (EHF)
that portion of the electromagnetic spectrum consisting of radiation frequencies between 30 GHz and 300 GHz

extremely low frequency (ELF)
that portion of the electromagnetic spectrum consisting of radiation frequencies below 300 Hz

extremity
see upper limb, lower limb

extrinsic
pertaining to a structure or mechanism which originates outside the structure on which it acts; *opp.* intrinsic

eye
the total of all structures and tissues enclosing and enclosed within the eyeball

eye blink
a brief closure and re-opening of both eyelids; *syn.* blink

eye blink rate
the number of occasions within a specified temporal interval that an individual executes an eye blink

eye dominance
see ocular dominance

eye height, sitting
the vertical distance from the upper seat surface to endocanthus
Comment: measured with the individual seated erect and looking straight ahead

eye height, standing
the vertical distance from the floor or other reference surface to endocanthus
Comment: measured with the individual standing erect, looking straight ahead, and his weight balanced evenly on both feet

eye movement
any active or passive, conscious or unconscious movement of the eyeball relative to the orbit

eye protection
any device which is capable of shielding the eye in an environment containing one or more possible eye hazards

eye scan
scan the visual field by eye movement alone, not allowing or using any head movements

eye sensitivity curve
see spectral luminous efficiency function

eyeball
the approximately spherical portion of the eye, including the sclera, cornea, pupil/iris, retina, intraocular fluids, lens, and blood vessels

eyebrow
the supraorbital ridge with its associated overlying tissues and hairs

eyeflush
the process of rinsing fluid over the conjunctiva and anterior eyeball with water or eyewash

eyelash
a short, curved hair embedded in the free edges of the eyelids, usually in two or three separate rows

eyelid
a thin, soft movable structure which overlies the anterior portion of the eye-

ball, is capable of closure to protect the eyeball from certain stimuli, is lined on its posterior surface by the conjunctiva, and contains various glands, a muscle, and the eyelashes

eyestrain

a visuo-motor fatigue resulting from a prolonged period of muscle tension to

focus or to overcome glare or any other vision-interfering condition; *syn.* visual strain

eyewash

a solution for flushing the eyes

eyewear

any type of eye covering, whether for eye protection or for improving vision

F

F
a variable obtained from computing the F ratio and used in tests of statistical significance

F distribution
that frequency distribution obtained by taking repeated random pairs of independent samples and calculating the F ratio

F ratio
the ratio of two chi squares divided by their respective degrees of freedom

F-test
the use of an obtained F value with the degrees of freedom for each of the mean squares in an F distribution to indicate the probability that the samples are from the same population; *syn.* variance ratio test

fabric softener
any of a class of cationic amine compounds of substituted fatty acids which act to reduce wrinkling and increase fluffiness while retaining moisture to reduce static electricity / cling; *syn.* textile softener

face
the anterior portion of the head, from crinion to menton, and from right otobasion to left otobasion; *adj.* facial

face shield
any transparent device intended to prevent hazardous substances from contacting the face

face validity
having an apparently relevant or appropriate measure, statement, or data

face velocity
the air velocity required or used at the frontal opening of a safety hood or booth to contain any possible contaminant

facet
a smooth, generally flat surface on a bone

facial angle
that angle formed by the intersection of a line connecting nasion and gnathion with the Frankfort plane of the head

facial breadth
see bizygomatic breadth

facial height
the vertical linear distance between crinion and menton in the midsagittal plane; *syn.* facial length; *see also facial height, total*

facial height, total
the sellion – menton length; *see facial height*
> *Comment:* uses a restricted definition of face

facial index
the ratio of the facial length to the face breadth

facial length
see facial height; see also facial height, total

facial nerve
a cranial nerve having both motor and sensory aspects, and which is involved in facial expressions, cutaneous sensations, and taste

factor [1]
a set of related variables as determined by factor analysis

factor [2]
see variable

factor analysis
a statistical data treatment in which

111

variable scores are analyzed and rotated to obtain orthogonality and achieve a summary in terms of a minimum number of factors

factor loading
a calculated measure of the degree of generalizability between variables and factors in a factor analysis

factorial design
a type of experimental design in which two or more independent variables are examined as part of the same process to permit the study of both their independent and interaction effects on a dependent variable

Fahrenheit (F)
an empirical temperature scale based on 32° as the freezing point of water, and 212° as the boiling point at STP

Fahrenheit degree (°F)
a division of the Fahrenheit temperature scale which divides the range between the freezing and boiling points of water into 180 equal intervals

fail operational
a design characteristic which allows continued operation of a (sub-)system despite a discrete failure

fail operational, fail safe
a fail operational design which also remains acceptably safe

fail-safe
a design aspect of a product or system such that when some portion of it fails, it will either (a) retain operating capability due to sufficient redundancy, or (b) fail in a non-hazardous manner and result in a non-hazardous condition, without injury or additional damage to itself or other systems; also failsafe, fail safe

failure
an inability to perform some intended function or achieve some specified level or range of performance in a task or system under a given set of conditions

failure analysis
see *failure mode and effects analysis*

failure assessment
the process in which the cause, effect, responsibility, and cost of a failure is determined and reported

failure management
the planning, decision-making, and policy implementation which attempt to identify and eliminate potential failures or apply corrective policies/procedures after a failure occurrence

failure mechanism
see *fault*

failure mode
the status in or process during which a piece of equipment failed

failure mode and effects analysis (FMEA)
a methodology which attempts to identify the possible failures of each system component, the mode in which it might fail, the effects of that immediate failure, and the overall system effects on performance; *syn.* failure analysis

failure tolerance
the ability of a system to experience one or more failures and still maintain some functional capability; *adj.* failure tolerant

faint
see *syncope*

fair day's work
a concept of the amount of daily work output expected by management from qualified employee(s), assuming no processing limitations; *syn.* expected attainment

Fair Labor Standards Act (FLSA)
a comprehensive federal employment regulation providing employer requirements such as equal pay, overtime, minimum wage, employment of minors, and record keeping; *syn.* Wage and Hour Law

false alarm
an indication of a problem when no operational problem exists other than in the sensing mechanism; *see also type I error*

false chokes
a choking sensation or cough due to

breathing 100% oxygen for an extended period of time, which results in dry lung tissues

falx

a sickle-shaped structure

far field

that part of a sound field in which spherical divergence occurs, resulting in a sound pressure level decrease of 6 dB for each doubling of distance

far infrared

that portion of the infrared radiation spectrum with wavelengths ranging from about 5000 nm to 1 mm; *syn.* long wavelength infrared; *opp.* near infrared

far ultraviolet

that portion of the ultraviolet radiation spectrum consisting of wavelengths from about 100 to 200 nm

far vision

the ability to see the distant physical environment

farad (F)

a unit of capacitance; that amount of capacitance between two conductors separated by a dielectric with a potential difference of one volt and charged by one coulomb

farmer's lung

see thresher's lung

farmer's skin

see sailor's skin

farsightedness

see hyperopia

fast twitch muscle

see white muscle; opp. slow twitch muscle

fat body mass

that portion of the body mass which is due to fat; *opp.* lean body mass

fat patterning

the distribution of subcutaneous fat throughout the body

fat-free body

a physical/metabolic state in which an individual has only the minimal amount of fat stored in his body

fat-free mass (FFM)

see lean body mass

fat-free weight (FFW)

see lean body weight

fatal accident

an accident causing the death of one or more persons in or as a direct result of that accident

fatality

a death due to any cause

fatfold

see skinfold

fatigue

a state characterized by lack of motivation, interest, and/or an inability to maintain normal, consistent productivity and quality due to recent physical or mental exertion; *see also muscular fatigue*

fatigue allowance

that additional time which is added to the normal time to permit a worker to rest

fatigue life

see cycle life

fatigue-decreased proficiency

a decrease in performance due to prolonged whole-body vibration exposure
Comment: an older term

fatigue-decreased proficiency boundary

those limits of human whole-body vibration exposure for certain time durations at specified frequencies which are intended to maintain a basic performance level
Comment: an older term

fauces

the opening between the posterior mouth and the oropharynx

fault

any condition which may or will cause a system to fail; *syn.* failure mechanism; *see failure*

fault tolerance

a measure of the ability of a system or set of systems to keep functioning after encountering a number of faults

fault tree analysis
 the development and examination of a tree-like structure in which all possible faults that can be thought to occur within a system are given, with linkage of the lower-level to higher level faults, and a description of the necessary and sufficient conditions for each failure

feces
 the collective excretions normally passing through the anus, including undigested and unabsorbed food and intestinal secretions; *adj.* fecal

Fechner's law
 a proposed logarithmic relationship between stimulus intensity and sensory strength, having the form

$$S = k \log I_s$$

 where:
 S = sensory strength
 k = constant depending on the units of measurement and modality
 I_s = stimulus intensity

Federal Standard 595a
 a color ordering system developed by the U.S. Government for standardizing colors used by federal agencies according to a 5-digit code and a gloss/luster criterion

feed
 a mechanism which introduces material to a machine for processing

feedback
 the return of meaningful information within a closed-loop system so that system performance can be appropriately modified; *syn.* knowledge of results

feedback control loop
 see closed loop

Fels index
 an estimate for the percentage of body fat and nutritional status of the body;

$$FI = \frac{(weight)^{1.2}}{(stature)^{3.3}}$$

femoral breadth
 see knee breadth

femur
 the long bone in the thigh; *adj.* femoral

fiber optics
 the use of fine flexible glass or plastic tubing or extruded filaments to carry signals using light pulses

fibrinogen
 a blood protein which precipitates out to form fibers during the clotting process

fibula
 the smaller, more lateral bone of the lower leg; *adj.* fibular

fibular height
 the vertical distance from the floor or other reference surface to the superior tip of the fibula
  Comment: measured with the individual standing erect and the weight distributed evenly on both feet

fidelity [1]
 the degree to which a system's input is reflected in its output

fidelity [2]
 the degree of realism in a simulation

field
 that portion of an interlaced display which is represented by every other horizontal scan line
  Comment: two fields make a frame on an interlaced video display

field of view
 the solid angle within the visual field for which the eye or other optical sensor provides useful data

field study
 an investigation in which subjects are observed or measured in their natural environments

figure
 any drawing, graphical display, photograph, or similar entity composed of more than just text in a document

file
 a collection of information or data which is stored as a single unit or within

a specified restricted location

fill up work

   see internal work

film analysis

   a systematic frame-by-frame study of an activity from motion picture film; *see also video analysis*

film analysis chart

   see film analysis record

film analysis record

   a record generated from a film analysis, containing sequential elemental motions or operations, the beginning and ending clock times, and some type of descriptive symbol; *syn.* film analysis chart

film loop analysis

   a film analysis with a cut and spliced segment of film to form a continuous loop for repeated viewing; *see also cassette loop analysis*

filter

   any device which removes undesired materials, noise, signal, or information

finger

   any of the structures on the hand composed of three phalanges and the surrounding tissues of a digit

finger dexterity

   the ability to make rapid, coordinated finger movements using one or both hands to manipulate small objects

finger diameter

   the maximum medial-lateral cross-sectional diameter of a finger

     Comment: measured by a determination of the smallest diameter hole into which the finger can be inserted; specify the digit involved

finger-shaping

   providing the alternating troughs and ridges on a handle or gripping structure to accommodate the fingers and the gaps between them

fingernail

   the harder elastic tissue covering the dorsal portion of the terminal phalanges of the hand; *see also toenail*

fingerprint

   the pattern of unique whorls and ridges on the pad of the distal phalanx of each finger; *syn.* digital dermatoglyph

fingertip height

   see dactylion height

finite element

   a small segment of a large object obtained by some standard division process

finite element analysis

   the use of finite elements to model force components on a large object or complex structure and draw conclusions about that object or structure as a whole

fire

   a rapid oxidation process involving some fuel and an oxidizing agent which emits heat, light, and often smoke

fire alarm

   any fire protection device or system which indicates the presence of a fire

fire classification

   a division of fires by the types of materials being burned:

   Class A: Ordinary combustibles (e.g., wood, paper)

   Class B: Flammable liquid or gas (e.g., oil, paint, grease)

   Class C: Energized electrical circuits (e.g., electrical wiring, equipment)

   Class D: Combustible metals (e.g., magnesium, sodium, lithium)

fire detection

   the use of any fire protection device or system intended to determine that a fire is present

     Comment: usually sensitive to heat, smoke, or flame

fire door

   any door which has been designed, tested, and rated for preventing the spread of fire

fire point

   the lowest temperature which will support continued combustion after ignition of a vapor- or gas-air mixture

fire prevention

the study and/or implementation of measures specifically designed to control ignition and fuel sources

fire protection

the implementation of measures for preventing, detecting, controlling, and extinguishing of fire to protect life and property

fire resistant

pertaining to a normally non-combustible material which will withstand the effects of a fire

fire retardant

any material or substance which slows the progress of a fire through reduced combustibility

fire wall

any self-supporting vertical structure designed to resist the horizontal spread of a fire from one enclosed region to another

first aid

any emergency care provided to an ill or injured person in order to relieve pain, counteract shock, or prevent death or further injury until better medical care becomes available

first phalanx length

the linear distance of the most proximal segment of a finger

Comment: measured across the surfaces from the distal tip of the third metacarpal to the proximal tip of the second phalanx while the hand is held in a fist; specify the digit involved

first piece time

the time permitted or required for the production of the first complete item in starting a sequence of several complete items

first shift

a day work shift of about 8 hours' duration, approximately between 7 A.M. and 5 P.M.; *syn.* day shift, A shift

first-class lever

a lever in which the fulcrum is located between the effort and resistance

first-order control

see rate control

fission

a type of nuclear reaction occurring in very heavy atoms in which the nucleus, following bombardment by neutrons or other atomic particles, splits into two nuclei of nearly comparable mass, accompanied by the release of energy; *syn.* atomic fission, nuclear fission; *adj.* fissile

fist

a hand posture consisting of a maximal flexion of the hand in which the phalanges of digits II – V (the fingers) are tightly collapsed into the palm with the metacarpals and phalanges of digit I (the thumb) flexed to overlie the fingers

fist circumference

the surface distance around the fist over the thumb and the knuckles

Comment: measured with the thumb lying across the end of the fist

fit [1]

the adequacy, suitability, and/or appropriateness of some individual, equipment, object, or structure with consideration of size, shape, conditioning, or other aspects to perform some function or fulfil a need or use; *syn.* fitness

fit [2]

a sudden, brief exhibition of emotion or motor activity

fit check

see fit test

fit factor

the value of the ratio of the outside concentration of a substance to the concentration of that substance inside a respirator/face mask during a fit test

fit test

the testing of a prototype item on either a sample or potentially the population as a whole to verify that a design is acceptable, appropriate, or the best option for the environment; also fit-test; *syn.* fit check

Comment: usually referring to clothing, personal protective equipment

Fitts' law

a rule for movement time prediction, in which the average movement time in a response is a function of the target separation distance and the width of the target; *see also **index of difficulty***

$$MT = a + b\log_2\left(\frac{2A}{W}\right)$$

where:

MT = movement time

A = distance to target

W = width of target

fixation [1]

the focusing and convergence of the eyes on some point or object at a distance

fixation [2]

having a particular attachment for one technique for performing some task

fixation disparity

a condition in which the visual axes intersect at some point other than in the desired fixation plane

fixation distance

that distance at which the visual axes intersect

fixation muscle

*see **fixator***

fixation plane

that fixation surface which is at such a distance from the observer that the arc may be assumed for practical purposes to be planar

fixation point

that location in a normal individual's line of sight at which the eyes' visual axes intersect; *syn.* point of fixation

fixation reflex

a ocular reflex mechanism which tends to orient the eyes toward a stationary light or object or to keep the eyes oriented toward a light or object which is in motion relative to the observer

fixation surface

that curved surface which is perpendicular to the observer's line of sight and which contains the fixation point of the eyes; occasional *syn.* fixation plane

fixator

a muscle which undergoes an isometric contraction to steady a body part or segment against some other muscle contraction or against an external force; *syn.* stabilizer, fixation muscle

fixed function key

a keyboard key which directs a computer to perform some unchangeable, specific function when pressed; *opp.* programmable function key

fixed linkage mechanism

*see **link***

fixed shift

a work shift in which the working hours remain the same over time

fixture [1]

any device at a workplace used for positioning or holding materials being assembled, worked on, or used

fixture [2]

*see **lighting fixture***

fixture hand

that hand being used to hold an object while the other hand performs some work on the object

flame

the electromagnetic radiation from a fire, typically referring to the visible range

flammable

pertaining to any substance which is readily ignited, burns rapidly, and is capable of rapid flame dissemination; *syn.* inflammable; *opp.* non-flammable; *see also **combustible***

flammable atmosphere

a surrounding gaseous environment which contains a mixture of gases or vapors within their flammable range(s)

flammable limit

*see **lower flammable limit, upper flammable limit**; *syn.* explosive limit

flammable liquid
a liquid which has a flash point below 100°F; *see also* **combustible liquid**

flammable mixture
any combination of flammable vapor or gas and an appropriate oxidizing agent within the flammable range

flammable range
the region of or difference in concentration of a flammable gas or flammable liquid vapor between the upper and lower flammable limits; *syn.* explosive range

flash [1]
a sudden, great increase in brightness for a short period of time

flash [2]
a highlighting technique in which a selected portion of a display momentarily increases in brightness

flash blindness
a temporary inability to see detail or objects having poor illumination following a brief exposure to very intense light

flash burn
any injury to tissue from sudden intense heat radiation

flash point
the lowest temperature at which a liquid gives off sufficient vapor to be ignitable in the surrounding atmosphere

flash rate
the number of times a highlighted portion of a display increases in brightness within a specified temporal interval

flat [1]
a smooth, level surface

flat [2]
having little or no gloss

flatulence
having gas in the gastrointestinal tract

flatus
gas or air expelled from the gastrointestinal tract

flex
move adjacent body segments connected by a common joint so that the

angle formed by the joint and the two segments is decreased; *n.* flexion; *opp.* extend

flexibility [1]
a measure of the mobility of a joint or a series of joints
Comment: quantified as the range of motion, reach

flexibility [2]
the capability for adjusting to varying conditions

flexibility of closure
the ability to discover a known pattern masked by the background material

flexible work schedule
see **flextime**

flexitime
see **flextime**

flexor
any muscle which causes joint flexion; *opp.* extensor

flexor retinaculum
the ligament which forms the carpal tunnel in the wrist through which the finger flexor tendons and the median nerve pass

flextime
a work schedule in which an employee has the freedom within certain limits to choose his work starting and stopping times, but which usually includes a period of time within a given shift during which the employee must be present; also flex-time; *syn.* flexitime, flexible work schedule

flicker
a perceptible temporal variation of brightness or movements occurring several times per second in a display or other source within the visual field

flicker fusion
the perception of a regular, intermittent visual stimulus as continuous or steady by the eye or video sensor; *syn.* fusion

flicker fusion frequency (fff)
the frequency at which flicker fusion occurs; *syn.* critical flicker frequency,

critical fusion frequency, critical flicker fusion frequency

flicker photometry

the use of a single field of view and rapidly alternating light sources of different colors to determine equal-appearing intensity

flicker-free display

a visual display unit with a refresh rate greater than 60 Hz

flight deck [1]

that region of an air- or spacecraft in which the flight controls and instrumentation, the pilot, and others involved in operating the vehicle are based

flight deck [2]

that region of an aircraft-carrying ship on which air- and ground-support operations, including launching and landing, take place

flight simulator

a flight trainer with computer-driven functional displays and controls, possibly including motion

flight trainer

a ground-based pilot training device containing a representation of an aircraft cockpit for familiarization and basic training purposes

float [1]

the amount of slack in a network

float [2]

see **bank**

flood

send multiple messages to the viewing screen, with clearing or scrolling off the screen before all can be read

floor reference plane

that plane through the floor reference point and perpendicular to the local vertical axis

floor reference point (FRP)

that point on a floor or other base surface which provides an origin for representing all other coordinates within the volume of interest

flow analysis

an examination of the progressive sequence of activities and locations of personnel, equipment, and materials involved in the performance of a particular task or operation

flow diagram

a scaled graphic/pictorial representation of the layout and locations of activities or operations and the flow paths of materials between activities in a process

flow line

see **flow path**

flow path

the route(s) taken by personnel, equipment, and materials involved in production as the manufacturing process continues; *syn.* flow line, line of flow

flow process chart

a graphic/symbolic representation using standardized symbols for the manipulations involved for an item through each of the various steps required; *see* **process chart, process chart symbols, person-type flow process chart, material-type flow process chart, equipment-type flow process chart**

flowchart

a diagram consisting of standardized symbols which enclose text and/or other symbols and are governed by specific layout rules for describing the steps involved in a given operation; also flow chart

fluence

the number of particles or photons passing per unit area, usually square centimeter; *syn.* radiation fluence

fluid balance

a physiological state in which water intake equals water loss; *syn.* water balance

fluorence

a hue similar to fluorescent materials

fluorescence

the process of absorbing short-wavelength light energy and emitting light energy with a longer wavelength; v. fluoresce; *adj.* fluorescent

fluorescent lamp
a light source which operates by passing an electrical current through a closed tube containing mercury vapor and one or more suitable fluorescing powders coating the interior surface of the tube

flush-mounted
pertaining to any piece of equipment which is embedded within a structure such that the exposed surface of the equipment is level with the structure surface

flutter 1
any deviation in frequency of the reproduced sound from the original sound

flutter 2
any low-frequency vibration of an object capable of such vibration

flux density threshold
see illuminance threshold

fly
control an aircraft or spacecraft in flight, generally including takeoff and landing

fly-by-wire
a technique for controlling aircraft in which a digital signal is carried by wire to hydraulic actuators in the wings and tail which move the flight control surfaces

flyback timing
see repetitive timing

flybar
a system which provides airspeed, turn, and bank indications via auditory signals, insteasd of the conventional visual flight instruments; *syn.* flying by auditory reference

foam
a fluid mixture of bubbles which floats on or flows over a surface

focused attention
see selective attention

follower
any selected object on a display which

is moved by manipulation of a control

font change
a highlighting technique in which a different font, a different pitch, point size, or representation of the same font, or some other alteration is used

food engineering
the implementation of food science and technology in the manufacturing, processing, and packaging of food items

foot 1
that bodily structure composed of the phalanges, metatarsal, cuneiform, navicular, cuboid, talus, and calcaneus bones with their associated, surrounding tissues; *pl.* feet

foot 2 (ft)
a unit of length in the English system; equal to 12 inches; *pl.* feet

foot breadth
the maximum width of the foot measured perpendicular to its longitudinal axis; *syn.* ball-of-foot breadth
   Comment: measured with the individual standing and his weight distributed evenly on both feet

foot control
any control device intended for normal operation using a foot

foot length
the maximum length of the foot measured parallel to its long axis, from the back of the heel to pternion
   Comment: measured without tissue compression, with the individual standing erect and his weight evenly distributed on both feet

foot restraint
a platform structure which serves to immobilize one or both feet to hold an individual in position for performing a task

foot-leg
involving both the foot and the leg, generally pertaining to sensory or other external influences on both the foot and the leg; *see also leg-foot*

foot-pound (ft-lb)
an English system measure of torque;

equal to one pound of force acting at a distance of one foot from the fulcrum

footbar
a rod or molded tube which serves as a footrest for a chair when the seat pan of the chair is too high for the feet to reach the floor or another surface

footcandle (fc)
a measure of the illuminance produced on a surface, all points of which are uniformly one foot from a point source of light with an intensity of 1 candela; a flux of 1 lumen per square foot of surface; also foot candle

footfall
the striking of the bottom of the foot or footwear on a surface in human gait

footlambert (fL)
a unit of luminous intensity; the luminance of a surface which receives 1.0 lumen per square foot; also foot lambert
Comment: an outdated measure

footrest
any structure on which the foot may rest, usually when seated

footrest angle
the angle between a footrest having a flat surface and the lower leg link

footring
a tube or rod attached in a circular pattern about the legs of a stool or chair as a footrest when the seat pan is too high for the feet to reach the floor or other base surface

footstool
a short structure which is easily portable and may be stood upon to improve one's vertical reach

footswitch
any type of switch which closes when the foot or some portion of the foot makes contact with the floor or ground

footwear
any type of material or covering worn over the foot

force
a physical influence exerted on an object which tends to cause a change in velocity

force arm
see effort arm

force feedback
any means of providing information to an operator about the forces involved on a remote or teleoperated end effector

force joystick
see isometric joystick

force plate
a system consisting of a cover plate and one or more transducers for measuring the forces or accelerations of an object either positioned on the cover or as the object strikes the cover plate; *syn.* force platform, reactance platform

force platform
see force plate

force reflection
providing an operator or system with tactile information about the forces/torques experienced by a remote device

force-velocity curve
a graphical plot showing a characteristic of concentric muscular contractions in which the velocity of a muscular contraction is inversely related to the force of the contraction; *syn.* force-velocity relationship

force-velocity relationship
see force-velocity curve

forced choice
an experimental methodology in which a subject must make a selection from one of the available choices

forced expiratory volume (FEV)
the volume of air from a maximally forceful and complete exhalation of air from the lungs after a maximal inhalation

forced grasping
a movement disorder in the adult in which the victim grasps any object which touches his hand, frequently with great strength

Comment: different from the normal grasp reflex

forearm
the radius, ulna, and all other organized tissues comprising that part of the arm from the elbow to the wrist

forearm circumference
the surface distance around the forearm at the level at which the maximum value is obtained
Comment: measured with the individual standing erect, the shoulder slightly abducted, and the hand relaxed with the fingers extended

forearm circumference, elbow flexed
the maximum surface distance around the forearm with the elbow flexed 90°
Comment: measured with the shoulder flexed 90° laterally (so that the upper arm is horizontal), and the fist clenched

forearm circumference, relaxed
the maximum surface distance around the forearm
Comment: measured with the elbow flexed 90° and the hand relaxed

forearm – forearm breadth, sitting
the horizontal linear distance from the most lateral point on the right forearm, across the body to the most lateral point on the left forearm
Comment: measured without tissue compression, with the individual seated erect, the upper arms hanging naturally at the sides, and the elbows flexed 90° while resting lightly against the torso

forearm – hand length
the distance from the posterior elbow to the tip of the longest finger; *syn.* elbow – fingertip length
Comment: measured with the individual seated erect, the upper arm vertical at the side, the forearm and hand horizontal, and the fingers maximally extended

forearm length
see ***radiale – stylion length***

forearm skinfold
the thickness of a vertical skinfold on the posterior midline of the forearm at the level of the forearm circumference
Comment: measured with the individual standing erect and the arms relaxed naturally at the sides

forefinger
see ***index finger***

forefinger length
see ***index finger length***
Comment: measured with the index finger fully extended

forefoot
the anterior portion of the foot, including the phalanges, metatarsals, cuneiform, and cuboidal bones and the soft tissues surrounding them

forehead
that superior portion of the face from the supraorbital ridges upward and between the maximum lateral bulges of the brow ridges near the ends of the eyebrow; *syn.* brow, frons

forehead breadth
see ***frontal breadth (maximum)***, ***frontal breadth (minimum)***

foreign element
a work element which is not normally part of the work cycle and provides an interruption to it, usually with a random/unpredictable frequency of occurrence

foreseeability
a concept in which an individual may be held liable for actions resulting in injury or damage only if he could be reasonably expected to foresee the risk or danger

forklift
a powered vehicle with two prongs projecting from a front, vertically adjustable platform

form
a display or hardcopy with organized categories for the user or operator to fill in; *pl.* forms

form analysis chart
see form process chart

form process chart
a flow process chart for one or more paperwork forms; *syn.* information process analysis, functional form analysis, form analysis chart, paperwork flow chart, procedure flow chart

formant
a resonance which is associated with vocal tract reflections in the production of sound

formed elements
the enclosed structures within blood, consisting of erythrocytes, leukocytes, and platelets

formulation time
the temporal period required for the end-user and manufacturer to determine what characteristics a desired system should have

forward chaining
a reasoning or control stategy in which the starting point is selected and all possible resulting states are derived from that point, *opp.* backward chaining

forward masking
a form of temporal masking in which the masking stimulus just precedes the test stimulus

fossa
a depression in the surface of a bone

foundation
a structural, knowledge, or economic base which enables further growth or development

foundation garment
underwear
Comment: an older term

Fourier analysis
the mathematical decomposition of a complex periodic waveform into its sinusoidal components
Comment: often used with non-periodic waveforms, however, to get frequency components

fovea
a depressed region within the macula lutea of the posterior retina at which cone density is highest and the greatest visual acuity occurs; *adj.* foveal; *syn.* fovea centralis

foveal blindness
the lack of visual capability in the center of the visual field, due to damage or other problem with the fovea or macula lutea, *syn.* central visual field blindness

foveal vision
that photopic sensory stimulation mediated by the fovea; *syn.* central vision; *opp.* peripheral vision

fracture
a sudden break or crack in a bone or other solid material

frame
one complete scan or image on a CRT, video tape, motion picture film, or other type of display

frame counter
any electrical, mechanical, or electromechanical device which is used to determine and/or display a count or the number of frames displayed on a film or video medium

frame rate
the number of frames recorded or displayed per unit time

Frankfort plane
an imaginary plane through the head established by the lateral extensions of a line between tragion and the lowest point of the orbit; also Frankfort horizontal plane, Frankfurt plane
Comment: for head orientation purposes

Frankfurt plane
see Frankfort plane

free field
see free sound field

free field room
an enclosed volume which provides essentially a free sound field

free float
that calculated additional time available for an activity from the earliest possible completion time of that activity and the earliest possible beginning of the next activity linked to it in a network

free radical
an atom or a chemically-combined group of atoms which have a free electron and are very chemically reactive; *syn.* radical

free sound field
a sound field in which the boundary effects are negligible over the frequencies of interest; *syn.* free field

free-running rhythm
a biological rhythm without the use of entrainment cues, often resulting in a slight change of period

frequency [1]
the number of occurrences of a periodic event within a given unitary time period; the reciprocal of the period

frequency [2]
the number of times a signal crosses a baseline per unit time

frequency [3]
the number of cases within a group or having a particular score or being within a range of scores

frequency distribution
the number of instances obtained or the probability of the occurrence of a score for a given variable value; *syn.* frequency function; *see also probability distribution function, histogram, frequency polygon*

frequency domain
the expression of a function in terms of frequency; *opp.* time domain

frequency function
see frequency distribution

frequency masking
see simultaneous masking

frequency of lift
the number of times a specified mass is raised and/or lowered within a unit time
  Comment: most common time interval is one minute

frequency of use principle
a rule that the most frequently used controls and displays should be placed in optimal locations

frequency polygon
a graphical representation in which the ordinal values corresponding to abscissal values are plotted in a coordinate system and connected by straight lines

frequency rate
see accident frequency rate

frequency response
that range of frequencies which a system is capable of producing or a sensor is capable of detecting

frequency response curve
a graph of the input frequency spectrum vs. output frequency spectrum for a system

frequency spectrum
a description of the frequency components and associated amplitudes of a time series waveform

frequency-time spectrum
see compressed spectral array

fricative
a consonant produced by the steady frictional or turbulent passage of air through a narrowing of a segment within the vocal tract; *syn.* spirant

friction
a force which opposes the motion of a body or tends to hold a stationary body in place; *see static friction, kinetic friction*

Friedman two-way analysis of variance
a non-parametric statistical test using matched sample rank data to test the null hypothesis; also Friedman two-way analysis of variance by ranks

fringe benefit
that compensation to an employee

which is not in the form of wages, salary, or bonuses

frivolous

pertaining to a lawsuit with no basis in fact, and which is based on nonsensical legal theory or intended to harass the defendant or grandstand in court; *syn.* non-meritorious

frons

see forehead

front-end analysis

the process of determining whether or not a problem exists; *syn.* needs assessment; discrepancy analysis

frontal

pertaining to the anterior portion of the body or of a body part, or the frontal plane

frontal arc, minimum

the minimum surface distance across the forehead to the temporal crests at their points of maximum indentation
    Comment: measured with the individual sitting or standing erect and the facial muscles relaxed

frontal bone

the flat bone making up the forehead and superior-frontal portion of the skull

frontal breadth, maximum

the horizontal linear distance between the maximum lateral bulges of the brow ridges near the ends of the eyebrow; *syn.* forehead breadth

frontal breadth, minimum

the horizontal linear distance across the forehead from the points of greatest indentation of the temporal crests; *syn.* forehead breadth

frontal lobe

the most anterior portion of the cerebral hemisphere, extending from the frontal pole to the central sulcus

frontal plane

any vertical plane at a right angle to the midsagittal and horizontal planes which divides the body into anterior and posterior portions; *syn.* coronal plane

frostbite

an injury to the skin and occasionally underlying tissues, depending on the degree, from exposure to extreme cold or windchills

fulcrum

a fixed point representing the axis about which a lever may operate

full-time employment

having a job consisting of about 35 or more hours per week on a regular basis

fumble

an unintentional sensory-motor error

fume

a collection of very small airborne solid particles formed by condensation of volatilized materials; *pl.* fumes

function [1]

that activity which a product or system is to carry out

function [2]

a software-supported capability to aid the user in performing a task or operation

function [3]

a rule defining the behavior of variables within a specified region of values

function area

a portion of a screen display reserved by a given application for a specific purpose

function key

a key which directs the computer to perform some specific function when pressed; *see fixed function key, programmable function key*

Functional Analysis Systems Technique (FAST)

a diagramming process which permits a hierarchy of two-word function definitions derived from a product's consequences and cause

functional anatomy

the study of the body and its component parts, relating them to biomechanical and/or physiological function

functional anthropometry
see *dynamic anthropometry*

functional capacity level rating scale
a seven-point classification for grouping individuals, especially the elderly, according to their ability to perform the activities of daily living

functional deafness
see *psychogenic deafness*

functional electrical stimulation
see *electrical stimulation*

functional flow logic diagram
a technique for determining what operations or processes are necessary to achieve certain objectives from a system

functional form analysis
see *form process chart*

functional impact
a purposeful impact in fulfilling a useful task; *syn.* beneficial impact

functional impairment
a reduced ability to perform certain functions; *syn.* functional limitation

functional injury
a form of trauma not readily detectable by visual examination, but which is indicated by one or more variables measuring a functional limitation

functional leg length
the linear distance from the back at waist level to the heel, measured along the longitudinal axis of the leg

Comment: measured with the individual sitting erect on the edge of a chair and the knee fully extended

functional limitation
see *functional impairment*

functional principle
see *functionality principle*

functional reach
see *thumb-tip reach*

functional residual capacity (FRC)
that volume of air which remains in the lungs after a normal exhalation; *syn.* lung functional residual capacity

functional vibration
an intentional vibration generated to accomplish some end

functionality principle
a rule that displays and controls having related functions should be grouped together; *syn.* functional principle

fundamental frequency
the lowest natural or resonant frequency in a vibrating wave system; *syn.* first harmonic

fusiform neuron
see *gamma motor neuron*

fusion
see *nuclear fusion, binocular fusion, flicker fusion*

fuzzy logic
the use of approximations in reasoning rather than exact, discrete data points or information

G

g
> that force experienced due to acceleration(s) on the body from vehicular or other motion; *see also* **positive g, negative g, transverse g**
>> *Comment:* usually expressed as some multiple or fraction of *g*

g force syndrome
> *see* **acceleration syndrome**
>> *Comment:* an older term

g-induced loss of consciousness
> *see* **gravity-induced loss of consciousness**

g-load
> that loading imposed on the body due to gravity or other accelerations

g-tolerance
> a measure of the ability to withstand positive acceleration(s) without a system failure or blackout

gain [1]
> the ratio of the output value to the input value

gain [2]
> the constant multiplier in the numerator of a transfer function

gain sharing
> any means through which an employee receives benefit in wages from his greater than standard production rates

gait
> the mobility style using an individual's or robotic legs; *see* **walk, run, jog, limp**
>> *Comment:* many clinical types of gaits

gait analysis
> the study of gait
>> *Comment:* usually with the intent to determine mechanisms or quantify disorders

galactic cosmic radiation
> that cosmic background radiation, consisting of extremely high energy particles, which comes from outside the solar system

galley
> that location on certain ships in which food is prepared for consumption; *see also* **kitchen**

gallon
> a U.S. unit of measure for liquid volumes; *see also* **Imperial gallon**

galoshes
> a type of waterproof footwear worn external to the shoes

galvanic current
> direct current from an electricity source, usually a battery
>> *Comment:* an outdated term

galvanic skin reflex
> *see* **galvanic skin response**

galvanic skin response (GSR)
> *see* **skin resistance response**; *syn.* galvanic skin reflex

galvanometer
> an electrical instrument for measuring small electric currents

galvo
> *see* **metal fume fever**

gamma
> a unit of magnetic field strength

gamma angle
> the angle formed by the intersection of the optical axis and the visual axis (line of sight), usually about 4°

gamma efferent
> *see* **gamma motor neuron**

gamma motor neuron
> an A-class motor neuron in the Erlanger-Gasser classification system

having a medium conduction velocity which innervates muscle spindle intrafusal fibers and is involved in regulating muscle activity; *syn.* gamma efferent, fusiform neuron

gamma radiation
the emission or presence of gamma rays in the environment

gamma ray
a very high energy and penetrating form of ionizing electromagnetic radiation having wavelengths between about 10^{-11} and 10^{-13} meters

gang chart
a multiple activity process chart used for coordinating work crews

ganged controls
a set of controls which are grouped or stacked on a single axis, usually having different outside diameters

gangrene
a condition in which a region of body tissues dies, usually due to blood supply failure from injury or disease

Gantt chart
a two-dimensional graphical representation of the planned activities and the dates/times at which each of those activities should be completed over the duration of a project or other activity

Gantt task and bonus plan
a wage incentive plan in which employees are rewarded with a percentage bonus for higher than normal performance

ganzfeld
a homogeneous, uniformly illuminated, formless visual field

garbage in, garbage out (GIGO)
(sl) a phrase indicating that if errors are made in computer input, errors will be present in the output, even if the programming and logic are correct

garment
any piece of clothing intended for wear over one or more body parts

garment design
the development of a garment, ideally

with consideration given to size, style, color, patterns, fabric types, layering, and insulation value

gas discharge lamp
a lamp which produces light at specific wavelengths of the spectrum by electrical excitation of the gas within the lamp; *syn.* gaseous discharge lamp

gas exchange
the diffusion of gases through a membrane or other porous material

gas mask
a transparent face covering device which protects the eyes and has a filter for removing toxic or noxious gases from the atmosphere being breathed

gas tension
the partial pressure of a gas

gastrocnemius muscle
the large voluntary skeletal muscle in the posterior lower leg
Comment: forms much of the calf

gastrointestinal tract
that portion of the digestive tract from the junction of the stomach with the esophagus to the anus

gauge pressure (psig)
the pressure by which a contained fluid exceeds ambient atmospheric pressure; also gage pressure

gauss
a unit of measure for magnetic induction; *syn.* abtesla
Comment: an older term

gaussian
pertaining to or having the appearance of a normal distribution; also Gaussian

Gaussian curve
see **normal distribution**

Gaussian distribution
see **normal distribution**

Gaussian noise
see **white noise**

gaze
look in one direction for an extended period of time

gender [1]

referring to feminine, masculine, or neuter terms in a language

gender [2]

a classification for the male or female of a species; *see sex*

Comment: preferred by some to the term sex when referring to the male and female

gene

a segment of DNA providing the basic unit for transmission of hereditary characteristics; *adj.* genetic

General Aptitude Test Battery (GATB)

a commonly-used test for determining general intelligence, numerical, verbal, and spatial skills, motor coordination, finger and manual dexterity, and form clerical perception

General Aviation Crashworthiness Project

an effort sponsored by the National Transportation Safety Board which was intended to improve the crashworthiness of small airplanes

General Duty Clause

a statement in the Occupational Health and Safety Act of 1970 requiring that an employee be provided with a workplace free of recognized hazards, whether those hazards are specified in the OSHA standards or not

general exhaust

see exhaust ventilation

general hearing

the ability to detect sound and/or discriminate between sounds over a wide range of pitch and loudness

Comment: an older term

General Industry Standard (GIS)

see OSHA General Industry Standard

general lighting

the approximately uniform background illumination within a specific area or volume

general ventilation

see exhaust ventilation

generality (r^2)

see coefficient of determination; opp. specificity

genetic

see gene

genetic mutation

a change in a gene which is reflected in body structure and/or function

genetics

the biological study of heredity

geodesic line

the shortest line which connects two points on a curved surface

geometric mean

the nth root of a product of n numbers; *syn.* geometric average

$$GM = \sqrt[n]{x_1 \cdot x_2 \cdots x_n}$$

geometric progression

a sequence of values corresponding to the form $a, ar^1, ar^2, ar^3, \ldots$

geometric series

an infinite series having the form $a + ar^1 + ar^2 + ar^3 + \ldots$

geometrical axis

see optical axis

geometry

the study of size and shape; *adj.* geometric, geometrical

geriatrics

the study of aging and any diseases associated with aging

germ

(sl) a microorganism

germicidal effectiveness

see bactericidal effectiveness

germicidal lamp

see bactericidal lamp

gerontology

the study of aging processes and their associated problems; *adj.* gerontological

get

pick up and acquire control of an object

Comment: may include several therbligs

giga- (G)

(prefix) 10^9 or 1 billion times the base unit

gimbal

a device with two mutually perpendicular and intersecting axes of rotation which permits orientation or motion in two directions

gingiva

the mucous membrane and other fibrous tissue covering the upper and lower jaws and bases of the teeth within the mouth; *adj.* gingival; *syn.* gum

gingival septum

that portion of the gingiva which lies between two teeth

gingival sulcus

the groove between the gingiva and the tooth surface; *syn.* gingival crevice

gingivitis

an inflammation of the gingiva

girth

the distance around an approximately circular object or cross-section of a structure

glabella

the most anterior point of the forehead between the brow ridges in the midsagittal plane

glabella – inion length

the horizontal linear distance from glabella to inion in the midsagittal plane; *syn.* glabella to back of head, head length; equivalent glabella to wall

Comment: measured with the individual standing erect and looking straight ahead

glabella to back of head

see glabella –inion length

Comment: measured with the individual standing erect and looking straight ahead

glabella to top of head

the vertical distance from the most anterior point of the forehead between the brow ridges to the level of the top of the head

Comment: measured with the individual standing erect

glabella to wall

the horizontal linear distance from a wall to the most anterior point of the forehead between the brow ridges; equivalent glabella – inion length

Comment: measured with the individual standing erect with his back and head against the wall and looking straight ahead

gland

a structure, ranging from a cell to an organ in size, which manufactures, stores, and/or secretes one or more substances for bodily use

glare

an excessive luminance or reflection in the field of view which is greater than that to which the eyes are adapted and which may cause discomfort or interfere with visual performance; *see also discomfort glare, disability glare*

glare sensitivity

the ability to see objects despite the presence of glare or strong ambient lighting

glare shield

any transparent structure which can be used to reduce glare

glass cockpit

(sl) an aircraft cockpit in which the use of multifunctional and computerized displays replaces many of the dedicated gauges and instruments

glassblower's cataract

a heat ray cataract due to prolonged exposure to intense heat from glass furnaces

glaucoma

an abnormally high pressure in the eyeball

glaze

a glossy coating or finish

glenoid cavity

the depression in the scapula inferior to acromion which articulates with the head of the humerus to comprise the shoulder joint

glide

a speech sound generally considered

as being between a vowel and a consonant, and which is produced by movement or gliding from an articulatory position to an adjacent sound

globe temperature
a thermal value representing the composite of the dry bulb temperature, radiational heating, and convection/ wind effects
Comment: measured with the thermometer in the center of a 6" sphere which is assumed to be a blackbody radiator or represent the material being tested

gloss
an attribute of a surface which results in a shiny appearance; *adj.* glossy

gloss trap
a cavity or other structure designed to absorb specular reflections from incident light

glossal
pertaining to the tongue

glossmeter
a photometer for measuring the gloss of a material in the general direction of specular reflection; also glossimeter

glossopharyngeal nerve
a nerve having both motor and sensory components, and generally involved in salivation, muscular control of the pharynx, and taste; *syn.* ninth cranial nerve

glottis
the opening between the vocal cords; *adj.* glottal

glove
an article of clothing which has separate appendages for covering the digits and the remainder of the hand, as well as possibly covering the wrist and some portion of the distal forearm
Comment: to protect tissue from some undesirable or hazardous environment

glove controller
a lightweight glove-like device which is equipped with transducers and can transmit information about arm, hand,

and finger position to a computer for controlling another device

glovebox
an enclosed structure which has openings to which gloves are attached or can be inserted for manipulation of materials and/or organisms, and which is generally sealed to prevent contamination of either the samples being manipulated or the worker and outside environment

glucose
a 6-carbon monosaccharide; *syn.* blood sugar
Comment: the most common type of sugar and the primary metabolic energy source

gluteal arc
that portion of the posterior body surface represented primarily by the curvature of the buttock

gluteal arc length
the surface distance over the buttock from the gluteal furrow to the posterior waist level
Comment: measured with the individual standing erect and the back/ hip/leg muscles relaxed except as necessary to maintain posture

gluteal furrow
the crease at the inferior junction of the buttock and superior portion of the posterior thigh

gluteal furrow height
the vertical distance from the floor or other reference surface to the gluteal furrow
Comment: measured with the individual standing erect and the back/ hip/leg muscles relaxed except as necessary to maintain posture

glycogen
a polysaccharide which is stored in various body tissues as a quick reserve source of sugar/energy; *syn.* animal starch
Comment: converted to glucose when additional energy required

glycolysis
: the breakdown of carbohydrates in bodily metabolism; *see aerobic metabolism, anaerobic metabolism*

go/no-go display
: (sl) a display which provides information from which the user can make only one of two opposing responses

go/no-go reaction
: (sl) one of a set of responses open to an individual in which he either responds (go) or withholds a response (no-go) depending on a stimulus, display, or other input

goal
: an objective for which some activity is initiated and sustained

goal gradient
: the influence of the nearness to reaching a goal on the energy expended toward achieving that goal

goals, operators, methods, and selection rules (GOMS)
: a method for analyzing and/or modeling the knowledge required for interface use

goggles
: any of a class of eye protection or vision enhancement objects consisting either of two transparent individual (yet connected) eye covers or a single rigid structure covering both eyes, but without covering the nose

going rate curve
: a relationship between the evaluation of jobs and their rates of pay in the labor market

Golgi tendon organ
: a stretch receptor located primarily near the tendon-muscle junction which measures muscle tension and provides feedback to the nervous system; *syn.* neurotendinous spindle

gonad
: a primary sex gland, consisting of an ovary in the female or testis in the male

gonial angle
: the point on the lower jaw at which the posterior lower portion of the ramus and lower body of the mandible meet

goniometer
: any instrument for measuring angles, for human joint angles consisting generally of two arms, at least one of which pivots about a center; *see electrogoniometer*

  Comment: the goniometer arms are normally aligned with the bones of adjacent body segments, and the angle read from the pivot point

goniophotometer
: an instrument for measuring the quantity of light emitted/reflected in various directions to determine the spatial distribution of light

goniophotometric curve
: a graph or function showing the light emitted/relected from an object at varying angles of view with a fixed angle of incidence

goodness of fit
: a measure of how well a sample or model approximates a prescribed curve

governing element
: a work element which requires a longer time than any other element being performed concurrently in a work cycle

grab bar
: a handhold or handrail placed in bathrooms, vehicles, or certain other locations for use as a safety or mobility aid; *syn.* grab rail

grab rail
: *see grab bar*

grab sample
: (v) collect an air sample for a short period of time to test for contaminants in a work or other environment

grade [1]
: (n) one level in a series of defined sequential levels according to a set of criteria

grade [2]
: (n) the angle of an incline, either up or down from horizontal

grade [3]

(v) segregate a quantity of some product by quality

gradient

the rate of increase or decrease in magnitude of a variable or response

gram (gm, g)

a unit of mass in the cgs system; *see also kilogram*

graph

a plot of some function or distribution using a coordinate system

graphic

a pictorial hardcopy or display representing an object or a dataset which involves more than simple straight or curved lines

graphic display

a graphic presented on a CRT, flat panel, or other graphics-capable monitor; also graphical display

Graphical User Interface (GUI)

the use of direct manipulation and icons or other graphical symbols on a display to interact with a computer

grapple

close a device on the end effector of a robotic or teleoperated arm to gain control of an object

grasp [1]

(v) position the required number of digits and/or the palm to enable an individual to move, pick up, or hold an object

grasp [2] (G)

(v) a therblig; flex the hand and fingers around an object to gain control of that object

grasp reflex

a grasping motion which occurs on stimulation of the palm or sole of the foot; also grasping reflex

graveyard shift

(sl) *see third shift*

gravitational field ($\vec{g}$)

that vector field due to gravity extending through space which would cause the source and any object entering that field to be mutually attracted to each other

Comment: one of the basic fields in nature

gravitational force

see gravity

gravitational physiology

the study of the effects different gravity levels on body structure and function

gravity (g, $g_{x,y,\text{ or }z}$)

a force which causes objects to attract each other as a function of their masses and the distance between them; *adj.* gravitational; *syn.* gravitational force; *see also g*

Comment: one of the basic forces in nature

gravity feed

the process of using gravitational force to pass materials from one location to another, lower location

gravity-induced loss of consciousness (g-LOC)

that loss of consciousness due to high positive g force maneuvers with the resulting reduction in cranial blood supply in high performance aircraft; *see also grayout, blackout*

gray [1]

an achromatic color between total white and total black

gray [2] (Gy)

the SI unit for the energy imparted to, or dosage absorbed by, a mass from ionizing radiation; *see also radiation absorbed dose*

gray scale

a series of achromatic shades with varying proportions of white and black, to give the full range between total whiteness and total blackness

graying of vision

see grayout

grayout

a condition in which the visual field begins to narrow and decrease in bright-

ness; *syn.* graying of vision; *see also*
gravity-induced loss of consciousness,
blackout

greater multiangular bone
see trapezium

greater trochanter
a large lateral projection of the proximal femur

green
a primary color, corresponding to that hue apparent to the normal eye when stimulated only with electromagnetic radiation approximately between 495 to 575 nm wavelength

Greenwich Mean Time (GMT)
a world time standard; the mean solar time at the Greenwich Meridian

grid
a flat section of a region which is subdivided into smaller, usually square, sections

grievance
any dissatisfaction with working conditions or pay which is expressed by one or more employees to management

grievance committee
a group of workers, usually in a union shop, who have been chosen by their fellow workers to represent employees to management

grievance procedure
any sequence of steps which should be followed in pursuing an employee's grievance through an organization in an attempt to obtain resolution

grind
process using an abrasive disk rotating at high speed

grinder's asthma
a form of pneumonitis caused by the inhalation of fine particles set free in grinding operations; *syn.* grinder's rot

grinder's rot
see grinder's asthma

grip [1]
(v) hold firmly; *see also* **grasp**

grip [2]
(n) that portion of a tool or other device which is normally held by the operator for carrying or operating

grip diameter, inside
the diameter of the widest level of a cone which an individual can grasp with his thumb and middle finger (digit III) touching; *see grip diameter, outside*
Comment: measured at the level of the thumb crotch

grip diameter, outside
the linear distance between the joint of the 1st and 2nd phalanges of the thumb and the metacarpal-phalangeal joint of the middle finger (digit III); *see grip diameter, inside*
Comment: measured with the hand held around a cone at the widest level at which the thumb and middle finger (digit III) can still touch

grip strength
the amount of force which may be applied when grasping or squeezing an object under specified conditions

groin
that region between the thighs at the apex of the pubic crotch

grooving
the practice of designing a tool with grooves to accommodate the user's fingers

gross adjustment
see primary positioning movement

gross anatomy
that portion of anatomy which involves the bodily features apparent to the naked eye

gross body coordination
the ability to integrate motion of the body segments while the entire body is in motion

gross body equilibrium
a measure of the ability to retain or acquire one's balance, regardless of bodily position or motion

gross metabolic cost

the total amount of energy expended to perform some specific activity; *syn.* total metabolic cost; *see also net metabolic cost*

ground [1]

(n) the surroundings of a figure or object which are perceived as behind or not belonging directly to the figure or object of interest

ground [2]

(n) *see electrical ground*

ground [3]

(n) the surface of the earth

ground [4]

(v) restrict from certain activities, especially flying

ground cover

those plants, generally except grass and trees, which cover the soil or terrain

ground current

any current passing to or through the earth from electrical equipment

ground fault interrupter (GFI)

a sensitive, rapidly responding circuit breaker designed to operate at very low current leakage levels to ground for limiting electrical shock below the current-time levels which might result in serious injury; also ground fault circuit interrupter

ground potential

see electrical ground

groundwater

that water which has been filtered through the soil or penetrated rock crevices and lies in underground layers

group

two or more persons having some common relationship or interest

group dynamics

the interactions between the members of a group or their functioning as a unit

group incentive plan

an incentive plan in which a number of workers are collectively rewarded based on the results of an entire group

group technology

a concept that the similarities of part geometric shapes or processes can be grouped to reduce manufacturing costs

growth [1]

an increase in the number of cells and / or cell size

growth [2]

an expansion in consciousness or value

growth curve

a graphic representation of the pattern of increase in some measure

growth rate

a measure of the rapidity in some aspect of individual or entity growth

guard [1]

a person whose primary function is to restrict entry to a certain facility and observe that facility for hazards or violations

guard [2]

any structure designed to restrict or limit entry into some hazardous region of a piece of equipment for preventing injuries

guardrail

any handrail-type structure, usually consisting of posts and interconnecting rail segments, intended to prevent people from contacting a hazard beyond that structure

guideline

a recommended practice, but one that is not mandatory

gum

see gingiva

gustation

the sense of taste; *adj.* gustatory

gut

(sl) the intestines

H

habilitate
bring to an initial state of fitness or capability, as in overcoming a congenital handicap; *see also* ***rehabilitate***

habit
an acquired, well-practiced behavior pattern which is carried out with minimal or no conscious direction

habitability
a measure of the interaction quality of an individual or group with their physical, social, and psychological environment to produce certain working and living conditions

habitable volume
that volume which is suitable for living, containing breatheable air and necessary or reasonable accommodations

habituation
a decline in response or conscious sensitivity to repeated or maintained exposure to one or more environmental stimuli

habituation error
the tendency to keep making the same response, even if the stimulus or conditions change

habutai
a soft, lightweight, plain weave silk

hacking
a massaging technique in which the medial edge of the open hand is brought repeatedly against the body surface

hair [1]
the collective hair shafts growing in various portions of the body, such as the scalp, face, or pubic region

hair [2]
a single keratinized shaft growing from a hair root within the skin

hair esthesiometer
a device developed by von Frey to determine skin touch sensitivity, consisting of a filament attached to some type of holder; *see also* ***von Frey filament***

hair follicle
that structure surrounding the root of a hair in the skin

half-life
see ***radioactive half-life***, ***biological half-life***

halfway-to-hip circumference
the surface distance around the torso at a level midway between the waist height and the trochanteric height levels
> *Comment:* measured with minimal tissue compression

halitosis
a condition in which one's breath is offensive to others

halo effect
a tendency for an evaluator to be overly influenced by an individual's ratings on one trait or due to some past outstanding achievement

halon
a heavier than air chlorofluorocarbon compound used to extinguish electrical fires

hamate bone
one of the distal group of bones in the wrist

hamstring
the tendon for the hamstring muscles

hamstring muscles
a group of muscles in the posterior thigh, consisting of the biceps femoris, semitendinosus, and semimembranosus muscles; occasional *syn.* hamstrings

hand
> the metacarpal and phalangeal bones and other associated tissues normally existing distal to the wrist

hand breadth, metacarpal
> the maximum linear width of the hand across the distal ends of the metacarpal bones; *syn.* hand breadth
>> *Comment:* measured with the fingers extended and adducted

hand breadth, thumb
> the maximum width of the hand at the level of the distal end of the first metacarpale of the thumb
>> *Comment:* measured with the fingers extended and adducted, and the thumb adducted to the side of the palm

hand circumference
> the surface distance around digits II – V at the metacarpal-phalangeal level
>> *Comment:* measured with the hand flat and the fingers extended

hand circumference, over thumb
> the surface distance around the hand, in a plane at right angles to the long axis of the hand, passing over the metacarpals and the metacarpal-phangeal joint of the thumb; also hand circumference including thumb
>> *Comment:* measured with the hand flat, the fingers extended, and the thumb aligned with the index finger

hand control
> any control on a panel or other structure which is used for controlling some process and is normally designed for positioning by the hand

hand feed
> that portion of a machine at which the materials or operating portion are fed for processing at a pace determined by the worker

hand hole
> a slot in the side or end of a container used for carrying items

hand length
> the linear distance from the plane where

the base of the hand/thumb joins the wrist to the fleshy tip of the middle finger (digit III) parallel to the longitudinal axis of the hand
>> *Comment:* measured with the fingers extended and adducted, the wrist rotated/supinated into the anatomical position

hand steadiness
> a measure of the ability to sustain a fixed position of the hand and/or finger with minimal tremor; *syn.* manual steadiness; *see also* **arm-hand steadiness**

hand thickness
> *see* **hand thickness (metacarpale III)**

hand thickness, metacarpale III
> the thickness of the metarcarpo-phalangeal joint of the middle finger (digit III); *syn.* hand thickness
>> *Comment:* measured with the hand flat, fingers extended and adducted

hand time
> *see* **manual time**

hand tool
> any small tool capable of being held and used easily by one or both hands for manufacturing, servicing, or other activities; also handtool

hand-arm
> involving both the hand and the arm, generally pertaining to sensory or other external influences on both the hand and the arm; *see also* **arm-hand**

handcontroller (H/C)
> a small device, usually grasped by or fitting the hand, which responds to axial and/or rotational movements for allowing an operator to control a larger/stronger/remote system; also hand-controller

handedness
> a preference for using one arm-hand or the other, or a combination of the two

handhold
> a structure consisting of a segment which normally is of elliptical- or rod-shaped cross-section and of suitable outside perimeter and length to permit

a hand to grasp it for carrying, for assistance in remaining in a desired position, or for mobility

handicap [1]

a compensating factor which attempts to equalize performance levels on one or more aspects in some activity

handicap [2]

a physical or mental condition which prevents an individual from functioning at a normal performance level, especially referring to those functions such as activities of daily living; n., adj. handicapped

handle [1]

(n) a structure designed for gripping an object

handle [2]

(v) move an object or material from one location to another, via a suitable combination of motions; ger. handling

handling aid

see **job aid**

handling time

the period of time required to move parts or materials to or from a work area or operation

handrail

a structure of much longer length than thickness/width, which is of suitable outer perimeter for hand grasping either to remain in a given position or to aid in mobility

handwear

any form of clothing worn over the hand

handwheel

a large control device intended for rotation when a mechanism requires a greater amount of torque than can be applied by a knob

happiness sheet

(sl) a written survey obtained from students at the end of a course or training session to provide feedback to the instructor regarding various aspects of the training

haptic

pertaining to the sensation of pressure;

occas. syn. tactile

hard hat

(sl) a safety helmet maintained in position on the head by straps, for protecting the wearer from injury by falling objects

hard light

a light source which causes objects to cast well-defined shadows

hard palate

the anterior portion of the roof of the mouth, backed by the maxilla and palatine bones and covered by mucous membranes; see also **soft palate**

hard soap

any soap made with sodium hydroxide and packaged in bar form; opp. soft soap

hard water

water having a relatively high concentration of dissolved calcium and magnesium salts; opp. soft water

hardcopy

a paper or other sheeted material display; also hard copy, hard-copy

hardware

equipment or components made of physical materials, often referring to the electronics and structural portion of a computer

harmonic

one member of a harmonic series; see **harmonic series**

harmonic mean

$$\overline{H} = \frac{n}{(\frac{1}{X_1}) + (\frac{1}{X_2}) + (\frac{1}{X_3}) + \ldots + (\frac{1}{X_n})}$$

harmonic motion

see **simple harmonic motion**

harmonic series

a set of overtones whose frequencies are separated by integral multiples of the fundamental frequency

harmonic vibration

see **simple harmonic motion**

harness

any combination of straps intended to hold an occupant of a vehicle in his seat, especially those straps holding the torso against the seatback

Harvard step test

a physical fitness test in which an individual repeatedly steps from a floor level to a bench, with the physiological variable being the heart rate; *syn.* step test

hat

any head covering made largely of soft materials, but having a rigid shape

hatch

a full-body or materials passageway through some solid structure which may be sealed to separate different fluids or pressures

Hawthorne effect

a phenomenon in which employee-perceived interest by the employer proved to be a factor in productivity and employee morale, as well a physical environmental variable being examined in an experiment, leading to a confounded experiment

> *Comment:* based on a study at the Western Electric Co. Hawthorne Works plant, Chicago; often generalized to apply to confounded results from unconsidered variables in experiments

hazard

any real or potential condition which either has previously caused or could reasonably be expected to cause personal injury or property damage; *syn.* unsafe condition; *adj.* hazardous

hazard analysis

an examination of one or more situations to identify and eliminate or control possible hazards

hazard elimination

the removal of a known, already existing hazard

hazard pay

see hazardous duty pay

hazard recognition

the detection and realization that a hazardous condition exists

hazardous atmosphere

any atmosphere which could cause injury or death to an individual remaining within it and not having suitable personal protective equipment

hazardous duty pay

the additional monetary compensation given to workers performing dangerous tasks; *syn.* hazard pay

hazardous material (HAZMAT)

any substance possessing a high risk of harmful effects to persons or equipment if exposed

haze

a cloudiness in a surface or coating

head [1]

that part of the human body superior to the neck when standing erect, including the skull and facial bones, skin, brain, and other associated tissues

head [2]

a point of origin, as in a muscle

Head And Neck Support (HANS™)

a head, neck, and upper torso restraint modeling system, consisting of a helmet and tethers, for minimizing neck injuries in a vehicular crash

head breadth

the maximum linear side-to-side width of the head superior to the auricles

> *Comment:* measured at whatever level provides the maximum, with minimal tissue compression

head circumference

the maximum surface distance around the head, including the hair, at a level just above, but not including, the brow ridges; *syn.* occipitofrontal circumference

> *Comment:* measured with hair compression

head diagonal, inion to pronasale

the linear distance from inion to pronasale

> *Comment:* measured with the face and scalp muscles relaxed, without tissue compression

head diagonal, maximum, menton to occiput

the maximum linear distance from menton to occiput

Comment: measured with the face and scalp muscles relaxed, without tissue compression

head diagonal, maximum, nuchale to pronasale

the maximum linear distance from nuchale to pronasale; *syn.* head diagonal, nuchale to pronasale

Comment: measured with the face and scalp muscles relaxed, without tissue compression

head height

the vertical distance between tragion or the lowest point on the inferior orbit and the horizontal plane which intersects the vertex in the midsagittal plane

Comment: this uses a restricted definition of "head"

Head Injury Criterion (HIC)

a measure for determining the tolerance to concussion in a head impact, based on the duration and acceleration involved:

$$HIC = \left[t_2 - t_1 \right] \left[\frac{1}{t_2 - t_1} \int_{t_1}^{t_2} a(t)dt \right]^{2.5}$$

where:

t_1 = start time of impact
t_2 = end time of impact
$a(t)$ = acceleration function (in g units)

Comment: an HIC value of 1000 with a duration of less than 15 msec is an acceptable tolerance level

head length

see glabella – inion length

head length, maximum

the horizontal linear distance between pronasale and inion in the midsagittal plane

head movement

any motion of the head as a unit, relative to the torso

head scan

scan through the visual environment using head movements, allowing for accompanying eye movements

head-down display

a display, generally located on a control panel, which requires the operator to lower his normal line of sight to obtain the desired information; *opp.* head-up display

head-mounted display (HMD)

any system which can be attached to the head, neck, and/or shoulders for enabling presentation of a head-up display

head-up display (HUD)

a display in which information is presented on a nearby transparent surface such that the operator is capable of viewing both the information and the external world with his normal line of sight; also heads-up display; *opp.* head-down display

headform

an object whose shape resenbles that of the human head for sizing, modeling, or simulation purposes

headgear

any protective structure worn on the head to protect the individual from possible injury due to hazards, usually from impacts; *see also* **headwear**

headgear retention

a measure of the ability of a piece of headgear to remain in place during an impact and any post-impact events

headgear retention assembly

any combination of chinstraps, napestraps, internal form fitting, or other techniques to aid in headgear retention

headrest

any padded structure which provides support to the head when sitting or reclining

headroom

that distance available to accommodate an individual's head, generally referring to that distance between the vertex of an individual's head and a roof, passageway, or other limiting environmental feature when standing, sitting, walking, or other motion/posture as the situation requires

headset
a device having one or a pair of transducers for converting electrical energy
to sound and having a spring mechanism or other device over the head,
under the jaw, or around the neck to
hold it/them in place

headward g
see negative g

headwear
any form of clothing worn only on or
around the head, such as a hat, cap, or
helmet; *see also headgear*

health
a state in which an individual's and/or
population's mental, physical, physiological, and social conditions are
within normal limits; *see also mental
health, physical health*

**Health And Nutrition Examination
Survey** (HANES)
*see National Health And Nutrition
Examination Survey*

health index
any qualitative or quantitative measure for describing the relative or absolute health of an individual or a population

health insurance
a program which includes some percentage of payment or reimbursement
for medical, dental, vision, counseling,
and/or other care beyond a specified
deductible limit
Comment: often a fringe benefit paid
at least in part by employers and
generally used to provide financial
protection in the event of a major
family health problem

health physicist
an individual with the appropriate
training, experience, and education in
health physics, and who is practicing
in that field

health physics
that field involved in the study and
implementation of radiation protection

hearing
that specialized sense through which
sound is perceived

hearing aid
a device which amplifies sound intensity or filters noise, typically for use by
persons with hearing impairments

hearing conservation
the prevention or minimization of hearing loss in noisy environments by using some combination of procedures,
exposure monitoring, hearing protection devices, noise control, or noise
reduction

hearing impairment
a reduction in one's hearing ability; the
deviation of an individual's absolute
auditory threshold in decibels using a
calibrated audiometer or by comparison to the absolute auditory threshold
of a person with normal hearing; *syn.*
hearing loss

hearing level
see hearing threshold

hearing loss
see hearing impairment

hearing protection device (HPD)
any physical structure covering the
auricle or inserted into the external
auditory canal which is intended to
prevent hearing loss in noisy environments

hearing scotoma
see tonal gap

hearing test
any method of evaluating hearing capabilities; *see audiometry, tuning fork
test*

hearing threshold
that sound intensity or pressure level
at which an individual reports hearing
test tones at a given frequency presented to the ear, and below which no
indication is given that sound is detected; *syn.* hearing level, hearing
threshold level

hearing threshold level
see hearing threshold

heart

the multi-chambered muscular organ within the thorax which pumps blood through the circulatory system

heart rate

the number of complete heart contraction cycles per minute; *syn.* pulse rate

heartburn

(sl); *see reflux esophagitis*

heat

a form of energy which can be transmitted from a system to its surroundings or to another system via radiation, convection, and/or conduction due only to temperature differentials

heat acclimatization

a physiological adjustment to living or working at higher external temperatures and/or humidities

heat balance

the difference between the heat produced by the body and that which is given off to the environment

heat capacity

that heat energy absorbed by an object under given conditions for each degree rise in temperature; *see also specific heat*

heat collapse

see heat exhaustion

heat conduction

heat transfer from one entity to another via direct contact

heat conservation

any mechanism such as peripheral vasoconstriction, piloerection, or reduction in sweating which may be used to retain heat within the body; *syn.* heat retention

heat convection

heat transfer from one entity to another or within an entity via a fluid capable of storing heat, such as air

heat cramp

a painful spasm of one or more voluntary muscles as a result of water and/or salt deficiency in heat stress; *syn.* miner's cramp, stoker's cramp

heat exhaustion

a condition consisting of muscular weakness/fatigue, reduced perspiration, vertigo, headache, and/or nausea, due to surface blood vessel collapse following salt and water loss; *syn.* heat prostration, heat collapse

heat index

see heat stress index

heat loss

the release of heat from the body to the environment via conduction, convection, radiation, or evaporation; *syn.* heat dissipation

heat prostration

see heat exhaustion

heat pyrexia

see heat stroke

heat radiation

the transfer of heat via electromagnetic radiation; *syn.* thermal radiation

heat ray cataract

an opacity in the lens of the eye which occurs in occupations requiring long exposures to high temperatures and glare; *see glassblower's cataract*

heat regulation

see thermoregulation

heat retention

see heat conservation

heat strain predictive system

a method for predicting heat stress based on variable clothing effects

heat stress

the excessive cooling load imposed on the body by ongoing activity or simple presence in a hot and humid environment; *opp.* cold stress

heat stress disorder

see heat cramps, heat exhaustion, heat stroke

heat stress index (HSI)

any of a number of estimators for body heat stress which may be based on temperature, humidity, air velocity, workload, clothing, and their interactions; *syn.* heat index; *see* Belding-Hatch heat stress index, effective temperature, index of physiological effect, index of relative strain, predicted four-hour sweat rate, wet bulb globe temperature, wet bulb globe temperature index, wet bulb temperature index, wet globe temperature, wet Kata thermometer

heat stroke
an excessive rise in body temperature, followed by prostration due to over-exertion for the working conditions and/or failure of the body thermoregulatory system; also heatstroke; *syn.* heat pyrexia

heat syncope
the fainting or collapse due to heat stroke

Heath-Carter somatotype
a body type classification system which uses a combination of anthropometric measures (such as stature, weight, skinfolds, girths, and breadths) for determining or modifying the basic classifications

heavy ion
an ion having a normal atomic mass equal or greater than that of carbon

heavy metal
any metal having a specific gravity of greater than or equal to about 5

heavy work
that level of work activity which involves the entire body and has a gross metabolic cost of 280 – 380 Calories per square meter of skin surface per hour

heel
the calcaneus and surrounding soft tissues of the inferior and posterior portion of the foot

heel – ankle circumference
the surface distance around the foot under the tip of the heel and over the instep at the junction of the foot and anterior lower leg
Comment: measured with minimal tissue compression, minimal weight on the foot being measured, and the foot muscles relaxed

heel bone
see calcaneus

heel breadth
the maximum medial to lateral linear width of the heel behind the vertical projection downward from the ankle bones
Comment: measured with the individual's weight equally distributed on both feet and with minimal tissue compression

height
the straight-line vertical distance from the floor or other reference surface to the level of the referenced body part or of the top an object; *see also* **stature**

height – breadth index of the nose
see nose height – breadth index

height velocity
the rate at which stature increases during physical maturation

helix [1]
the rolled outer portion of the auricle

helix [2]
a spiraling geometrical pattern

helmet
a piece of headgear with a hard exterior covering and internal cushioning designed to fit over the top of or enclose the entire head to protect the head from impacts or other hazards

helmet-mounted display (HMD)
a display projected within or on the visor of a user's helmet such that both the information presented and the external environment are simultaneously within the line of sight

Helmholtz resonator
a passive acoustical filter consisting of a cavity with a narrow neck and an enlarged interior

Helmholtz-Kohlrausch effect
a tendency for apparent brightness to increase as color saturation increases

help
an on-line software user assistance feature

hematocrit
the percentage of the red blood cell volume in the total blood volume; *syn.* hematocrit reading

hematoma
an enclosed volume of blood in tissue external to the circulatory system, from whatever cause; *see also* **bruise**

hematopoiesis
see hemopoiesis; adj. **hematopoietic**

hematopoietic radiation syndrome
see hemopoietic radiation syndrome

hemi-
(prefix) half; pertaining to one side of the body

hemianopsia
a unilateral or bilateral blindness in one half of the visual field

hemiballismus
a unilateral form of ballismus

hemiplegia
a condition in which one side of the body (especially both limbs) is affected by paralysis; *adj.* hemiplegic

hemodynamics
the study of the physical principles of blood and its circulation

hemoglobin (Hb)
the iron-containing protein which serves as the respiratory pigment of erythrocytes, and for carrying oxygen from the lungs to bodily tissues and carbon dioxide to the lungs

hemorrhage
bleeding; the loss of blood from the cardiovascular system

hemorrhoid
an enlarged blood vessel in the anal or rectal wall; *syn.* pile

hemostasis
the stoppage of blood flow or loss

hemothorax
a collection of blood in the pleural cavity

henry (H)
the inductance of a closed circuit in which a potential of one volt is produced when the electric current in the circuit is uniform at one ampere per second

herb
a plant which may be used as food flavoring or for medicinal purposes

herbicide
a chemical which kills plants or inhibits plant growth

heredity
the characteristics passed from parent to child via genes; *adj.* hereditary

Hering opponent process theory
see **opponent process theory**; *syn.* **Hering theory**

hernia
a failure/rupture or weakness in the wall of a bodily structure, usually a rupture of the abdominal wall or an intervertebral disk which results in the protrusion of part of an organ or tissue through the failure; *adj.* hernial

herniate
create a hernia

herniated disk
a protrusion of the nucleus pulposus of an intervertebral disk into or through the annulus; *syn.* slipped disk

Hertz (Hz)
a measure of frequency in cycles per second (cps); *syn.* cycle per second

hetero-
(prefix) relating to combinations of different entities; *also* heter-

heterochromatic
combining or pertaining to two or more different colors

heterochromatic flicker photometry
a technique used for measuring an observer's relative sensitivity to light of different wavelengths by comparing a fixed-luminance reference light alternating with a light of different wavelength to minimize the sensation of flicker

heteromodal
see **multisensory**

heterophoria

a tendency to position the eyes such that binocular vision can not be used

heuristic

pertaining to a learning or problem solving technique which uses certain empirical rules or guidelines to ultimately reach a solution

heuristic program

a set of instructions which directs a computer to use a heuristic approach to problem solving

hexadecimal

pertaining to a numbering system based on 16 using the alphanumerics zero through nine and A through F

Hick-Hyman law

a rule that the choice reaction time is linearly related to the logarithmic transformation of the amount of stimulus information presented

$$CRT = d + t_b H$$

where:

CRT = average choice reaction time

d = summed time required for all nondecision-making activities, e.g., stimulus transmission time plus motor response time, assumed to be a constant

t_b = time required to process one bit of information, assumed to be a constant

H = amount of information in bits (= $\log_2 N$), often taken as the number of available choices

hidden digit test

(sl); *see Stilling test*

hidden line

a graphic line not displayed on a model, especially a wire-frame model, which would not be visible from a particular view if the model were solid

hidden window

a display window partially or completely covered by another

hierarchical decomposition

the breakdown of a high-level task into smaller, lower level steps

hierarchical menu

a menu structure or format in which each item on a given menu has another menu consisting of a subset of additional selections until the lowest level menu is reached

high density lipoprotein

a substance present in blood which functions to return cholesterol to the liver for reprocessing and elimination; *see also low density lipoprotein*

high fidelity

pertaining to an audio or graphic/photographic reproduction which is comparable with the original; *syn.* hi-fi

high frequency (HF)

that portion of the electromagnetic spectrum consisting of radiation frequencies between 3 MHz and 30 MHz

high task

see incentive pace

high-definition television (HDTV)

a video medium with a resolution of approximately 1200 lines

high-energy heavy ion (HZE)

a high-velocity particle consisting of an ionized heavy atom

high-speed cinematography

the sampling of activities using motion picture film with a frame rate much higher than the normal projection rate; *syn.* high-speed photography

high-speed photography

see high-speed cinematography

highlight

use some feature different from the background to attract a user's attention to some portion of a display

highpass filter

a device which allows frequencies higher than the cutoff frequency to exit from the device unattenuated, while the intensity of frequencies lower than the cutoff frequency are attenuated;

also high-pass filter; *opp.* lowpass filter

highway
a public roadway, usually one which interconnects urban areas and allows high speed vehicular travel between such areas

hinge joint
the type of joint which permits only a single degree of freedom, as rotation about a pivot point within a plane
Comment: e.g., elbow, knee

hip
the coxal bone, its joints with the sacrum and femurs, and all the associated surrounding tissues

hip bone
see coxal bone

hip breadth, sitting
the maximum horizontal linear distance across the widest portion of the hips
Comment: measured with the individual sitting erect, knees flexed at 90°, knees and thighs together, and feet flat on the floor

hip breadth, standing
the maximum horizontal linear distance across the lower torso in the hip region
Comment: measured with the individual standing erect, feet together, and his weight distributed evenly on both feet

hip circumference at trochanterion
the surface distance around the hip at the trochanteric height
Comment: measured with the individual standing erect and his weight equally distributed on both feet

hip height
see trochanteric height

hip joint
the joint composed of the junction of the femur head and the coxal bone

histogram
a graphical representation of two or more amplitude measures using rectangular shapes along either a discrete or continuous dimension; *syn.* bar graph, bar chart

historical data
that data which have been previously collected in a given work situation and serve as a standard reference for performance
Comment: typically refers to historical time, but not necessarily restricted to that

historical time
see statistical standard time

histotoxic hypoxia
an inability of the tissues to use oxygen, even though it is present in amounts equal to or greater than normal

hold (H)
a therblig; a work element in which an object is held in a fixed orientation and location by the hand or other body member

homatropine
a chemical which dilates the pupil and paralyzes accommodation when applied to the eye surface
Comment: usually applied during an eye examination to permit viewing of the eyeball interior

home row
the row of letters in a typewriter or computer-style keyboard on which the fingertips normally rest when typing in a standard mode

homeostasis
the concept of a baseline steady state of approximately constant physiological conditions within the body; *adj.* homeostatic

homeothermy
the ability of some species to regulate body temperature within narrow limits, despite large temperature fluctuations in the environment; also homoiothermy; *adj.* homeothermic; *syn.* warm-blooded

homesickness
a strong desire to be home such that one becomes sluggish and performance is affected, with possible psychoso-

matic or other symptoms if prolonged; *adj.* homesick

homogeneous menu hierarchy
a menu hierarchy having the same number of options in each menu

homograph
a word which is spelled the same as another, but which has a different origin, pronunciation, and/or meaning

homoiothermy
see homeothermy

homologous
having the same structural relationship

homologous motion
a movement which can be achieved in more than one way

homoscedasticity
a condition in which each distribution has the same variance; *adj.* homoscedastic

homunculus
a representation of the human body mapped onto the surface of the brain cortex; *see sensory homunculus, motor homunculus*

hook
a point within software at which additional steps or code can be easily added at a later time

hook grip
a type of grip where only the fingers flex around an object, with the thumb not being used

horizon
the apparent boundary line between the earth's surface and the sky

horizontal axis of Helmholtz
the horizontal axis connecting the centers of rotation of the two eyes

horizontal disparity
see binocular disparity

horizontal job enlargement
see job enlargement

horizontal leg room
see knee well width, knee well depth

horizontal plane
any plane parallel to the floor, ground,

or other reference surface

horizontal scroll
move the cursor sufficiently to the left or right under operator control such that the display changes to present information not visible before; *opp.* vertical scroll

hormone
a chemical released from certain glands or cells which are absorbed into the bloodstream for transportation to a target tissue in a distant part of the body where its action is initiated

horology
the study of time measurement, including the principles and technologies involved in time-measuring devices

horopter
the locus of points in space which produce images falling on the corresponding points of both eyes with a constant amount of convergence such that a single image is seen

horsepower (hp)
an English unit of power

hostility
an outwardly directed expression of anger, animosity, or antagonism toward another entity; *adj.* hostile

hot-wire anemometer
a device which is used to measure low air movement levels by the rate of heat loss from a resistive wire

hour (hr)
a unit of time, corresponding to 1/24 of the time required for the earth to rotate about its axis

housekeeping
the process of cleaning or maintaining a state of cleanliness, neatness, and order of an occupied or occupiable space

hue
a perceptual attribute of color determined primarily by the wavelength of the light entering the eye

hue composition (H_C)
an expression of hue as percentages of

the components

hue contrast
see *chromatic contrast*

human describing function
see *human transfer function*

human ecology
the study of the relationships of individuals with each other and with their community's environment

human engineering
see *human factors engineering*

human error
any error due to a human behavior or response; *syn.* operator error; *see also* error [1]

human error probability (HEP)
a measure of the likelihood of occurrence of a human error under specified conditions

$$HEP = \frac{error\ count}{number\ of\ possibilities}$$

human error rate prediction analysis
see *Technique for Human Error Rate Prediction*

human factors
that field which is involved in conducting research regarding human psychological, social, physical, and biological characteristics, maintaining the information obtained from that research, and working to apply that information with respect to the design, operation, or use of products or systems for optimizing human performance, health, safety, and/or habitability; *syn.* ergonomics

human factors analysis
a systematic study of those elements involving a human-machine interface or other situation with the intent of improving working conditions, operations, or an individual's well-being; *syn.* ergonomics analysis

human factors engineer
one who has the appropriate education, training, and experience to be capable of properly performing human factors engineering activities

human factors engineering
the use of information derived from human factors research, theory, and modeling for the specification, design, development, testing, analysis, and evaluation of products or systems for human use

human factors specialist
an individual who has the necessary educational, training, and experiential background to have a working understanding of human factors principles and be capable of research or other work toward achieving human factors' goals

human modeling
the use of any system which is capable of modeling one or more human structures or other characteristics for education, research, or engineering purposes

human operator
an individual who is involved in the routine control, function, or support of a system or subsystem, but is specifically not involved in any maintenance on that system

human performance [1]
the degree to which an individual's skill or ability is implemented in a specific task

human performance [2]
any result from the measurement of human activity under specified conditions

human performance technology
the use of people, systems, and/or programs to influence behavior and accomplishment

human reliability
the probability that an individual or group will adequately perform a given task at the appropriate time

human resources engineering
the process of using human skill resources as factors in design tradeoffs

human tolerance
the ability of the human body and/or psyche to withstand physical and/or mental stresses without permanent injury or damage

human transfer function
a mathematical description of what output(s) the human operator would produce as a function of specific input(s); *syn.* human describing function

human-computer dialogue
the interchange of data, commands, or information in those activities between a human and computer; also human-computer dialog; *syn.* man computer dialogue

human-computer interaction (HCI)
the total of the relationship and activities occurring between a human operator and a computer or terminal; *syn.* man-computer interaction, user-computer interaction

human-computer interface (HCI)
the point(s) of exchange between a human operator and a computer; *syn.* computer-human interface, man-computer interface, user-computer interface

human-environment interface
any region of contact between man and his surroundings; *syn.* man-environment interface

human-machine interface (HMI)
any region or point at which a person interacts with a machine; *syn.* man-machine interface

human-machine system
a system in which the functions of both man and machine are interrelated, both being necessary for proper system operation; *syn.* man-machine interface

humanization of work
see **work humanization**

Humanscale
a manual modeling system for estimating body link, strength, postures, and other aspects for use in human factors engineering
Comment: an older system

humectant
any chemical which absorbs and helps retain moisture

humeral breadth
see **elbow breadth**

humerus
the bone in the upper arm; *adj.* humeral

humidify
increase the water vapor content of the atmosphere; *n.* humidification

humidistat
a device for measuring and/or controlling humidity levels

humidity
a measure of the water vapor content of the surrounding atmosphere

hunger
the feeling of a need for food to satisfy an empty feeling in the stomach

Hunter Lab color system
a color ordering system which is defined from a simple relationship to the CIE X, Y, and Z tristimulus values and is specified by lightness (L), redness or greenness (a), and yellowness or blueness (b)

Huntington's chorea
a hereditary type of chorea which develops in adults and is accompanied by mental deterioration; *syn.* adult chorea; *see also* **Sydenham's chorea**

hurricane
a storm originating in warm or tropical waters of the Atlantic Ocean, Caribbean Sea, or Gulf of Mexico which has circulation about a center and sustained wind speeds of 117 km/hr (74 or greater miles per hour); *see also* **Saffir/Simpson Hurricane Scale, typhoon**

hurricane scale
see **Saffir/Simpson Hurricane Scale**

HYBRID
one of a series of anthropomorphic dummies developed for use in automotive and aircraft crash testing

HYBRID II
an instrumented anthropomorphic dummy used in automobile head-on collision research

HYBRID III
an instrumented anthropomorphic

dummy used in automobile head-on collision research; *syn.* Part 572 dummy

hydraulic
pertaining to the use or flow of water or other liquid, typically referring to the use of fluids in transferring force or motion

hydrocarbon
any compound consisting solely of chemically combined hydrogen and carbon atoms

hydrodynamic element
a modeling fluid which is governed by pressure and volume laws

hydrostatic weighing
a part of one technique for estimating body composition by weighing an individual completely submerged under water to determine body volume

hygiene
the study and implementation of healthful practices

hyoid bone
a U-shaped bone in the neck which is connected by ligaments to the temporal bone and which supports the tongue
   Comment: unique in that it doesn't articulate directly with any other bone

hyper-
(prefix) greater than normal, excessive

hyperabduct
abduct a joint beyond the normal joint range of motion limits, with or without injury; *n.* hyperabduction; *syn.* superabduct

hyperactivity
a disorder characterized by prolonged generally excessive movement, but which may be voluntarily controlled; *adj.* hyperactive; *syn.* hyperkinesis, hyperkinesia, hyperkinetic syndrome

hyperbaric
pertaining to ambient pressures greater than the standard atmosphere

hyperbaric oxygen therapy
a treatment using pure oxygen at greater than atmospheric pressures in a pressure chamber to treat decom-

pression sickness, lesions or sores that resist healing, and other pathologies; *syn.* hyperbaric oxygenation

hypercapnia
an excessive amount of CO_2 in the blood

hyperextend
extend a joint beyond its normal range of motion or comfortable working limits, with or without injury; *n.* hyperextension

hyperflex
flex a joint beyond its normal range of motion or comfortable working limits, with or without injury

hyperkinesis
see hyperactivity

hyperkinetic syndrome
see hyperactivity

hypermetropia
see hyperopia

hyperopia
a refraction disorder in the eye, in which the focal point of parallel light rays from a distant object come to a focus posterior to the retina under relaxed accommodation; *adj.* hyperopic; *syn.* farsightedness; *opp.* myopia

hyperoxia
a condition in which the partial pressure of oxygen is greater than that found in a standard atmosphere; *adj.* hyperoxic; *opp.* hypoxia

hyperpnea
an increased depth and rate of respiration

hypersonic
travelling at or pertaining to a velocity equal to or greater than five times the velocity of sound

hypertension [1]
a state in which an individual chronically maintains an arterial blood pressure higher than optimal levels, generally ≥ 90 mm Hg diastolic and/or ≥ 140 mm Hg systolic

hypertension [2]
a state in which a muscle is overly tensed; *opp.* hypotension

hyperthermia
a condition in which the core body temperature of a homeothermic species greatly exceeds the norm; *opp.* hypothermia

hypertonia
having an above normal muscle tension; *adj.* hypertonic; *opp.* hypotonia

hyperventilation
an abnormally rapid, deep breathing when the body's respiratory needs do not require such breathing

hypo-
(prefix) less than normal

hypoallergenic
having a low probability of stimulating allergic reactions

hypobaric
pertaining to lower than normal ambient pressure

hypodynamia
the lack of gravitational loading on the skeleton

hypoglossal nerve
a cranial nerve which regulates part of the motor activity of the tongue

hypogravity
a state in a which significantly reduced gravitational force is experienced, generally with reference to the accepted standard gravitational force of the earth at its surface; *syn.* subgravity; *see also* *microgravity*

hypokinesia
an abnormally-reduced capacity for voluntary muscular movement while having full, normal consciousness

hypokinetic hypoxia
a hypoxic condition due to the reduced flow of blood

hypothalamus
a neural structure at the base of the anterior brain which has a major role in body development, physiological drives, temperature regulation, and maintaining hormone levels

hypothenar
pertaining to the fleshy mass on the medial/ulnar side of the palm

hypothenar eminence
the fleshy protrusion on the medial/ulnar side of the palm

hypothermia
a condition in which the core body temperature of a homeothermic species has fallen significantly below the average; *opp.* hyperthermia

hypothesis
an initial proposal or assumption which is testable and presumed to be true for purposes of discussion and/or appears to show a causal relationship between two or more factors

hypothesis testing
the conducting of a properly-controlled experiment, including any supporting statistical analyses, to determine the likelihood of a hypothesis being true

hypotonia
a condition involving decreased muscle tone; *adj.* hypotonic; *opp.* hypertonia

hypotonic hypoxemia
a hypoxemic condition due to a low partial pressure of oxygen

hypoxemia
a reduced amount of oxygen carried by blood, usually referring to arterial blood; *adj.* hypoxemic; *see also* ***hypotonic hypoxemia, isotonic hypoxemia***

hypoxia
any condition in which insufficient oxygen is available for consumption by body tissues, regardless of the cause; *adj.* hypoxic

hysteresis
a condition in which the current state of a system depends not only on the present forces acting on it, but on its history; *adj.* hysteretic

hysteresis error
the difference in response output when increasing a variable as opposed to decreasing that variable

hysteretic damping
damping due to the internal mechanical properties of materials

hysteria
a highly emotional state; *adj.* hysterical

I

ICAO word list
a standard word list in which the first letter of each word represents the corresponding sequence of letters in the alphabet
Comment: e.g., alpha, bravo, charlie

icon
a graphical, nonlinguistic representation of an object or action; *adj.* iconic

iconic memory
a sensory memory associated with the visual system

ideal blackbody
see blackbody

ideal radiator
see blackbody

ideal spectrum
a frequency distribution in which a pure tone appears as a vertical line due to perfectly sharp filtering

ideation
the mental process(es) through which ideas are formed

ideational fluency
the ability to generate a number of ideas on a given topic; *syn.* fluency of ideas

idiopathy
a disease or other dysfunction of unknown cause; *adj.* idiopathic

idiosyncratic error
a type of human error due to peculiarities of an individual's characteristics, such as attitudes, social problems, or emotional state

idle machine time
see machine idle time

idle time
a temporal interval, excluding standby time, during which a worker, a piece of equipment, or a system is at the workplace, but not producing output, regardless of the cause; *see also delay time*

ignition temperature
the lowest temperature at which sustained combustion for a volatile substance will occur when heated in air or another specified oxidizing environment; *syn.* ignition point

iliac crest
the lateral, superior rim of the coxal bone

iliac spine
a projection from the coxal bone at the anterior portion of the iliac crest

iliocristale height
the vertical distance from the floor or other reference surface to the highest point of the iliac crest in the midaxillary plane
Comment: measured with the individual standing erect and his weight equally balanced on both feet

iliospinale
the most anterior point on the iliac spine

iliospinale height
the vertical distance from the floor or other reference surface to iliospinale
Comment: measured with the individual standing erect and his weight evenly distributed between both feet

illiteracy
having no ability to read and write; *syn.* true illiteracy

illness incidence rate
the number of annual occupational illnesses experienced by a company in one year, based on 100 full-time employees

$$IIR = \frac{No.\ of\ illnesses \times 200,000}{No.\ of\ man\text{-}hours\ worked}$$

illuminance (E)

the luminous flux density incident per unit area on a surface

illuminance category

an alphabetic character, ranging from A through H, representing illumination ranges for various types of work such that the further the letter is from A, the brighter the light

illuminance meter

a device, composed of a photodetector, filter, and electronic circuitry, for measuring the luminous flux incident on a plane; *syn.* light meter

illuminance threshold

that lowest luminance level which the eye or other image sensor is capable of detecting, given a specified luminance contrast, position within the field of view, dark adaptation, flicker rate, source dimensions, and color; *syn.* threshold illuminance, flux density threshold

illuminant

any light source or combination of light sources

Illuminant A

a standard CIE illuminant corresponding to a typical tungsten filament incandescent lamp

Illuminant B

a standard CIE illuminant corresponding to direct sunlight

Illuminant C

a standard CIE illuminant corresponding to average daylight

Illuminant D

a series of standard CIE illuminants corresponding to daylight which measures beyond the normal visible spectrum

illuminate

distribute or provide light to an area or a region; *n.* illumination

illumination level

see illuminance

illusion

a perceptual misinterpretation of a stimulus

image [1]

the sum of the perceptions by an individual, group, or population about itself or another entity

image [2]

an electronic or photographic representation of one or more entities

image [3]

a subjective sensory experience, especially in the visual modality

image analysis

any computer or other electronic processing to quantify an image, usually with the intent of deriving some statistically-based conclusions

image enhancement

that electronic or other processing to improve the resolution, features, or other quality of an electronic or photographic image

image processing

any type of computer-based alteration of the data representing an image, including enhancement, analysis, and reconstruction

image reconstruction

the process of re-working data for image enhancement

immersion foot

that damage to the skin, blood vessels, and nerves of the feet resulting from prolonged exposure to water at temperatures between freezing and about 60°F

impact

a rapid transmission of physical momentum from one object to another in a mechanical system; *syn.* impulsive force

impact acceleration

an acceleration lasting less than one second

impact acceleration profile
a graphical display or plot of the deceleration sequence experienced by a vehicle in a crash

impact analysis
a subjective technique for attempting to quantify the positive and negative aspects of a system or plan

impact attenuation
the reduction in impulsive forces due to cushioning or other means of spreading out the forces in space or time

impact biomechanics
see **biodynamics**

impact load
a force implemented by a rapid blow

impact noise
see **impulse noise**

impact strength
the impulse energy required to fracture a material

impact velocity
the velocity at which one object strikes another

impactor
an object which makes contact with another body or structure

impairment
any dysfunction in which one or more body systems or subsystems is not capable of functioning to the degree considered normal

impedance (Z)
see **electrical impedance, mechanical impedance**

Imperial gallon
a British gallon, slightly larger than the U.S. gallon
    Comment: an older term

implementation allowance
that time allowance provided for workers in beginning new techniques or changing to a different method to prevent them from losing income during the change; see also **changeover allowance**

implosion
a very rapid inward burst of the surrounding medium due to some change or the rupture of some structure; *opp.* explosion

importance
a subjective rating of greater worth, necessity, or regard relative to other items or functions

importance principle
a rule that displays and controls with the greatest operational importance should be placed in optimum locations with regard to convenient access and visibility

imprecision
that variance due to measurement error from repeated measurements within a short period of time, and which are attributed to measurement process only; see also **unreliability, undependability**

impulse [1]
a human urge based more on emotional than cognitive factors and without significant consideration of possible consequences

impulse [2] (J)
the area under the curve of a force for the brief time duration of the force application

impulsive force
see **impact**

in phase
pertaining to waveforms having the same frequency and which are at the same point in their respective cycles at the same time; *opp.* out of phase

in situ
in a structure or object's normal location

in vivo
within the living body

inaccessible
incapable of being reached or entered by a human, a human body part, a remotely-operated system, or a tool for retrieval or repair of a system or subsystem

inactive window
a open, perceptually- and functionally-available window which must be activated before the user may work within it; *opp.* active window

incandescence
the emission of light and other forms of electromagnetic energy due solely to heating a source material; *adj.* incandescent

incandescent lamp
a light source derived from incandescence, usually from electrical heating of a filament within a sealed bulb

incentive
any condition which motivates behavior to obtain a reward or avoid punishment

incentive operators
those employees whose wages are determined either entirely or in part by the quality and/or quantity of their output

incentive pace
the performance level of a worker under incentive conditions and without excess fatigue

incentive plan
any procedure by which an organization attempts to promote increased productivity; *syn.* incentive scheme; *see wage incentive plan*

incentive scheme
see incentive plan

incentive wage plan
see wage incentive plan

incentive wage system
see wage incentive plan

inch
a unit of length in the English system; equals 2.54 cm in the metric system

incidence
see morbidity

incidence rate [1]
any of a number of OSHA-reportable work-related illness, injury, or death rates based on 200,000 man-hours in the workplace; *see injury/illness inci-*dence rate, fatalities incidence rate, lost workday case incidence rate, no lost workday case incidence rate

incidence rate [2]
see morbidity rate

incident
any occurrence or near-occurrence of an event which is recorded

incidental element
see irregular element

incidental learning
the acquisition of information or skills as a byproduct of one's simple presence or through other, unrelated activities

incidental vibration
any unintended vibration
Comment: an older term

inclusion
any unintended or undesirable foreign particle in a finished object

incombustible
see noncombustible

incompetence
an inadequacy for performing a certain function, regardless of cause

incomplete diffusion
see partial diffusion

incontinence
an inability to control the elimination of feces and/or urine

incorporation by reference
the inclusion of specifications, requirements, regulations, or other information into a given document simply by referring to a second document which already contains the desired information; *syn.* specification by reference

incremental threshold
see difference threshold

incubation time
the temporal interval between infection by an agent and the appearance of disease symptom(s)

incus
the middle bone of the auditory ossicles in the middle ear

independent psychomotor abilities
a set of movement capabilities reportedly determined by factor analysis to be independent of one another and which may be used for task and job analyses, performance measurement, etc.

independent variable
a variable which can be either set to a desired value or controlled by the experimenter, or either matched or observed as it occurs naturally; *opp.* dependent variable

index finger
the three phalanges and surrounding tissues of digit II of the hand; *syn.* forefinger

index finger length
the linear distance from the thumb crotch to the tip of the index finger; *syn.* forefinger length
 Comment: measured with the index finger fully extended; this definition is not consistent with other finger or finger segment lengths, since it includes a portion of the metacarpal length

index of difficulty
an indication of the amount of information required to generate a movement; *see Fitts' law*

$$ID = \log_2\left(\frac{2A}{W}\right)$$

where:
A = distance to the target
W = width of the target

index of forecasting efficiency (E)
that reduction in prediction error obtained by using the correlation between two variables

$$E = 1 - \sqrt{1 - r^2}$$

where:
r = the correlation between the variables

index of physiological effect
a measure of heat stress; *syn.* heat stress

index

index of refraction (n)
the value of the ratio of the velocity of electromagnetic radiation in one medium relative to another medium; *syn.* refractive index, refraction index
 Comment: a constant for a given pair of media and a given wavelength

index of relative strain
a measure of heat stress based on clothing insulation and clothing effects on evaporation; *syn.* heat stress index

index of skin wettedness
see skin wettedness

index of thermal stress
an indicator of the degree of heat stress which predicts the sweating rate required to cool the body based on the heat load combined with the effects of clothing and humidity levels; *see also heat stress index*

indicator
any device for displaying information

indirect anthropometric measurement
a bodily measurement obtained by remote or non-contact techniques, such as stereometric anthropometry; *opp.* direct anthropometric measurement

indirect cause
a contributing causal factor other than the direct cause associated with an incident

indirect cost
any of the expenses incurred from indirect operations; *opp.* direct cost

indirect labor
that work which is a part of indirect operations; *opp.* direct labor

indirect lighting
that illuminated environment in which approximately 90% or more of the luminous flux is directed toward a continuous solid structure away from a task; *opp.* direct lighting

indirect material
any of the materials not used in direct operations

indirect operations

those administrative, management, or other functions within an organization necessary to support the manufacture or output of a product but which are not directly involved in producing a product or service for sale in the marketplace and which does not add value to that product; *opp.* direct operations

indirect radiation effect

any of those cellular effects causing damage to DNA by first creating radicals in other body or cellular materials, which in turn affect the DNA; *opp.* direct radiation effect

indirect viewing

the use of video or other aids to view a scene or object being manipulated when direct viewing is not practical or possible; *opp.* direct viewing

indirect vision

peripheral vision

indirect worker

an employee involved in indirect operations; *opp.* direct worker

individual incentive plan

an incentive plan in which each worker is rewarded based on his own efforts

indoor air pollution

that pollution found within living, working, or other structures, generally consisting of airborne hazardous or offensive substances such as dust, tobacco smoke, fungi, and solvent or cleaning material fumes

induced environment

that environment imposed upon an object or system from man-made conditions; *opp.* natural environment

induction [1]

the generation of an electrical current by a change in magnetic flux in a conductor

induction [2]

see **inductive reasoning**

induction [3]

the alteration of a perception by indirect stimulation

inductive reasoning

the ability to integrate specific, diverse bits of information to arrive at a general conclusion

industrial anthropometry

the use of anthropometry for designing and constructing equipment for human use in the industrial environment; *see* **human factors**

industrial cost control

see **cost control**

industrial disease

see **occupational disease**

industrial engineer

one who is qualified by education, training, and experience to practice the discipline of industrial engineering

industrial engineering

that engineering discipline concerned with the design, development, installation, and improvement of integrated systems of people, materials, equipment, and energy in the industrial environment; *syn.* production engineering

industrial ergonomics

human factors applied to an industrial setting

industrial hygiene

that aspect of medicine generally promoting healthy conditions in the working environment, including attempting to prevent occupational diseases and accidents, and developing or obtaining proper measures for any emergency treatment which may be required

industrial hygienist

a person with the ability, training, and experience to practice industrial hygiene

industrial medicine

see **occupational medicine**

industrial psychology

that field of study and practice involving the testing, development of criteria and predictors for personnel selection and human performance in the workplace

industrial robot
a programmable manipulator for moving or operating on materials, components, products, or other objects in the industrial environment

industrial safety
see occupational safety

ineffective time
that part of the elapsed time spent on any activity which is not a specified part of the task or job, excluding check time

inert dirt
any form of dirt which has no inherent attraction to any surface except through gravitation

inert gas narcosis
a toxic effect of the diluting or carrier gas in a breathing mixture at increased pressures, characterized by euphoria, diminished cognitive function, and impaired coordination; *syn.* diluent gas narcosis; *see **nitrogen narcosis***

inertia
the resistance to a change in momentum

inertial frame
a reference frame to which the law of inertia applies

infant
a child less than two years of chronological age

infant mortality
the failure of a system in the early portion of its projected useful life

infant mortality phase
*see **run-in phase***

infant mortality rate
the reported death rate for infants under one year of age per 1000 reported live births in a calendar year for a specified region

inference
the conclusion resulting from the inductive reasoning process

inference space
those limits within which the results of an experiment may be applied

inferential statistics
the use of statistical methods to derive population parameters from sample data; *syn.* statistical inference; *opp.* descriptive statistics

inferior [1]
of less than acceptable quality or performance

inferior [2]
lower than or beneath some reference structure in position

inferior angle of the scapula
the thick lowermost portion of the scapula

inferior nasal concha
a bone forming part of the lateral wall of the nasal cavity

inferior oblique muscle
a voluntary extraocular muscle extending beneath the eyeball
Comment: principally for rotation of the upper part of the eye laterally about the optical axis

inferior rectus muscle
a voluntary extraocular muscle parallel to the optical axis beneath the eyeball
Comment: involved in the anterior downward pitch/rotation of the eye

inflammable
*see **flammable***

inflammation
a reaction of living tissue to some injury, usually characterized by one or more of the following symptoms: turning red or warm, becoming swollen and/or painful, and the release of fluid

inflection point
a point on a curve such that the following are true: (a) the curve changes from concave to convex, (b) the mathematical derivative of the curve is increasing on one side of the point and decreasing on the other side, and (c) the second derivative changes sign; *syn.* point of inflection

information [1]
a meaningful collection of facts, figures, and/or data

information ² (H)

that which reduces uncertainty
Comment: typical unit is the bit

information aid

any work aid which provides the worker with text, numbers, figures, or other details appropriate for performing in the working environment

information area

any region of a display containing useful general purpose information

information ordering

the ability to correctly follow a set of rules in arranging items

information process analysis

see form process chart

information theory

that aspect of communications dealing with the coding of messages and with the content and amount of information conveyed; *see information, bit*

informed consent

see voluntary informed consent

infra-

(prefix) under, below, or less than

infradian

having less than one cycle per day, or a period longer than 24 hours; *opp.* ultradian

infradian rhythm

a biological rhythm having less than one cycle per day, or a period longer than one day

infrared (IR)

pertaining to that radiational energy in the portion of the electromagnetic spectrum with wavelengths longer than those visible as red light to those wavelengths of the microwave region, generally from about 750 nm to about 1 mm

infrared lamp

a lamp which emits its primary radiation in the infrared portion of the electromagnetic spectrum, and any radiation in the visible portion of the spectrum is not normally of interest

infrared radiation

see infrared

infrared touchscreen

a display having a frame with embedded opposing infrared transmitters and receivers which uses blockage of the infrared beam to indicate a touch location

infrasonic

see infrasound

infrasound

a mechanical vibration at frequencies below those normally heard by the human ear, generally below about 16 20 Hz; *adj.* infrasonic

Ingersoll glarimeter

an early instrument for measuring gloss using polarized light

ingestion

the taking in of substances, especially via the mouth

ingress

(v) enter a region or space; *opp.* egress

ingress point

(n) the location for entering a region or space

inguinal crease

the groove at the junction of the anterior-medial thigh and the torso

inhalant

a substance which is inhaled

inhale

see inspire; n. inhalation

inherent delay

see delay time

inhibitor

a substance or condition which slows or prevents a response, usually via competition

inion

the most posterior protruberance of the occipital bone; *syn.* external occipital protuberance

initial luminance

that luminance reaching the work surface from a given luminaire when new

initiate

a mental activity preceding a psychomotor task

injury
any physical disturbance to, damage to, or destruction of one or more body structures which prevents/impairs normal functioning or appearance

injury incidence rate
the number of injuries experienced by a company based on a year's work for 100 full-time employees

$$IIR = \frac{No.\ injuries \times 200,000}{No.\ man\text{-}hours\ worked}$$

injury potential
a potential difference across a membrane, generally of about 30–40 mv, between regions of normal and injured tissue

innate
due to one's genetic make-up; *opp.* learned

inner ear
that portion of the ear embedded in the temporal bone and consisting of the vestibule, cochlea, and semicircular canals; *syn.* internal ear, labyrinth

innervation
the distribution of nerves or neurons to all or some portion of the body

innominate bone
the coxal bone

input
that information, signal, or form of energy which enters a system

input device
any piece of equipment or instrumentation used to provide the human with an interface for providing input to a system

input/output (I/O)
any activity which inputs to or receives output from a computer

input point
the physical location at which some information or signal can enter some system

input storage
the temporary placement of data in a computer file until time for processing

inrolling nip point
a system in which two or more rollers rotate parallel to each other, but in opposite directions, and which can grab and pull on such items as loose clothing, and ties; *syn.* inrunning nip point

inrunning nip point
see **inrolling nip point**

inseam
see **leg inseam, sleeve inseam**

insert
(v) place one object inside another; *n.* insertion

insert mode
a data entry mode in which text or information entered by the user is placed in front of any existing text or information, shifting that existing text

insertion
the distal point of attachment of a muscle tendon to bone; *syn.* point of insertion; *opp.* origin

inside work [1]
see **internal work**

inside work [2]
that work performed inside some structure which shields the worker at least in part from the atmospheric elements; *syn.* indoors work; *opp.* outside work

inside-out display
any display which uses the vehicle as the frame of reference such that the display reflects the way the operator would see the external environment from inside; *opp.* outside-in display

insomnia
an extended period in which sleep is disturbed, not resulting from immediate external stimuli

inspect (I)
a therblig which may involve testing, observation, or other processes to provide quality control; *n.* inspection

inspection error
any incorrect reading, action, or other error of either omission or comission in the inspection process

inspiratory capacity (IC)
the maximum volume of air which can

be inhaled after a normal expiration

inspiratory reserve volume
that additional volume of air which can be inspired beyond the normal tidal volume

inspire
breathe or take air into the lungs using one's own resources; *n.* inspiration; *syn.* inhale

instant
any given point in time; *adj.* instantaneous

instant start fluorescent lamp
a fluorescent lamp which does not require pre-heating of the electrodes; *syn.* cold start fluorescent lamp

instantaneous acceleration ($\vec{a}$)
the rate of velocity change with time at any instant

$$\vec{a} = \frac{d\vec{v}}{dt}$$

instantaneous velocity ($\vec{v}$)
the rate of change of displacement with time at any instant

$$\vec{v} = \frac{d\vec{x}}{dt}$$

instep
the arch on the medial side of the foot

instep circumference
the surface distance around the foot in a coronal/frontal plane at the anterior junction of the leg and foot
> Comment: measured with the individual standing erect, having his weight distributed equally on both feet on the floor, and without any unnecessary leg or foot muscle tension

instep length
the linear distance from the plane of the most posterior aspect of the heel to the point of maximum medial protuberance of the foot
> Comment: measured with the individual standing erect and the body weight equally distributed between

both feet on the floor

instinct
a genetically-based or natural motivation or behavior; *adj.* instinctive, instinctual

instruction
one item of a set of procedures, standard practices, or steps for accomplishing a given task or job

instruction aid
a job aid containing written instructions on a card or sheet of paper; *syn.* instruction card, instruction sheet

instruction card
see **instruction aid**

instruction sheet
see **instruction aid**

instrument
any device for measuring, recording, and/or controlling the value of one or more variables

instrument error
any error made by an instrument

instrument flight rules (IFR)
those rules governing flight in which instruments must be or are used for controlling an aircraft; *opp.* visual flight rules

instrumental activities of daily living (IADL)
those functions likely to be carried out on a daily or near daily basis which involve the use of equipment or instrumentation for sustenance of the individual or a normally habitable environment; *see also* **activities of daily living, daily living tasks**

instrumental conditioning
see **operant conditioning**

insulation value of air
see **thermal insulation value of air**

insulation value of clothing
see **thermal insulation value of clothing**

integral-mode controller
a type of controller whose output signal is proportional to the integral of the error signal

integrate [1]

compute the area under a curve; *opp.* differentiate

integrate [2]

combine activities, information, or objects in a meaningful way for some purpose

integrated controller

a device which coordinates the control of more than one aspect of some operation

integrated electromyogram (IEMG)

the computed area under the curve of an electromyographic signal; also integrated EMG

integrated error

the sum of the errors accumulated over a given task

Integrated Noise Model (INM)

a computer modeling system used by the FAA to develop noise contours for airports and surrounding areas

integrated surface electromyogram

see integrated electromyogram

intellect

the capacity for understanding and reasoning

intelligence

the ability to recognize, learn, understand, reason, create, and react appropriately to a given set of living conditions

intelligence quotient (IQ)

a numerical score attributed to be one's intelligence level, typically the value of the ratio of mental age to chronological age, multiplied by 100

$$IQ = \frac{mental\ age}{chronological\ age} \times 100$$

Comment: generally of limited value

intelligence test

any of a set of standardized tests which purport to measure an individual's intelligence

Intelligent Vehicle Highway System

a planned passenger car-highway system in which the routine driving, safety, and navigation functions are assumed by integrated computer systems

intelligibility

see speech intelligibility

intelligibility score

see speech intelligibility score

intensity

a measure of the strength or amount of some entity or sensation

intensity-duration relationship

see load-endurance curve

inter-

between

inter-individual variation

the differences between individuals on the same or equivalent aspect or variable

Inter-Society Color Council-National Bureau of Standards color system (ISCC-NBS)

a color ordering system which specifies color by boundaries rather than points

interaction

the result from a particular combination of events, due solely to the combination and not any particular individual event

interaction effect

that experimental or statistical result attributable solely to a particular combination of variables and beyond that which can be predicted from the variables independently

interactive

having the capability for one or more cycles of human input with rapid display feedback

interactive window

an active window which is receptive to user input; *opp.* non-interactive window

interaural phase

the apparent relative phase difference of a tone between the left and right ears

interdigital crotch

that region of soft tissue between each

pair of digits on the hand or foot; *see also* **thumb crotch**

> *Comment:* which pair of digits being referred to should be specified

interface

a common boundary or point of connection between two or more parts of a system or between systems, whether physical or perceptual

interference allowance

that time compensation given a worker for lost production due to interference time; *syn.* synchronization allowance

interference time [1]

that machine idle time which results from an operator's inability to service one or more machines due to other assignments

interference time [2]

that worker idle time when working as a member of a team in which one or members of the team is required to wait while some task is carried out by another member

interflection

the multiple reflections of light from an enclosed volume other than the luminaire prior to reaching the surface of interest

interlace

scan across a display screen or other medium such that the distance from line to line in a field is approximately twice the line width and adjacent lines belong to different fields

interlaced display

a display which uses an interlaced scanning format such that two fields must be written to completely update the display; *opp.* non-interlaced display

interlock

a mechanism which interacts with another mechanism to prevent operation of a device under certain circumstances for safety or other reasons

intermediate cuneiform bone

one of the distal group of foot bones in the tarsus; *syn.* second cuneiform bone

intermediate infrared

that portion of the infrared spectrum from about 1400 to 5000 nm; also intermediate IR

intermittent work

that work, often physically-demanding, which is performed only at certain points in time, not on a continuous basis

internal [1]

see **medial**

internal [2]

within or beneath the surface of a body part or other structure

internal ankle height

see **medial malleolus height**

internal biomechanical environment

the mechanical forces to which bodily tissues, particularly the musculoskeletal system, are subjected when executing motions or being acted upon by outside forces; *syn.* internal mechanical environment

internal canthus

see **endocanthus**

internal clock

a hypothetical internal bodily mechanism responsible for maintaining biological rhythms; *see also* **circadian pacemaker**

internal consistency

having data within an experiment, analysis, or test which are repeatable across subjects or which have logical relationships within a subject

internal desynchronization

the loss of normal phase relationships between biological rhythms within a single entity

internal ear

see **inner ear**

internal injury

any injury to organs lying within the thoracic or abdominal cavities

internal naris

the junction of the posterior nasal cavity with the nasopharynx; *opp.* external naris

internal pacing
    pertaining to self-paced work

internal work
    that manual work done by an operator
    during the operation of a machine or
    process he is supervising; *syn.* fill up
    work, inside work; *opp.* external work

internally-paced work
    see self-paced work

international candle
    see candle

International System of Units
    see Système International d'Unites

interoceptor
    any sensory receptor sensitive to
    changes within the viscera and blood
    vessels

interocular breadth
    see endocanthic breadth

interocular distance
    see interpupillary breadth

interphalangeal
    between the phalanges of the hand or
    foot

interpolate
    estimate one or more unknown values
    within a range of known values using
    some predictor; *n.* interpolation; *opp.*
    extrapolate

interpupillary
    pertaining to the region between the
    eye pupils

interpupillary breadth
    the horizontal linear distance between
    the centers of the pupils of the eyes;
    syn. interpupillary distance, interocular
    distance
        Comment: measured with the indi-
        vidual's scalp muscles relaxed, the
        eyes open, and looking straight
        ahead

interpupillary distance
    see interpupillary breadth

interscapulae
    pertaining to the region of the back
    between the two scapular bones

interscye, bent torso
    the surface distance across the back

between the scye points; *syn.* interscye,
maximum, II
        Comment: measured with the indi-
        vidual standing, the torso bent for-
        ward from the waist at an angle of
        about 90°, and the arms hanging
        relaxed

interscye, seated forward reach
    the surface distance across the back
    between the scye points; *syn.* interscye,
    maximum
        Comment: measured with the indi-
        vidual sitting erect with his arms
        extended forward horizontally

interscye, seated leaning
    the surface distance across the back
    between the posterior axillary folds at
    the lower level of the armpits; *syn.*
    interscye II
        Comment: measured with the indi-
        vidual seated and leaning forward
        with his hands on his knees

interscye, standing erect
    the surface distance across the back
    between the posterior axillary folds at
    the lower level of the armpits; *syn.*
    interscye, back
        Comment: measured with the indi-
        vidual standing erect and his body
        weight distributed equally between
        the two feet

interstimulus-onset interval
    the length of time between the onset of
    one stimulus and the onset of a second
    stimulus; *syn.* stimulus-onset interval

interval scale
    *see equal-interval scale, logarithmic
    interval scale*

intervertebral disk
    a circular mass of fibrous cartilage lo-
    cated between adjacent vertebrae in
    the spine; also intervertebral disc

interview
    a spontaneous or organized sequence
    of questions and discussion to exchange
    information relevant to a particular
    situation between two or more indi-
    viduals

intervocalic
occurring between vowels

intort
rotate a structure toward the midline, especially the eye; *n.* intorsion; *opp.* extort

intorter
a muscle which intorts

intoxication [1]
a state of having been poisoned by any toxic substance, whether unknowingly or due to one's own voluntary actions

intoxication [2]
a state of intense mental or emotional excitement

intra-abdominal pressure (IAP)
that pressure exerted on the internal walls by gravity, the abdominal viscera, arterial supply, and the musculature

intra-individual variation
that variation which occurs within a single person over time on the same or similar testing or observation

intra-ocular muscle
an involuntary intrinsic smooth muscle within the eye, specifically the ciliary and pupillary muscles

intra-ocular pressure
that fluid pressure within the eyeball

intracellular water (ICW)
that water contained within the cells of the body; *opp.* extracellular water
  Comment: one of two components of total body water

intracranial
within the skull

intrafusal fiber
the small muscle fibers within a muscle spindle which are involved in sensing length changes; *opp.* extrafusal fiber

intramodal matching
see **intrasensory matching**

intrasensory matching
a procedure in which a subject matches the magnitude of a stimulus in a sensory modality with the magnitude of another stimulus using the same modality; *syn.* intramodal matching

intravehicular activity (IVA)
any activity occuring within a vehicle, especially referring to a space vehicle; also intra-vehicular activity; *opp.* extravehicular activity

intrinsic
pertaining to a structure or mechanism which originates within a structure and acts on itself; *opp.* extrinsic

intrinsic muscle
any muscle having both its origin and insertion located within a given structure and which is involved in the function of that structure

invasive
pertaining to a procedure object which requires breaking the skin, insertion of any object into any body cavity except the mouth, or which causes extreme discomfort; *opp.* non invasive

inventory
those materials, supplies, products, or capabilities available in-house

inventory control
see *inventory management*

inventory management
those techniques involved in maintaining the desired inventory levels, including planning, tracking, distribution, providing storage, and purchasing; *syn.* inventory control

inverse power function
an exponential mathematical relationship involving a negative exponent or where the variable would be represented in the denominator with a positive exponent, such as

$$y = ax^{-3} = \frac{a}{x^3}$$

inverse square law
a law stating that the force or intensity of some physical entity emitted from a source varies inversely as the square of the distance from that source

$$I = \frac{I_0}{d^2}$$

invert
turn inward; *n.* inversion

inverted image
an image which has been rotated within its plane by 180°

inverted U function
see concave function; opp. U function

invertor
any muscle which turns the sole of the foot inward; *opp.* evertor

involuntary muscle
those muscles not normally under conscious control, such as the smooth muscles; *opp.* voluntary muscle

involuntary reflex
see reflex

involuntary response
see reflex

involution
the process of decline or decay in human processes later in life

ion
any atom or molecule which has a net positive or negative charge

ion range
see range

ionization track
the detectable path of an ionizing photon or particle following passage through tissue or another substance

ionizing radiation
that electromagnetic or particulate radiation which causes ion formation as it passes through materials or tissues

iontophoresis
the process of transferring ions across some barrier with direct current

iridescent
pertaining to the optical interference effects in thin films or of reflected diffracted light from ribbed surfaces; *n.* iridescence

iris ¹
the colored circular structure in the aqueous humor of the eye which encircles the pupil between the cornea and the lens and regulates the amount

of light reaching the retina; *pl.* irides

iris ²
an arrangement of flat leaf-like structures which provides an approximately circular opening on retraction

iris reflex
the adjustment of muscle fiber length in the iris to accommodate light levels to which the eye is exposed

irradiance (E)
the density of radiant flux per unit area on a specified surface

irradiate
expose to some form of directed energy; *n.* irradiation

irregular element
a work element occurring at other than regular intervals, but which may be statistically predicted; *syn.* incidental element

irregular shift
a variable work schedule, set by the employer for his convenience, usually to accommodate anticipated workloads

irritant
any agent which produces an active response by a living organism or system within that organism

ischemia
a lack of blood flow to a tissue, regardless of cause

ischemic hypoxia
a form of hypokinetic hypoxia in which arterial blood flow is reduced

ischial tuberosity
a projection at the base of the ischium which can become a pressure point when sitting on a hard surface; *syn.* sitting bone

ischium
the inferior and posterior portion of each coxal bone; *adj.* ischial

Ishihara test
a commonly used color deficiency test using plates on which numbers of a given color are embedded in a variety of hues
Comment: one number is seen by

those with normal color vision, another number by those with a color vision deficiency or color blindness

isoacceleration

having a constant acceleration

isocandela diagram

a set of plotted isocandela lines on a coordinate system to show lighting intensity spatial relationships

isocandela line

a contour line representing an area of equal lighting intensity

isoforce

see isotonic

isoinertial

pertaining to the force applied to a constant, moving mass

isoinertial action

that dynamic muscle action involved in moving a constant mass; *syn. isoinertial contraction*

isoinertial contraction

see isoinertial action

isokinetic

pertaining to movement at a constant velocity

isokinetic action

a dynamic muscle action in which muscle contraction occurs at a constant velocity and maximal tension is maintained during the entire movement sequence; *syn.* isokinetic contraction

isokinetic contraction

see isokinetic action

isolated word recognition

see word recognition

isolation

any spatial or physical separation from other humans or certain individuals

isomer

one of a pair of color stimuli which are physically the same; *see metamer*

isometric

having or maintaining the same dimension; *see isometric contraction, isometric view*

Comment: may refer to either a

muscle length or an engineering display/drawing

isometric action

a muscular process in which tension increases in one or more muscles, but the muscles retain approximately the same length and essentially no movement of the body link(s) occurs for prolonged periods of time; *syn.* isometric contraction, static contraction, static muscle contraction

isometric contraction

see isometric action

isometric exercise

that exercise in which muscle actions are counteracted by equal and opposing forces

isometric joystick

a non-moving joystick which provides a directional output proportional to the force applied by the user; *syn.* force joystick, pressure joystick

isometric muscle contraction

see isometric action

isometric muscle work

see static work

isometric strength test

a test to determine the safe static load handling capabilities for workers

isometric view

a three-dimensional appearing view of an object on a display or drawing which has been constructed so that perspective has been ignored

isometric work

see static work

isotonic

having uniform tension or force; *syn.* isoforce

isotonic action

a dynamic muscle action in which the muscle length of one or more muscles shortens and movement of one or more body links occurs, with constant muscle tension throughout the movement

isotonic contraction

see isotonic action

isotonic hypoxemia

a hypoxemic condition with a normal partial pressure of oxygen, generally due to decreased hemoglobin or toxin/drug effects

isotonic joystick

a joystick whose output is proportional to and in the same direction as the displacement of the joystick from its null point; *syn.* displacement joystick

isotonic muscle work

see dynamic work

isotope

one of a set of nuclides having the same atomic number as another atom, but a different atomic mass and neutron number

isotropic

having an equal spatial distribution or growth in all directions

iterate

calculate a desired result using repeated operations; *n.* iteration

J

jacket
a short, lightweight coat

jackhammer
a hand-controlled chisel device operated by compressed air

jackknifing
the restriction of blood flow to the distal portion of a limb due to a large angle of flexion in a proximal joint

jaw
the two bones forming the skeletal framework for the mouth, the maxilla for the upper jaw bone, the mandible the lower

jaw height
see menton-subnasale length

jawbone
see maxilla (upper), mandible (lower)

jerk [1]
the rate of change of acceleration with time; *syn.* jolt

jerk [2]
a sudden, spasmodic or reflex body movement

jet lag
(sl) the state of general discomfort from crossing one or more time zones rapidly, due to circadian rhythm desynchronization; *syn.* desynchronosis

jig
any precision mechanical device used to support or hold parts in position or act as a guide

jitter
a periodic jumping of a target or small structure on a display

job [1]
the sum of all the tasks and duties assigned to and carried out by one or more workers toward the completion of some goal

job [2]
that work specified in a contract work order, usually to be performed by several people

job [3]
a position within a company serving as one's employment

job aid
see work aid

job analysis
an evaluation of job requirements through an evaluation of the duties and tasks, facilities and working conditions, and worker qualifications and responsibilities necessary to perform a job; *syn.* job study; *see also* **task analysis**

job breakdown
a division of a job into its elements; a listing of the elements comprising a job

job characteristic
see job factor

job class
a job classification level in which jobs involve similar types of work, difficulty, and/or pay; *syn.* job family, job grade, labor grade

job classification
the arrangement of jobs by job class

job content
the total makeup of a job, including the physical tasks and the psychological factors of challenge, variety, and feeling of worth

job costing
a cost determination in which manufacturing costs are attributed to individual items

job demand
the combined physiological, sensory-perceptual, and psychological require-

ments for or loads experienced by a worker performing a particular job

job description
a written general statement of the scope, duties, and responsibilities of a particular job

job design
the process of determining what the job content should be for a set of tasks, how the tasks should be organized, and what linkage should exist between jobs; *see also* **work design**

job dimension
any of the primary quantifiable aspects of a job for evaluation purposes

job element
some distinct portion of a specified job

job enlargement
an increase in job scope with the intent to make jobs more interesting, through the addition of more tasks of a similar nature to the duties or tasks currently being performed; *syn.* horizontal job enlargement

job enrichment
an increase in the scope of a worker's job, with the intent to increase variety and significance by adding additional duties such as planning, greater control over operations, and more interaction with others; *syn.* vertical job enlargement

job evaluation
the process of determining the relative worth or utility of a job

job factor
an essential element of a job which gives management some basis for setting a wage range for the job, as well as the selection and training of workers; *syn.* job characteristic

job hazard analysis
see **job safety analysis**

job modification
a minor change to a job; *see also* **job redesign**

job performance aid
see **work aid**

job plan
an organized approach or document by management showing detailed procedures for each job

job ranking
see **job evaluation**

job redesign
a significant, intentional change in job design

job restriction
a condition in which an individual returning to the workforce following an illness or occupational injury is not permitted to perform certain tasks which might aggravate that illness or injury

job rotation
the assignment to or performance of different activities by a group of workers on a periodic basis

job safety analysis
the division of any job into its components for the purpose of: (a) identifying possible hazards and the accidents which might be caused by them, (b) developing solutions intended eliminate accidents or counteract such hazards, and (c) determining the qualifications or requirements for the operators involved; *syn.* job hazard analysis, safety job analysis

job safety training
see **safety training**

job satisfaction
the degree to which the work environment provides such qualities as variety, comfort, compensation, and social expression to make a job meaningful in meeting an individual's goals

Job Severity Index (JSI)
a guideline for matching job design and employee placement such that an acceptable risk of injury potential is present

job sharing
a work schedule in which two part-time workers perform the duties which would normally be assigned to one full-time person; *syn.* shared time

job shop
a company whose primary function is to produce small quantities of specialized parts or components for customers who will integrate them into larger products

job skill
the combination of physical and mental abilities, experience, and training, which enable a worker to perform a given task

job specification
see *work specification*

job standardization
having or implementing a standard practice or method for some job

job study
see *job analysis*

jockeying
that customer or user behavior in which he has the option of using several queues or lines, possibly even changing lines while waiting

jog
an intermediate gait between walking and running, or an alternating combination of continuous walking and running, which is used as a form of exercise; see also *gait*

joint [1]
(n) a region of articulation between two bones

joint [2]
(adj) pertaining to a coordinated action between two or more groups

joint capsule
the connective tissue and membrane surrounding a synovial joint cavity

joint mobility
see *flexibility*

joint range of motion (JROM)
the angle through which a single joint can be flexed, extended, rotated, or otherwise normally moved without discomfort, pain, or injury; see also *range of motion*

joint stability
a measure of the rigidity of a joint; *opp.* joint mobility, flexibility

jolt
see *jerk*

joule (J)
a unit of work or energy in the MKS system; the amount of work done when a force of one newton displaces an object a distance of one meter in the direction of the applied force

journaling
the recording and storage within a computer of the keystrokes input by a user

joystick
a lever control or computer input device having at least 2 degrees of freedom and with which an operator may control an electromechanical system or a cursor or other activity on a display; *syn.* stick

jugular notch
see *suprasternal notch*

Julian date
a number representing the current day within a given year
> Comment: the range is from 1 through 365 (366 in a leap year)

jumper's knee
pain at proximal end of the patellar tendon; *syn.* Sinding-Larsen-Johansson disease

just cause
having good and fair reason(s) for taking disciplinary action

just noticeable difference (JND)
the smallest amount of change from a reference stimulus at which an observer will report a difference on a given trial; *syn.* just perceptible difference, least noticeable difference, minimal change; see also *difference threshold*

just tolerable limit
the maximal level of short term exposure to an agent which will prevent the average individual from developing

either acute or chronic symptoms caused by that agent

justify

arrange text, graphics, or other material to be formatted such it is aligned along the left and/or right margins

juxta nipple skinfold

the thickness of a skinfold just superior to the nipple and parallel to the lateral margin of the pectoral muscle

K

Kan-sei engineering
a system for developing consumer products in which the product is designed to have sensory and emotional appeal

Kata thermometer
an alcohol-based thermometer used for determining low air currents/velocities, in which the time required to cool from 100°F to 95°F corresponds to air velocity at that location; also kata-thermometer, catathermometer; *see also wet Kata thermometer*

Kelvin (K)
a temperature scale whose zero is absolute zero and whose units are separated by Kelvin degrees; *syn.* Kelvin temperature scale, absolute temperature scale
> *Comment:* the basic reference temperature scale

Kelvin degree (°K)
a unit of temperature equal to 1/273.16 of the absolute triple point temperature of water

Kelvin temperature scale
see Kelvin

Kendall rank-order correlation
a statistical test which requires at least ordinal data on two variables for ranking purposes and provides a determination of the significance of any association between the sets of ranks

Kendall rank-order correlation coefficient (τ)
the correlation coefficient computed for the Kendall's rank correlation test

Kendall's coefficient of concordance (W)
a measure of the degree of similarity in the rankings of a set of entities across two or more independent rank orderings of that set; *syn.* coefficient of concordance

keratitis
an inflammation or infection of the cornea, from any cause

key click
an audible click which is presented whenever a keystroke is performed on a keyboard or keypad
> *Comment:* provides feedback to the user that a keystroke was made

key event
one incident which is primarily responsible for the time, place, and severity of an accident or other significant happening

key grasp
see pinch grasp

key job
a job which has been evaluated itself and may be used as a benchmark for evaluating other similar, non-key jobs or work classes for evaluation, classification, and/or wage establishment purposes in the same company or industry

key repeat rate
the repeat rate with a keyboard key when continuously depressed, providing the number of characters input per second

keyboard
a computer input device or typewriter keying mechanism consisting of a panel containing alphanumeric, grammatical, function, and/or other keys for typing or information/data entry to a computer or onto hardcopy

keypad
a data entry pad consisting of the numeric keys 0–9, simple arithmetic

function keys, a decimal key, and an enter key

keypunch

a electromechanical device with a keyboard which punches holes in a card or onto tape; *syn.* cardpunch

    Comment: an older device

keystone distortion

an inequality in the shape of a projected image due to the film or projection lens plane not being parallel with the screen plane; *syn.* keystone effect

    Comment: usually presents a trapezoidal image

keystroke

the depression of a key on a keyboard with a force greater than the actuation force for that key

kickback

the reaction of a piece of material back toward the operator as it being fed into a mechanical processing device and it meets the cutting or processing tool

kickplate

any vertical structure or covering device on or near the floor which protects the surface it covers from impacts or which prevents accidental entry of the shoe/toes into a region which might be hazardous

kilo- (K, k)

(prefix) 1000 or 10^3 times a base unit

kilocalorie

    see **Calorie**

kilogram (kg)

an international standard unit of mass in the SI/MKS system, corresponding to a specific platinum-iridium alloy mass

kilogram force (kgf)

a force equivalent to that which the earth's gravity exerts on a one kilogram mass at the earth's surface

kilopond

    see **kilogram force**

kinanthropometry

the study of human nutrition, growth/development/maturation, size/shape/proportion, body function, and body composition to understand and improve upon health and performance; *syn.* nutritional anthropometry

kinematic chain

an open series of links or body segments, where the dimensions of each link are determined by the linear distance from one joint axis of rotation to another, with muscle mass and the type of articulation generally ignored

kinematics [1]

the study of the geometry of motion without consideration of causal factors; *adj.* kinematic

kinematics [2]

a technique which allows a computer graphics system to simulate the movement of part or all of the image

kinesimeter

a device which makes it possible to obtain quantitative measures of body motion, including displacement, velocity, and acceleration; *syn.* kinesiometer

kinesiology

the study of the anatomical, physiological, and mechanical bases for voluntary human motion

kinesiometer

    see **kinesimeter**

kinesis

objective physical body movement

kinesthesia

that sense which originates in the stimulation of mechanoreceptors in joints, muscles, and/or tendons and leads to awareness of position, movement, weight, and/or resistance of the limbs or other body parts; *syn.* kinesthesis; *adj.* kinesthetic

kinesthesiometer

a device for measuring an individual's ability to sense body part position or movement

kinesthesis

    see **kinesthesia**

kinetic

pertaining to movement or motion

kinetic art

the use of objects in motion as an expression of creativity

kinetic energy (KE)

that portion of the energy of an object resulting from its motion; *opp.* potential energy

$$KE = \frac{mv^2}{2}$$

kinetic friction

that friction between two surfaces in contact where there is relative motion between them; *opp.* static friction

kinetic momentum

*see **momentum***

kinetics

the study or use of the effects of mechanical forces and moments on material objects in motion or to produce motion, especially of the human body

kinetosphere

a reach envelope for the hand/arm combination or the leg/foot combination in which only translational motion of the limb is permitted, with the terminal segment (hand or foot) held in a constant position; *see also **strophosphere***

kitchen

a location in restaurants, homes, and some vehicles in which food is prepared for consumption; *see also **galley***

knee

the junction of the femur, tibia, fibula, and patella, including all surrounding tissues

knee breadth

the horizontal linear distance between the most medial and lateral projections of the femoral epicondyles; *syn.* femoral breadth

Comment: measured using firm pressure with the individual standing erect and with no excessive leg muscle tension

knee cap

*see **patella***; also kneecap

knee cap height

*see **knee height (standing)***

knee circumference, fully bent

the distance around the maximum knee prominence and through the crease behind the knee

Comment: measured with the individual in a squatting position with the knee joint maximally flexed

knee circumference, sitting

the maximum surface distance around the knee, under the popliteal area and over the kneecap at an angle of 45° to the floor

Comment: measured with the individual sitting erect, the upper leg horizontal, the lower leg vertical, and the foot flat on the floor

knee circumference, standing

the surface distance around the knee measured at the level of the midpoint of the patella

Comment: measured with the individual standing erect and his weight evenly distributed on both feet

knee height, recumbent

the horizontal linear distance from the base of the heel to the anterior surface of the thigh at the femoral condyles

Comment: measured with the knee flexed 90° and the longitudinal axis of the foot perpendicular to longitudinal axis of the lower leg

knee height, sitting

the vertical distance from the floor or other reference surface to the most superior part of the quadriceps musculature above the knee

Comment: measured with the individual sitting, the knee flexed 90°, the foot flat on the floor/reference surface, and the lower leg vertical

knee height, standing

*see **midpatella height, patella top height, patella bottom height**; syn.* knee cap height

Comment: measured with the individual standing erect, but with relaxed leg musculature, and his weight evenly distributed between both feet

knee – knee breadth, sitting

the maximum horizontal linear distance between the lateral surface from one knee to the lateral surface of the other knee

Comment: measured with the individual sitting erect, the knees flexed at right angles, and both knees touching but without significant tissue compression

knee pad

a cushion for placement over the patella, usually having a strap or other attachment device around the knee, to protect against injury from kneeling or impacts between the anterior knee and other objects

knee switch

a uniaxial control device which is operated by a lateral movement of the knee

knee well

that region from the edge and extending under a table, desk, or other seated workstation which accommodates the legs in a seated posture, usually with the knees flexed

knee well depth

the horizontal distance from the user's edge of a table, desk, or other seated workstation platform to a terminus against a wall or vertical panel on the opposite side

knee well height

the vertical height from the floor or other reference level to the underside or lower surface of the structure forming a knee well; *syn.* leg clearance, vertical leg room

knee well width

the horizontal distance from one side of a knee well to the other

knob

a cover for placement on a rotational device or mechanism which normally protrudes from a surface for easier gripping and turning

knot

a measure of velocity; one nautical mile per hour

knowledge engineering

the process of identifying what information must be gathered, obtaining that information from one or more recognized experts and organizing it into a rule structure to be used in decision-making for a specific problem

knowledge of results

see feedback

knowledge-based behavior

a cognitive operating mode in which the individual attempts to achieve a goal in a situation with no clearly pre-established rules

knuckle

the protuberance of the heads of the metacarpals when the hand is clenched into a fist, or the protuberance at the interphalangeal joints when the fingers are flexed

knuckle height

the vertical distance from the floor to the point of maximum protrusion of the metacarpal III knuckle; *syn.* metacarpal III height

Comment: measured with the individual standing erect, the arm adducted to the side of the body, the palm flat against the side of the thigh, and the fingers extended

knurled

pertaining to a surface texture with small ridges, generally for providing a more firm grip

konimeter

a device for sampling airborne dust

Kretschmer somatotype

a body structure classification system developed by Ernst Kretschmer, supposedly to represent human character traits, in which men are divided into three basic groups: pyknic, athletic, and asthenic

Comment: an old system no longer used

kurtosis

a measure of the peakedness of a distribution, based on the fourth moment about the mean; *see also* **leptokurtic, platykurtic**

kymograph

an electromechanical device consisting of a rotating smoked drum or paper-covered cylinder with one or more styli for recording time-based events

kyphosis

an abnormal curvature of the thoracic region of the spine with a posterior convexity; *adj.* kyphotic; *syn.* humpback, hunchback

L

L-1 maneuver

an anti-*g* straining maneuver for preventing *g*-induced loss of consciousness during high positive acceleration forces in high performance aircraft, in which the crewmember strains his skeletal body muscles and closes the glottis for a few seconds, then inhales and exhales rapidly before repeating the process; *syn.* Leverett technique; *see anti-g straining maneuver*

Laban notation

a systematic method for describing body position in the field of dance

label [1]

a descriptor of the contents of some container, which may include such information as the product name, manufacturer, amount present, instructions, and any warning(s)

label [2]

a descriptor which helps to identify displayed screen or control structures

label coding

the use of text, numerals, symbols, or other means to identify a control, device, or system

labiodental

articulated with the lower lip touching the upper central incisors

labium

a lip or lip-shaped structure; *pl.* labia; *adj.* labial

labor [1]

the process of doing work, especially that involving physical effort

labor [2]

a group of individuals consisting of or representing those working for hourly wages

labor cost

the portion of an employer's total cost of doing business which is attributable to wages and salaries, benefits, and other aspects of employment practices

labor standard

see direct labor standard, indirect labor standard

labor turnover

a measure of how many employees enter and leave a particular workplace within a specified interval

laboratory study

an experimental study conducted in an environment in which the experimenter(s) have some degree of control over the variables involved in the phenomenon of interest

labyrinth

see inner ear

labyrinthine nystagmus

see vestibular nystagmus

laceration

a wound caused by the cutting or tearing of tissues

lacrimal bone

a small bone making up part of the medial orbit of the skull

lacrimate

secrete tears; *n.* lacrimation

lactic acid

a three-carbon organic acid product of anerobic metabolism in tissue, especially muscle tissue

lag [1]

the period of time by which a second event trails leading event

lag [2]

that distance at which a second moving object trails a leading object

lambert (L)

a unit of luminance; equals $1/\pi$ candela per cm^2

Comment: an older unit

lambert surface

a reflecting or emitting surface whose brightness appears to be the same regardless of the angle of observation

Lambert's cosine law

a law providing that the luminous intensity from a perfectly diffusing surface varies with the cosine of the angle between the perpendicular and the direction of interest

lamp

an assembled, artificial light source, including the bulb, attached electrical wiring, any shading, and other accessories

lamp burnout (LBO)

the cessation of light output from an artificial source

Comment: a recoverable light loss factor

lamp burnout factor

the proportional loss of illuminance from the non-replacement of burned out lamps

lamp lumen depreciation (LLD)

see lumen depreciation

landing

the level region at the bottom of a stair

landmark

an easily located position on or near the body surface; *syn.* anatomical reference point; *see also pointmark*

Landolt C

see Landolt ring

Landolt ring

a ring having a small gap at some orientation, both the width of the gap and the ring thickness being one-fifth the outer diameter of the ring; *syn.* Landolt C, Landolt C-ring

Comment: for use in vision testing, in which the observer is expected to report the orientation of the gap

lap

that region formed by the upper thighs to the junction of the lower abdomen with the body in an erect sitting posture

lap belt

see seat belt

laryngopharynx

the lowest portion of the pharynx, which extends from the level of the hyoid bone to the junction of the esophagus and larynx

larynx

a tube-like structure, consisting of many pieces of cartilage interconnected by ligaments and muscles, which contains the vocal apparatus and lies in the airway between the laryngopharynx and the trachea in the anterior portion of the neck; *adj.* laryngeal; *pl.* larynges; *syn.* voice box

Comment: the anterior protruding part forms the Adam's Apple

larynx to wall

the horizontal linear distance from a wall to the most anterior portion of the tissue overlying the thyroid cartilage

Comment: measured with the individual standing erect, and with buttocks, shoulders, and occiput against the wall

laser

a device for emitting coherent monochromatic electromagnetic radiation in the wavelength region within or near that perceived as visible light

Comment: originally an acronym for Light Amplification by Stimulated Emission of Radiation

laser burn

a tissue injury caused by exposure to a beam of laser radiation

late radiation effects

those ionizing radiation effects which have a long latency

latency

the time period from exposure to some event and the manifestation of an individual's response to that exposure

lateral [1]

(adj) pertaining to, near, or toward the sides of the body or a symmetrical structure; *opp.* medial

lateral [2]

(n) a consonant produced by closing off the midline of the mouth with the tongue, but allowing passage of air around one or both sides

lateral bending moments

those torques acting on the spine which result from sideways motion

lateral canthus

see ectocanthus

lateral cricoarytenoid

a skeletal muscle in the larynx which, on contraction, causes the glottis to close; *opp.* posterior cricoarytenoid

lateral cuneiform bone

one of the distal group of foot bones making up the tarsus; *syn.* third cuneiform bone

lateral disparity

see binocular disparity

lateral displacement

see abduct

lateral inhibition

a phenomenon in which neurons in the vicinity of a stimulation point, especially in sensory pathways, show reduced reactivity compared to those at the stimulation point; *syn.* surround inhibition

lateral malleolus

the lateral protrusion of the fibula at the ankle

lateral malleolus height

the vertical linear distance from the floor or other reference surface to the most lateral point of the lateral malleolus

Comment: measured with the individual standing erect and his weight distributed equally on both feet

lateral rectus muscle

a voluntary extraocular muscle with an anterior-posterior extent parallel to the optical axis along the lateral eyeball for rotating the anterior portion of the eyeball to the side

lateral retinal image disparity

see binocular disparity

lateral transfer

a personnel reassignment to another position at the same or approximately the same level of salary or responsibility

laterality

a concept that different functions and modes of operation are allocated to different sides of the brain

lateralization

the localization of a dichotically-presented sound via earphones in apparent space along an imaginary line connecting the two ears

laundry booster

any substance or combination of substances intended to aid detergents in the removal of certain stains from fabrics

law of inertia

see Newton's first law

law of reflection

a physical law that an energy wave is reflected from a surface at equal angle from the perpendicular as the incident wave, and both are in the same plane; *syn.* reflection law

layer

a cross-section through a three-dimensional object or computer model

layering

the use of multiple display windows, allowing them to overlap and partially or completely hide the contents of the covered windows

lazy foot rule

a workplace design guideline that guards and lock out switches should be easily removable and replaceable so that workers will replace them

Le Système International d'Unites (SI)

see Système International d'Unites

lead

(n) a toxic, heavy chemical element

lead intoxication

exhibiting of any of the neural, anemic, or colic symptoms resulting from lead absorption into body tissues; *syn.* lead poisoning

lead poisoning
see *lead intoxication*

lean body mass (LBM)
that mass of the body, including bones, muscles, and other tissues except for body fat; *syn.* fat-free mass; *opp.* fat body mass

lean body weight
the lean body mass acted on by the acceleration due to gravitational or other forces according to Newton's second law

leaning
pertaining to a posture in which the body longitudinal axis is away from vertical

learn
change behavior as a result of formal education, training, practice, or experience; *ger.* learning

learner's allowance
see *learning allowance*

learner's curve
see *learning curve*

learning allowance
that time allowance given to a trainee or new worker while their skills are developed on a new job or task; *syn.* learner's allowance

learning control
having a control system with adequate memory and computing power to be able to modify its own operation in concert with newly acquired knowledge

learning curve
a concept, mathematical function, or graphical representation of performance versus time in which performance improves with time as a result of learning/feedback; *syn.* learner's curve; see *start-up curve, progress curve*

learning hierarchy
a set of behavioral objectives, concepts, and principles arranged in the order in which they should be learned for optimum performance

learning hierarchy analysis
a determination of the order in which

the learning hierarchy should be taught

least squares method
a mathematical technique for fitting a straight line or curve to a set of data points where the sum of the squares of the perpendicular distances from each data point to the line or curve is minimized

leg
the femur, tibia, fibula, and their surrounding associated and supporting soft tissues

leg clearance
see *knee well height*

leg inseam
the inside length of a trouser leg from the pubic crotch to approximately the dorsal/superior surface of the foot; *syn.* inseam

leg room
a measure of that usable volume beneath some table, platform, or other structure which the legs would normally occupy when in a seated posture

leg-foot
involving both the leg and the foot, generally referring to internally-generated or motor activities; see also *foot-leg*

legend
an explanatory symbol on a display or control

legend switch
a labeled switch

length
an open anthropometric measurement from one point on the body to another which contains as a major portion a relatively straight line, but may also contain some brief curvature, such as around a flexed joint

length-tension curve
an inverted-U-shaped function which indicates that muscle tension capability falls off to either side of an optimum length

lens
a transparent device for refracting or

otherwise directing electromagnetic radiation

leptokurtic
pertaining to a highly peaked normal distribution; *opp.* platykurtic; *see also kurtosis*

lesion
any wound or unnatural alteration in tissue

lesser multiangular bone
see trapezoid bone

lesser trochanter
a rounded projection on the medial proximal femur; *opp.* greater trochanter

lethal dose 50
see median lethal dose

letter of intent
a written promise to carry out a specified action at some point in the future

leukocyte
a white blood cell

leukopenia
having a below normal leukocyte count

levator
any muscle producing an upward movement; *opp.* depressor

level above threshold
see sensation level

level of effort [1]
a type of contract or agreement in which a certain number of people are supported to do specified tasks

level of effort [2]
the amount of physical or mental activity exerted or required to perform at a certain level

level of illumination
see illumination level

leveled element time
see normal element time

leveled time
see normal time

leveling
a performance rating method in which an observer adjusts a worker's time to compare with normal time; *see also performance rating*

lever
a rigid linear structure which is capable of movement and exerting force about a fulcrum; *see also fulcrum, effort, resistance*

lever arm [1]
the distance from a joint axis to the point of a muscle attachment

lever arm [2]
the distance from the fulcrum to the point of effort or resistance on a lever

lever switch
a type of toggle switch in which the activating mechanism is a manually-operated lever

leverage
that mechanical advantage achieved by using a lever

Leverett technique
see L-1 maneuver

lexical decision task
the process in which a judgment is made as to whether or not a letter string is a word

life characteristic curve
a life cycle curve describing the expected phases over the lifetime of a machine or electromechanical system; consists of a steeply declining initial segment (run-in, or infant mortality phase), a relatively flat middle segment (the useful life phase), and a moderately increasing terminal segment (the wear-out phase); *syn.* machine life curve, bathtub curve

life cycle cost
the total cost of an item over its useful life, including purchase, maintenance, and operations

life expectancy
the number of years a person may be expected to live from a given age, based on the mean length of life of persons of a similar age

life jacket
a personal flotation device worn about the torso and normally secured with straps across the torso and through the pelvic crotch; *syn.* life preserver

life performance curve
a functional relationship between some particular characteristic of a lamp and its age

life support
that function which addresses the sustenance, health promotion, and protection of personnel under all reasonably expected conditions for a specified activity

life support system
any system which provides life support

lifeline
a rope or other type of cable intended to save an individual's life should an accident occur under hazardous working conditions
Comment: may function to break fall before striking an object, to keep from drifting off, to keep from being washed overboard, etc.

lifting task
any task which involves manually changing the location of an object without external mechanical assistance and applies a force and/or torque to the vertebral column

lifting technique
a procedure recommended or used by an individual to perform a particular lifting task

lifting torque
the product of the load and distance of the load from a fulcrum in the vertebral column which is created by a lifting task

ligament
a band of dense fibrous connective tissue which interconnects the articular aspects of bones

light
(n) radiation from that region of the electromagnetic spectrum of which an organism becomes aware through stimulation of the retina or other visual receptor; that stimulation which excites visual receptors; *see visible spectrum*

light
(adj) not heavy

light aberration
see chromatic aberration, spherical aberration

light activity
that level of physical activity which requires/consumes 60–100 calories per square meter of skin surface per hour, including the basal metabolic rate

light adaptation
an adjustment within the visual system making it more or less sensitive to light by adjusting the threshold; *syn.* photopic adaptation; *opp.* dark adaptation

light duty
a work classification in which an individual is not permitted to do heavy lifting for health or other reasons

light effort
that level of physical work which can be maintained for a work shift without undue fatigue

light flux
see luminous flux

light intensity
see luminous intensity

light loss factor (LLF)
any of a set of possible variables which may contribute to a decrease in available luminance for a given location; *see recoverable light loss factor, non-recoverable light loss factor, total light loss factor*

light meter
see illuminance meter

light pen
a pen-shaped interactive device which emits a light beam for striking a certain region of a display to initiate a certain system action

light quantity (Q)
a measure of the amount of light used, equal to the product of the luminous flux and the time duration for which it is sustained; *syn.* quantity of light

light scatter fraction
the ratio of scattered light to specularly reflected light

Light Amplification by Stimulated Emission of Radiation (LASER)
see laser

light-emitting diode (LED)
any semiconductor diode which emits light when current is applied

light-emitting diode display
any display using LEDs as a radiant source

lighting
the collective sensation or description of the light being input to the visual environment

lighting effectiveness factor (LEF)
the ratio of equivalent sphere illumination to calculated illumination or illumination measured with a meter

lighting fixture
any structure designed and built specifically for the installation of light-producing devices and to direct illumination

lighting quality
a measure involving some combination of the following variables in a given environment: the distribution of luminance, the color temperature, the spectral makeup, the luminous intensity, the ability to see, visual comfort, and esthetics; *syn.* quality of lighting

lightness [1]
a judgement as to the weight of an object, on a scale from light to heavy

lightness [2]
that apparent degree to which something is judged as lighter or darker compared to a similarly reflecting or transmitting white or achromatic reference

Likert scale
a technique for rating surveys on a discrete, integer-based scale having an odd number of discrete options and consisting of a range, generally from 1 to 5, from strongly disagree to strongly agree, respectively

Comment: occasionally see scales to 7 or 9 options

limb coordination
a measure of the degree of integrated functioning of the limbs in performing some activity

limb movement velocity
the rate at which a single movement of a limb can be accomplished, without regard for accuracy, coordination

limb-load aggregate
the combined mass/torque from the working load plus the mass/torque from the limb(s) involved in a lifting or movement task

limen
see threshold

liminal contrast
see contrast threshold

limit stop
any device or mechanism which prevents further movement of a control, door, drawer, or other object at a certain point when motion beyond that point might have undesirable consequences; *syn.* stop
Comment: may be accompanied by audible click or tactile sensation

limit switch
an electrical switch which is capable of cutting the power supply if the device being monitored goes beyond a specified range

limp [1]
a type of gait in which steps are halting and the time spent on one leg is shorter than the other; *see also gait*

limp [2]
flaccid; having less than normal tonicity

line of flow
see flow path

line of sight (LOS)
that path from the lateral and vertical center of the eye pupil to an object being fixated or direction being viewed; *syn.* primary line of sight; *see also preferred line of sight*

line spectrum

a frequency spectrum in which the components are shown as lines at discrete frequencies

line width

the width of a line on a display or hardcopy

lineal

see linear

linear [1]

pertaining to a linear function

linear [2]

measured in a straight line; *syn.* lineal

linear algebra

the study and/or use of simultaneous linear equations, as used in vectors and linear transformations

linear correlation

a relationship between two variables which may be represented graphically by a straight line or by a linear function

linear energy transfer (LET)

a measure of the radiation energy loss per unit of ionization track length

linear equation

see linear function

linear function

a mathematical function which may be represented by a straight line, having an equation of the form below; *opp.* non-linear function

$$y = mx + b$$

linear momentum $(\vec{p})$

the tendency for an object to continue moving in a straight line

$$\vec{p} = m\vec{v}$$

linear motion

see rectilinear motion

    Comment: an older usage

linear movement control

a control device which moves in a straight line when force is applied

linear programming (LP)

a technique for determining a solution to a problem using the assumptions (a)

that the function is linear and (b) that the process involved can be represented as a set of linear equations or inequalities

linear regression

a statistical technique for estimating the value of one variable from the value of another when the two variables are known or assumed to be linearly related; *syn.* simple linear regression

linear system

a system in which output varies according to some proportionality constant and the input

linearity [1]

the straightness of a line, or column or row on a display

linearity [2]

that property between two variables in which a change in one variable results in a directly proportional change in the other

linguadental

articulated with the tip of the tongue placed on the upper front teeth

link [1]

any interface between between the human operator and a machine, at which movement in one produces movement in the other; *syn.* linkage

link [2]

a straight line representing a body segment, terminating at pivot points on the body

link [3]

any interface, interaction, or bond between individuals; *syn.* linkage

link analysis [1]

an examination and study of the biomechanical link actions or positions of the body

link analysis [2]

an identification and examination of the sensorimotor and mechanical/electrical interfaces between individuals, machines, or human and machine in a system

linkage

see link

lip breadth
the maximum horizontal linear distance between the most lateral point of the junction of the upper and lower lips on each side of the mouth opening; *syn.* lip length, mouth breadth
Comment: measured with the facial muscles relaxed

lip breadth, smiling
the maximum horizontal linear distance between the corners of the mouth opening; *syn.* lip length (smiling), mouth breadth (smiling)
Comment: measured with the individual smiling broadly

lip length
see lip breadth

lip length, smiling
see lip breadth, smiling

lip – lip length
the vertical distance, in the midsagittal plane, from the lower margin of the lower lip to the upper margin of the upper lip
Comment: measured with the facial muscles relaxed and the lips together

lip prominence
see lip protrusion

lip protrusion
the most anterior point of either the upper or lower lip, whichever is more anterior
Comment: specify which lip if different

lip protrusion to back of head
the horizontal linear distance from inion to the most anterior point of the lips; *equiv.* lip protrusion to wall
Comment: measured with the individual standing or sitting erect, facing straight ahead

lip protrusion to wall
the horizontal linear distance from a wall to the most anterior point of the lips; *equiv.* lip protrusion to back of head
Comment: measured with the individual standing or sitting erect with the back of the head against the wall

liquid crystal display (LCD)
any display constructed from a material whose reflectance or transmittance varies on application of an electric field

liter (l)
a unit of volume in the SI/MKS systems

little league elbow
an overuse injury caused by stress on the muscles, tendons, epiphyses, and articular surfaces of the elbow joint

little league shoulder
a condition of tendonitis or metaphyseal fracture causing pain from excessive internal and rotational stresses around the shoulder

load
the performance demands required from a system or individual at any given time

load cell
a strain gauge-based device for measuring the amount of force applied to an object

load factor
that proportion of the work cycle time required for a worker to perform the necessary work at standard performance during a machine-paced cycle

load limit
the maximum weight or stress which an individual, floor, vehicle, or other structure can safely support; *syn.* allowable load

load stress
that type of sensory overload caused by having too many channels of information to process effectively

load weight
the maximum weight which a vehicle can safely carry; *syn.* allowable load

load-endurance curve
a graphical curve illustrating the relationship between the percentage of maximum load and the length of time which that load will be voluntarily held; *syn.* intensity-duration curve, intensity-duration relationship

loading secondary task

a secondary task which must be constantly attended to; *opp.* non-loading secondary task

Local Area Network (LAN)

a communication link over which computers and peripherals may be connected within a limited geographical region

local definition

an elaboration on a more generic definition by providing additional detail to suit the purpose of a specialized condition or location

local exhaust

a system for removing contaminated air from the workplace by collecting and redirecting it from near its source to the outside, rather than letting it escape into the general enclosed workplace

local horizontal

pertaining to a region within a larger coordinate system in which a secondary, smaller coordinate system defines a horizontal axis or plane

local lighting

that lighting intended to provide illumination only for a small region

local minimum

the smallest value within a restricted range of values

local vertical

pertaining to a region within a larger coordinate system in which a secondary, smaller coordinate system defines a vertical axis

localize

determine the source of a stimulus or signal in space and/or time; *n.* localization

location coding

the identification of controls, devices, or systems through their placement on some panel or other structure

locator

a landmark on the body whose distance from another point or plane is being measured

lockjaw

see ***tetanus***

locomotion

the active movement of the body from one place to another; *adj.* locomotor

locomotor system

the various bodily systems, structures, and tissues used in locomotion

loft

the trapped air in clothing

logarithm

a function represented by the real-valued exponent of some base number; *adj.* logarithmic; *opp.* antilogarithm

logarithmic interval scale

an alternative to the basic measurement scales in which the magnitudes corresponding to points are given by $\log x_n - \log x_{n+1} = \log x_{n+1} - \log x_{n+2}$, etc.

long bone

any bone whose length greatly exceeds its width; *syn.* tubular bone

long ton

a unit of mass in the English system; equal to 2240 pounds; *see also* ***ton***

long-waisted

(sl) having a longer than normal trunk for the total stature

long wavelength infrared

see ***far infrared***

long wavelength ultraviolet

see ***near ultraviolet***

long-term

pertaining to events or conditions which develop or are maintained for an extended period of time, typically on the order of years

long-term memory

a coded form of memory which apparently exists indefinitely

longitudinal axis

an approximate center line of a body segment which is parallel to the length dimension of that segment

longitudinal design

any research methodology in which data is collected from the same individual(s) over a long period of time

longitudinal study

an experiment or observation using a longitudinal design

longitudinal wave

a waveform in which the direction of propagation and displacement are the same; *syn.* compression wave, acoustic wave

lordosis

a curving of the cervical-lumbar regions of the spine in the sagittal plane to yield an anterior convexity; *adj.* lordotactic; *opp.* kyphosis

lost time [1]

that time for which an individual would normally be at his workplace but is not due to an occupational illness or injury

lost time [2]

see delay time

lost time accident

an accident which results in a significant period of time away from the job

lost time illness

an occupational illness which results in more than one day off from work, usually referring to something more serious than a minor illness

lost time injury

an occupational injury due to which a worker misses at least one day of work

loudness

a subjective attribute of auditory perception whereby sounds may be rank ordered on a scale extending from inaudible to loud

loudness contour

a curve of sound pressure level values plotted against frequency which are required to produce a given loudness sensation for a normal listener

loudness level

the median sound pressure level of a free progressive 1 KHz frequency wave presented to listeners facing the source relative to 20 μN, in which the listeners judge the sound to be equally loud across multiple trials

low density lipoprotein (LDL)

a substance present in the blood which carries high levels of cholesterol, occasionally depositing it on arterial walls as plaque; *see high density lipoprotein*

low frequency (LF)

that portion of the electromagnetic spectrum consisting of radiation frequencies between 30 KHz and 300 KHz

lower arm length

see radiale –stylion length

lower explosive limit (lel)

see lower flammable limit

lower flammable limit (lfl)

the concentration of a flammable gas or flammable liquid vapor below which a flame will not be propagated despite the presence of an ignition source and an appropriate oxidizing agent; *syn.* lower explosive limit; *opp.* upper flammable limit

lower thigh circumference

see thigh circumference, distal

lowpass filter

a device which allows frequencies lower than the cutoff frequency to exit from the device unattenuated, while the intensity of frequencies higher than the cutoff frequency are attenuated; also low-pass filter; *opp.* highpass filter

lumbago

a low level of pain in the lumbar region of the back

lumbar disk

an intervertebral disk separating the lumbar vertebrae in the spine

lumbar spine

that region of the spine comprised of the five lumbar vertebrae

lumbar vertebra

any of the vertebral bones in the lumbar spine, L1–L5; *pl.* lumbar vertebrae

lumbosacral angle

the angle between the posterior of the lumbar spine and the sacrum

lumen [1]

a hole or passage in a tube-like structure within the body

lumen [2] (lm)

an SI unit for that luminous flux passing through one steradian from a point light source of one candela intensity

lumen depreciation

that decrease in luminous flux emitted by certain types of light sources over time; *syn.* lamp lumen depreciation

lumen-hour (lm-hr)

a unit for that amount of light delivered by a luminous flux in one hour

lumen-second (lm-sec)

see talbot

luminaire

a complete lighting system, consisting of the lamp(s), positioner(s), protector(s), and any required connections to a power supply

luminaire dirt depreciation (LDD)

the loss of luminous flux from lighting due to dirt collection on the luminaire or particulates in the atmosphere

Comment: a recoverable lighting loss factor

luminaire surface depreciation

any reduction in luminous output due to physical or chemical changes in materials associated with a luminaire, such as transmittance through or reflections from enclosing materials

Comment: a non-recoverable light loss factor

luminance

a physical measure of the luminous flux per unit solid angle incident on a surface; old *syn.* photometric brightness; *see also brightness*

luminance contrast

a measure of the physical relationship in luminance between two adjacent, non-specular surfaces under the same general illumination and immediate surroundings, generally defined by an equation similar to the form below; *see also brightness contrast*

$$C_L = \frac{\Delta L}{L}$$

luminance ratio

the value of the ratio between the luminances of any two surfaces or objects in the visual field

luminescence

the emitting of light due to some mechanism other than high temperatures; *adj.* luminescent

luminosity

a measure of the relative efficiency of various wavelengths of visible light for exciting the retina

luminosity function (V_λ)

see spectral luminous efficiency function

luminous efficacy, flux

the value of the ratio of the total luminous flux to the total radiant flux encompassing all wavelengths

luminous efficacy, source

the value of the ratio of the total luminous flux emitted by a lamp to the total electrical power input; old *syn.* luminous efficiency

luminous efficiency

see luminous efficacy (source), luminous efficacy (flux)

luminous efficiency

see spectral luminous efficiency

luminous efficiency function

see spectral luminous efficiency function

luminous environment

that portion of the visual environment generated by the luminaire type, luminous intensity, direction, and hues

luminous flux (Φ)

the rate of visible light energy emitted from a source over time; *syn.* light flux

luminous intensity

a measure of the power of a light source in terms of luminous flux per unit solid angle; *syn.* light intensity

luminous reflectance

see reflectance

luminous transmittance

see transmittance

lunate bone
one of the proximal bones of the wrist

lung
one of the asymmetrical bilateral organs within the chest which is involved in gaseous respiration

lung diffusing capacity
a measure of the amount of gas at STP which diffuses across the pulmonary membrane in the alveolus

lung expiratory reserve volume
see *expiratory reserve volume*

lung functional residual capacity
see *functional residual capacity*

lung inspiratory capacity
see *inspiratory capacity*

lung vital capacity
see *vital capacity*

lung volume
the volume of measurable gas in the lungs under specified conditions

lunula
the lighter-colored portion of the nail body near the nail root

Lustermeter
a device developed by Hunter to measure contrast gloss and compute luster

lux (lx)
the SI/metric unit for illuminance

luxon
see *troland*
Comment: an older term

lying
pertaining to a posture in which an individual's torso is horizontal to a reference surface, but not prone, with possible flexion of the hips and knees

lymph
the water and various dissolved substances and particulates which enter the lymphatic system from the interstitial fluid

lymph gland
see *lymph node*

lymph node
an ovoid-shaped structure occurring in lymph vessels which serves as a collection and filtration point for lymph in fighting infection; *syn.* lymph gland

lymph vessel
any of a range of diameters of tubular structures from capillary size to those resembling moderate-sized veins in the cardiovascular system which carry lymph

lymphatic
pertaining to lymph, lymph vessels, or lymph nodes

lymphatic system
the fluid and the various structures involved in collecting interstitial fluid, removing foreign particles, and returning the fluid ultimately to the cardiovascular system

M

M-1 maneuver
a technique for aircrewmen to prevent gravity-induced loss of consciousness due to high positive acceleration maneuvers in aircraft, in which the crewmember generally grunts with the glottis partially closed to increase intrathoracic pressure, thereby increasing blood pressure and blood flow to the brain; *see* **anti-g straining maneuver**

mach
a unit representing the velocity of sound, usually in air

mach indicator
an aerospacecraft display which provides the vehicle's velocity as a ratio to the velocity of sound; *syn.*

machine
a mechanically- or electromechanically-powered device consisting of both fixed and moving parts and having one or more specific functions

machine ancillary time
that time during which a machine is unavailable for use due to calibration, changeover, cleaning, or other related causes

machine assignment
that function or operator to which a machine is assigned

machine attention time
that time during which an operator must observe a machine's operations in the event intervention or servicing is required
   Comment: does not involve actually operating for production or servicing the machine

machine available time
that time during which a machine is performing or could perform work

machine capability
see **machine capacity**

machine capacity
some measure of the normally-expected output from a machine; *syn.* machine capability

machine cycle time
that time required for a machine to perform one complete cycle of a process

machine downtime
that amount of time during which a machine is not able to perform its designated function due to a breakdown, routine servicing, or a materials shortage; also machine down time

machine effective utilization index
the value of the ratio of the time which a machine is running under standard conditions compared to the time which the machine is available

$$MEUI = \frac{machine\ standard\ running\ time}{machine\ available\ time}$$

machine efficiency index
the value of the ratio of the machine standard running time to the machine running time

$$MEI = \frac{machine\ standard\ running\ time}{machine\ running\ time}$$

machine element
a work element performed entirely by a machine

machine guard
any piece of equipment or device on a machine intended to reduce or eliminate the chance of injury through the use of that machine

machine hour
a unit of measure for the utilization of machines, corresponding to one machine working for one hour; also machine-hour

machine idle time
the amount of time a machine is available but not productive due to the operator performing other work, due to a shortage of materials; *syn.* idle machine time

machine interference
a situation in which a demand for simultaneous operator attention by two or more machines results in machine idle time

machine interference time
that amount of time lost by an operator due to machine interference

machine life curve
see *life characteristic curve*

machine load
the proportion or percentage of scheduled or actual usage of machine available time during a given time interval

machine maximum time
the total time in a day, week, or other time period which one or more machines could work

machine pacing
see *machine-paced work*

machine running time
the actual operating, productive time by a machine

machine running time at standard
see *machine standard running time*

machine standard running time
that time at which a machine operates at optimum capacity

machine time
see *machine-controlled time*

machine time allowance
see *machine-controlled time allowance*

machine utilization index
the value of the ratio of the amount of time a machine is running compared to the time it is available

$$MUI = \frac{machine\ running\ time}{machine\ available\ time}$$

machine-controlled time
that time in a given work cycle which a machine requires to perform its portion of a task or process, independent of an operator; also machine controlled time; *syn.* machine time

machine-controlled time allowance
the expected or scheduled time given a worker for a machine to perform its portion of a given task; *syn.* machine time allowance

machine-controlled work
see *machine-paced work*

machine-paced work
that restricted or externally-paced work in which machinery controls the rate at which the work cycle progresses; *syn.* machine pacing, machine-controlled work; *see also self-paced work*

machmeter
see *mach indicator*

macro
a set of keystrokes or computer instructions which may be executed with a single command

macro command
that command which initiates a macro execution

macro-
(prefix) large, large-scale, or long-length

macroelement
a work element which is of sufficiently long duration to permit observation and timing with a manually-operated stopwatch or stopclock; *opp.* microelement

macroskelic
having long legs relative to the torso length

macula
see *macula lutea, utricular macula*

macula lutea
the yellow-colored central region of

the fovea, at which visual acuity is greatest; *adj.* macular; *syn.* yellow spot

macula utriculus

see utricular macula

magnetic field ($\vec{B}$)

that vector field generated by a magnetic substance or which exists in conjunction with an electric field via current conduction or electromagnetic radiation

Magnetic Resonance Imaging (MRI)

the use of a combined static and radio frequency electromagnetic field to measure energy absorption by certain atoms, which can be processed and presented as an image cross-section of the body, a body segment, or any other object transparent to the electromagnetic field; *syn.* Nuclear Magnetic Resonance Imaging

magnitude

the numerical absolute value of a vector

main menu

the top-level menu within a software package

maintainability

the probability of restoring a failed system to a specified operating condition within a given period of time using prescribed procedures

maintained illuminance

that proportion of initial illuminance which a light or luminaire retains over some specified period of time

maintenance

the performance of those functions necessary to keep a machine, process, or system in or return it to a proper state of repair for safe and/or efficient operation; *see routine maintenance, scheduled maintenance, planned maintenance, corrective maintenance, preventive maintenance, predictive maintenance, composite maintenance, overhaul*

maintenance management

the processes of deciding what type of

maintenance will be used for systems under an individual's or organization's control, which may include: (a) the conducting of tradeoff studies, (b) a decision as to what risks acceptable and what are not, (c) the scheduling and implementation of maintenance, and (d) the development of maintenance procedures

maintenance time

that time estimated, allowed, used, or required to perform some act of maintenance on a system

major axis

the longer axis in defining an ellipse; *opp.* minor axis

major defect

a defect which results in a serious malfunction of a product; *opp.* minor defect

major injury

an occupational or other injury which results in a loss of time to the injured person and a medical expense; *opp.* minor injury

make-or-buy analysis

a study to determine whether it is more advantageous to develop and produce an item in-house or purchase the item from outside sources

make-ready allowance

see setup allowance

makeup air

that outdoor air supplied to a workplace as replacement for the air removed by exhaust ventilation

makroskelic

see macroskelic

Malcolm Baldrige National Quality Award

a nationally-based award which is presented annually to an organization judged best in several categories such as human resource utilization, quality assurance, and leadership; *syn.* Baldrige Award

malignant

pertaining to continuing abnormal tissue growth, possibly with eventual metastasis, culminating in death unless successfully treated

malingerer
an individual who pretends illness, physical disability, or other inability to perform certain duties or service to avoid work or gain compensation

malleolus
a rounded bony projection at the ankle; *see lateral malleolus, medial malleolus*

malleus
the auditory ossicle interfacing with the tympanic membrane and the incus

malnutrition
a condition in which there is an inadequate nutritional intake or an inability to utilize ingested nutrients

malodorant
any odorant having a strong or offensive odor

malpractice
the intentional or careless misconduct or the lack of competence by an individual involved in professional work

mammalian diving response
a physiological response to high environmental pressure in which the peripheral arteries contract and the heart rate slows due to the body's attempt to preserve oxygen flow to the brain and other vital organs

man modeling
see human modeling

man process chart (MPC)
see operator process chart; also man-process chart

man-amplifier
the concept of a human using an exoskeleton or other device which enables him to perform feats requiring much greater strength or other capabilities than would be normally humanly possible without such a device

man-computer dialogue
see human-computer dialogue

man-computer interaction
see human-computer interaction

man-computer interface
see human-computer interface

man-environment interface
see human-environment interface

man-hour
a unit measure of work equivalent to the utilization, scheduling, or availability of one person working for one hour; also man hour

man-in-the-loop
see human-machine system

man-machine chart
a multiple activity process chart in which both personnel and machines are used

man-machine interface
see human-machine interface

man-machine system
see human-machine system

man-made fiber
any textile fiber made from synthetic or natural chemical substances

man-made noise
any electrical or acoustic noise having a human source or resulting from man-made equipment

man-minute
a unit of measure of work equivalent to the utilization, scheduling, or availability of one person working for one minute; also man minute

man-multiplier
a concept in which one person controls many machines, all performing the same tasks; *syn*. doppelgang

man-paced work
see self-paced work

man-tool interface
any portion of a tool where a person might grasp, carry, and/or hold a tool for performing manipulations on other objects

man-type flow process chart
see worker-type flow process chart

manage
organize and direct human, economic, and material resources toward devel-

oping and accomplishing one or more specified objectives; *n*. management

management

the group of people within an organization who manage

mandatory standard

a procedural, performance, or other type of standard which is regulated by law via one or more governmental agencies; *see also* **standard**

mandible

the lower jawbone of the skull

manikin

see **mannikin, mannequin**

Comment: incorrect or stylized spelling of mannikin

manipulate

handle, move, or operate on one or more objects or controls using the hands or other dexterous controlling device(s) in conjunction with a vision or other sensory system; *n*. manipulation; *adj*. manipulative

manipulative dexterity

a measure of the skill which an individual or robotic device possesses for the coordinated use of fingers / hands / wrists or their robotic analogues for fine tasks

manipulative grasp

see **tripodal grasp**

manipulative skill

a measure of the ability to operate or control with minimal placement errors and time duration

manipulator

any non-mobile mechanical device for handling, moving, or controlling operations at a distance

Mann-Whitney U test

a non-parametric statistical test using rank-ordered data for comparing two independent groups

mannequin

an anthropomorphic figure which has joints or other superficial human physical characteristics which is used in

modeling, clothing display, training, or art; *see also* **mannikin**

mannikin

an anthropomorphic dummy which has joints or other simulated human physical characteristics and which may be used for a variety of biodynamic functions; also manikin; *syn*. dummy; *see also* **mannequin**

manoptoscope

a device for determining which eye is dominant

Manpower and Personnel Integration (MANPRINT)

a U.S. Army management and technical human factors program for improving weapon-soldier system performance

manual [1]

pertaining to an operation or set of operations performed solely by humans, rather than by machines or with machine-assistance

manual [2]

a document which provides instructions or other information for operation of some equipment

manual control [1]

any control mechanism intended for manipulation by humans

Comment: the individual is the feedback element

manual control [2]

a discipline which studies and incorporates the human operator as a feedback element within a closed-loop system; *syn*. manual feedback control

manual dexterity

a measure of the ability to make rapid, coordinated, fine or gross movements movements of the fingers, hand(s), and/or arm(s) for handling independent objects

manual element

a work element performed by a worker using no more than simple tools, and not involving machines; *syn*. unrestricted element

manual feedback control
> see *manual control*

manual input
> the use of a human operator to input data to a computer via some computer input device

manual materials handling (MMH)
> the non-equipment-aided human act of relocating an object, consisting of approximately the following stages: approach, grasp, pickup, move/carry, putdown, adjust

manual steadiness
> see *hand steadiness*

manual time
> the amount of time required to execute a manual element; *syn.* hand time

manually-controlled work
> see *self-paced work*

manubrium
> the triangular-shaped superior segment of the sternum

Manufacturing Automation Protocol (MAP)
> a set of communication standards for use in automated manufacturing

manufacturing cost
> the total cost of manufacturing an item, including materials, direct labor, overhead, and depreciation

manufacturing engineering
> that field of engineering specializing in the research, planning, design, integration, and development of the methods, facilities, tools, and processes involved in the production of goods

manufacturing progress function
> the improvement in production efficiency with time

manuometer
> a spring device for measuring static strength of the finger flexor muscles

margin [1]
> a distance, setting, or other limit which should not be exceeded under normal circumstances

margin [2]
> that region, typically without printing, which separates printed text and/or graphics from the paper or other material edge on a hardcopy

margin of safety
> the value of the ratio of that stressor which would result in failure of a system to the level of the stressor planned to be or currently imposed; *syn.* safety margin

marginal cost
> that cost incurred for an additional unit of output

marginal product
> that additional unit of output which is obtained by adding an extra unit of some factor

marginal revenue
> that additional income realized by selling one additional product unit

marine vessel accident incidence rate (VIR)

$$VIR = \frac{number\ of\ vessel\ accidents}{number\ hours\ of\ operation} \times 100,000 hours$$

market analysis
> a study involving the collection of data to determine information for a product or service such as who potential consumers would be, trends in the marketplace, why a consumer might purchase it, etc.

market research
> the process of gathering and analyzing data regarding the potential sale of goods or services to the consumer

marketing policy
> that guideline which determines what products will be offered, what types of markets will be approached, what selling and promotional techniques will be used, what prices will be charged, etc.

marrow
> a soft tissue material in the interior of many bones; see *red marrow, yellow marrow*

marstochron

see **chronograph**; also marsto-chron; syn. marstograph

marstograph

see **chronograph**

mask

(n) any covering for the face and/or head, usually for a protective function

mask

(v) increase the threshold level of a stimulus or condition by presenting a second (masking) stimulus simultaneously or in close time or space proximity; ger. masking

masked differential threshold

see **masking level**

masking level

that difference in original stimulus intensity required to reach a reported threshold due to a masking stimulus

mass

that measure of an object's resistance to acceleration, compared to a standard

mass media

those forms of the media which typically reach large numbers of people, especially newspapers, television, and radio

massage

rub, stroke, knead, or impact the superficial muscles of the body, either by hand or with some instrument for therapeutic or other purposes

massed practice

continuous, repeated, or extended training, without time for rest periods; opp. distributed practice

Master Standard Data (MSD)

a universal predetermined motion time system

Master's two-step test

the simple exercise of repeatedly ascending over two nine-inch steps to test cardiovascular function; syn. two-step test

master-slave manipulator

pertaining to any device in which the remote operator is intended to follow either exactly or proportionately the motions and forces of the input controller; also master/slave manipulator

mastoid process

the bony projection on the inferior lateral surface of the temporal bone

matched groups design

an experimental design in which group selection is made by matching individuals across those groups based on one or more variables which are to be manipulated or controlled during the experiment; syn. equivalent groups method

matched pairs design

an experimental methodology in which assignment to groups is not strictly random, but based on one or more pairing criteria on which individuals are paired

matching individual

an individual acting as a control for another individual in a matched pair

material damping

sound attenuation due to energy loss in the substance through which the energy is being transmitted

material requirements planning (MRP)

the process of reducing each final product to its elementary parts, forecasting the product output required, and coordinating the production quantities of elementary parts

material-type flow process chart

a flow process chart which indicates materials usage

materials handling

the study or actual movement of any substances, components, or products, whether in or out of containers or packaging

maternity leave

that leave, granted either with or without pay, for a female employee to give birth to a child and recover before returning to work

mathematical reasoning

the ability to understand and organize

a mathematical problem, then select a method to find a solution to the problem

> *Comment:* excludes the actual numerical manipulation

matrix

a rectangular array of numbers with a designated rows-by-columns structure

matt

see matte

matte

having or pertaining to a surface with a dull appearance, exhibiting primarily or only diffuse reflections; also matt; *opp.* glossy

maxilla

a bilaterally-fused bone making up much of the anterior portion of the face, including the upper part of the mouth/jaw, part of the nasal cavities, and the floor of the orbits; *adj.* maxillary

maximal aerobic capacity ($\dot{V}_{O_2 max}$)

the level at which oxygen uptake during performance of a task reaches a steady state and no additional oxygen can be used by the muscles involved in the task; *syn.* aerobic capacity, aerobic endurance capacity, aerobic work capacity, maximal oxygen uptake/consumption, maximal aerobic power, maximum aerobic work capacity, maximum oxygen uptake

maximal isometric force

the maximum force generated during an isometric contraction for a specified muscle or muscle group

maximal oxygen consumption

see maximal aerobic capacity

maximal oxygen uptake

see maximal aerobic capacity

maximal voluntary contraction (MVC)

the greatest force which the muscle or muscle groups involved can develop under voluntary control when contracting against a resistance under specified conditions; also maximum voluntary contraction

maximum

the largest measured, existing, or permissible value of a set; *adj.* maximal; *opp.* minimum

maximum abdominal depth

see abdominal extension depth

maximum acceptable concentration

see permissible exposure limit

maximum aerobic capacity

see maximal aerobic capacity

maximum aerobic work capacity

see maximal aerobic capacity

maximum allowable concentration (MAC)

see permissible exposure limit; *syn.* maximum acceptable concentration

maximum allowable flight duty period

the greatest number of hours an air crew can fly in an aircraft in any 24-hour period

maximum allowable slope

that ratio of the horizontal distance from the edge of an excavation to the depth which must be provided for the existing soil or rock conditions

maximum breathing capacity

see maximum voluntary ventilation

maximum metabolic rate

the highest metabolic rate consistent with sustained aerobic metabolism

maximum operating pressure

the highest pressure which a pressure vessel is expected to experience during use

maximum oxygen uptake

see maximal aerobic capacity

maximum performance

the performance level which results in the highest possible production

maximum permissible concentration (MPC)

that maximal amount of radioactive material in air, water, or food which will not result in dangerous accumulations in humans

maximum permissible dose (MPD)

that amount of ionizing radiation which

can be absorbed per unit mass of irradiated material at a specific location without being expected to cause radiation injury to a person during one's lifetime

maximum permissible limit (MPL)
a NIOSH guideline for manual lifting under specified conditions, above which musculoskeletal injury is a high probability; *see also action limit*

maximum voluntary contraction
see maximal voluntary contraction

maximum voluntary ventilation (MVV)
the greatest volume which an individual can force himself to inhale and exhale per minute for a brief period of time; *syn.* maximum breathing capacity

maximum working area
that portion of the working surface which is easily accessible to the operator's hands with the elbow and shoulder fully extended in the normal working posture

maximum working volume
that maximal region within which an operator can be expected to reach via any combination of shoulder, elbow, and wrist motions

Maynard Operation Sequence Technique (MOST)
a predetermined motion time system

meal break
that segment of a work shift, typically about mid-shift, which an employee is allotted for eating a meal
Comment: may be compensable or not, depending on whether primarily for benefit of employer or employee

mean
a number representing the expected value of a set; *syn.* average; *see also arithmetic mean, geometric mean, measure of central tendency*

mean body temperature ($\overline{T}_b$)
an estimated value of the average body temperature based on skin and core

temperature measurements, usually as a function of the weighted mean skin temperature and the rectal temperature

mean deviation
the average of the absolute deviations of values in a distribution from the mean

$$MD = \frac{\sum |X - \overline{X}|}{N}$$

mean radiant temperature
a calculated estimate of the amount of heat transfer via radiation using the dry bulb temperature, the wet bulb temperature, air velocity, and the globe temperature

mean skin temperature
a measure intended to represent the average temperature of the skin over its total body surface; *see also weighted mean skin temperature*

mean time between failures (MTBF)
the average time expected between failures of a system or piece of equipment

mean time to failure (MTTF)
the average time to the first failure of a component or system

measurable engineering parameter
any attribute of physical behavior of a system which is detectable by the appropriate instrumentation capable of describing the external environment applied to, or the structural response of, the system

measure
(v) read or otherwise obtain one or more numerical values from observations for analysis according to certain rules; *n.* measurement

measure
(n) an aspect or dimension

measure of availability
see availability

measure of central tendency
any variable or value which is used to represent the central tendency of a dis-

tribution, such as the mean, mode, or median; also central tendency measure

measure of dispersion
any value which is an indicator of the spread of a distribution, such as the range, variance, or standard deviation; also dispersion measure; *syn.* measure of variability

measure of variability
see measure of dispersion

measured daywork (MDW)
that work performed at standard levels for an established hourly, non-incentive wage; also measured day work

measured work
that work for which performance standards have been set using some form of work measurement technique

measurement
the taking of data or the data resulting from a measure

measurement error standard deviation
the square root of the within-subject variance when a group of individuals have each been measured more than once; *syn.* technical error of measurement

mechanical advantage
the value of the ratio of force output by a mechanical device to the force applied to it

mechanical analogue
see simulator

mechanical efficiency (ME)
the value of the ratio of external work performed to physiological energy production

mechanical hazard
any unsafe situation due to machinery, equipment, tools, and/or physical structures

mechanical impedance (Z_m)
the complex ratio of force to velocity during simple harmonic motion

mechanical ohm
a unit for mechanical resistance, reactance, and impedance

mechanical reactance (X_m)
the imaginary portion of mechanical impedance

mechanical resistance [1]
a qualitative indication of the mechanical forces which must be overcome to move an object, control, or other mechanism

mechanical resistance [2] (R_m)
the real portion of the mechanical impedance; the opposition of a structure or object to a mechanical force either to change motion or to deform the structure; *syn.* resistance

mechanical shock
a relatively rapid transmission of mechanical energy into or out of a system; *syn.* shock

mechanics
that field which studies the mechanical environmental effects on physical systems

mechanize
introduce machinery to carry out certain functions previously performed by humans; *n.* mechanization

mechanoceptor
see mechanoreceptor

mechanoreceptor
any sensory receptor which is stimulated by a local change in mechanical pressure, force, or tension due to some type of movement; also mechanoceptor

medial
lying near or toward the midsagittal plane of the body or other approximately symmetrical structure; *syn.* mesial; *opp.* lateral

medial calf skinfold
the thickness of a vertical skinfold on the medial surface of the calf at the level of the calf circumference point midway along the antero-posterior direction

> *Comment:* measured with the individual standing, the knee flexed 90°, and the foot resting flat on an elevated platform

medial canthus
see *endocanthus*

medial cuneiform bone
one of the distal group of foot bones of the tarsus; *syn.* first cuneiform bone

medial malleolus
the medial projection from the tibia at the ankle joint

medial malleolus height
the vertical distance from the floor or other reference surface to the most medially projecting point of the medial ankle bone; *syn.* internal ankle height
Comment: measured with the individual standing erect and his weight evely distributed on both feet

medial plane
see *midsagittal plane*

medial rectus muscle
a voluntary extraocular muscle located parallel to the optical axis along the medial side of the eyeball
Comment: involved in rotating the anterior portion of the eyeball toward the body midline

median
the middle value in a sorted quantitative data set which divides a distribution into two equal parts; *syn.* 50th percentile, second quartile

median lethal dose (LD$_{50}$)
the acute dose of some drug, biological agent, or radiation which can be expected to be fatal to 50% of the population within a specified period of time; *syn.* lethal dose 50

median lethal time
that time required for 50% of the organisms to die following a given dose of a drug, radiation, biological agent, or other agent

median nerve
a spinal nerve innervating generally the forearm and volar-thumb region of the hand

median plane
see *midsagittal plane*

medical expert
any licensed physician found qualified to give testimony as an expert witness by a court

medical radiation
any ultrasound, electromagnetic, or particulate radiation emitted by or received from diagnostic or therapeutic radiological procedures

medium frequency (MF)
that portion of the electromagnetic spectrum consisting of radiation frequencies between 300 KHz and 3 MHz

mega- (M)
(prefix) one million or 10^6 times the basic unit

mel
a subjective unit of pitch with reference to a pure 1 KHz tone as 1000 mels at a loudness 40 dB above a listener's threshold

melatonin
a hormone produced in the pineal gland with a circadian cycle, and believed to have a relationship to circadian rhythms

membrane potential
that voltage difference measured across the membrane between the interior and exterior of a cell or across an artificial membrane

membranous labyrinth
a collection of soft-tissue ducts containing endolymph within the osseous labyrinth of the inner ear comprising the semicircular ducts, the cochlear duct, saccule, and utricle

memo motion
a method of visually sampling work activities at specified periods of time using time-lapse photography or videography

memomotion study
the use of memo motion for the analysis of long-duration events or processes; *syn.* camera study

memorize
absorb information with perfect recall,

usually in text, numeric, or pictorial form; *n.* memorization

memory [1]
the capacity for mental storage of feelings, sensations, information, movement patterns, and events

memory [2]
any of several types of storage means for bits in a computer

menarche
that phase in a female's life when menstruation begins

meninx
a layer of tissue which covers the brain and spinal cord; *pl.* meninges; *adj.* meningeal

menopause
that phase of a woman's life at which the menstrual cycle terminates; *adj.* menopausal

menses
the time of menstruation; *syn.* menstrual period

menstrual cycle
that hormonally-regulated period of about 28 days during which a woman normally undergoes ovulation and menses

menstruate
pass blood and other tissues from the uterus via the vaginal orifice during part of the menstrual cycle; *n.* menstruation; *adj.* menstrual

mental
of or pertaining to the mind or intellectual/cognitive activities or functions

mental age
the mental competence of an individual relative to the chronological age of an average individual with equivalent mental competence; *see also chronological age, developmental age*

mental basic element
any work element which involves some form of mental activity

mental health
a state in which an individual or population has accomplished a high degree of self-realization and integrated its own desires while successfully adapting to its environment; *see also health*

mental hygiene
that field of study and practice for the development and/or preservation of mental and emotional health

mental load
see mental workload

mental retardation
a mental handicap in which less than normal intellectual functioning is exhibited

mental set
see set

mental work
any work done by an individual primarily using perceptual and cognitive abilities, especially those involving such activities as calculating, reasoning, monitoring, decision-making, and verbal/image processing

mental workload
any measure of the amount of mental effort required to perform a task; *syn.* mental load; *opp.* physical workload

menton
the point at the tip of the chin in the midsagittal plane
Comment: convention: represented by the most anterior point in anterior-posterior measures, by the most inferior point in vertical measures

menton – crinion length
the vertical distance from the bottom surface of the tip of the chin to the hairline in the midsagittal plane
Comment: measured with the individual standing or sitting erect, looking straight ahead with the facial muscles relaxed; not applicable on the bald or balding

menton projection
the horizontal linear distance in the midsagittal plane from the most anterior point of the chin to the juncture of the neck and the bottom of the jaw
Comment: measured with the facial musculature relaxed

menton – sellion length

the vertical linear distance from the inferior tip of the chin to the deepest point of the nasal root depression

Comment: measured with the individual sitting or standing erect, looking straight ahead with the facial muscles relaxed

menton – subnasale length

the vertical linear distance between the junction of the base of the nasal septum and the superior philtrum to the base of the chin in the midsagittal plane; syn. jaw height

Comment: measured with the individual sitting or standing erect, looking straight ahead with the facial muscles relaxed

menton to back of head

the horizontal linear distance from inion to the most anterior portion of the chin; equiv. menton to wall

Comment: measured with the individual standing erect and looking straight ahead

menton to top of head

see menton to vertex

menton to vertex

the vertical linear distance from the inferior tip of the chin to the vertex plane level; syn. menton to top of head

Comment: measured with the individual standing erect and looking straight ahead with the facial muscles relaxed

menton to wall

the horizontal linear distance from a wall to the most anterior portion of the chin; equiv. menton to back of head

Comment: measured with the individual standing erect with his back and head against the wall

mentum

the chin; adj. mental

menu

a display of the possible options available to the user from a given command location in a software package

menu bar

a function area within a screen display which contains a menu

menu hierarchy

the structure with which a menu is organized, generally with higher level menus providing access to other comparable level menus as well as lower level menus under the given level

menu navigation

the process of a user finding his way through the menu structure of a software package to locate a desired function or operation

mercury

a heavy metal, existing in a liquid state under standard conditions, which can have toxic effects on the nervous system, primarily from inhalation of it or its salts or other compounds

mercury lamp

an illumination source which operates by passing an electrical current between two electrodes in a ionized mercury vapor atmosphere, giving off a bluish-green light with a significant amount of ultraviolet light; syn. mercury vapor lamp

mercury-fluorescent lamp

a high-intensity discharge lamp using high pressure mercury enclosed within a tube whose interior is coated with phosphors to convert the ultraviolet light into visible light; syn. phosphor mercury lamp

merit rating [1]

the process of assessing, or the result of an assessment of, performance regarding an employee in a job, usually according to some periodic interval and some specified group of factors such as dependability and work quality/quantity

merit rating [2]

the process of determining or the resulting determination of tax or insurance premium rates based on an employer's record for disabling injuries and layoffs; syn. experience rating

mesh

that latticework in computer modeling which divides a large object into finite elements

mesial

see medial

mesomorph

a Sheldon somatotype denoted by prominent muscular tissue, heavy bones, broad shoulders, and a flat abdomen; *adj.* mesomorphic

mesopic vision

that vision using both the rods and cones at moderate luminous intensities; an intermediary between photopic and scotopic vision; *syn.* twilight vision

message area

a function area for the system or other users to communicate with a user

message line

a single line within a message area

met

a unit of physiological workload; the metabolic thermal output of an average, sitting-resting individual under conditions of thermal comfort

metabolic cost

see gross metabolic cost, net metabolic cost

metabolic gradient

a difference in degree of metabolic activity from one region of the body to another

metabolic heat production

the transformation of chemical energy into heat energy by the body

metabolic rate

the amount of chemical energy transformed into heat and work for the performance of bodily activities per unit time; *see also basal metabolic rate*

metabolic reserves

the potential chemical energy source, stored primarily as glycogen, which can be rapidly mobilized for use by the body, especially for muscular activity involving effort beyond one's normal level of activity

metabolism

the sum of all physico-chemical processes by which energy is produced and substances are organized and maintained within the body; *adj.* metabolic; *see also catabolism, anabolism*

metacarpal bone

one of the bones in the hand between the wrist bones and the phalanges which make the rigid structure of the palm and back of the hand; *adj.* metacarpal

metacarpale III height

see knuckle height

metal fume fever

an influenza-type occupational illness believed due to inhalation of the fumes of certain metal oxides; *syn.* galvo, zinc chills

metal halide lamp

a high-intensity discharge lamp in which the primary light is produced from metal halide radiations and their dissociation products

metallic

possessing a brilliant luster, characteristic of most metals

metamer

a visual stimulus which is perceptually indistinguishable from another visual stimulus under one given type of illumination, but which has a different spectral composition and may be distinguishable under another type of illumination; *adj.* metameric

metameric pair

two colored visual stimuli which appear identical to the eye, but which consist of different spectral compositions

metamerism

a condition in which two colored stimuli appear the same under one illuminant but different under a different illuminant

metaphysis

the region of bone growth near the ends of a long bone; *adj.* metaphyseal

metastasis

the transfer of disease from some initial location to one or more distant locations via causal agents or the cardiovascular or lymphatic systems; v. metastasize; *adj.* metastatic; *pl.* metastases

metatarsal bone

one of the bones of the foot anterior to the tarsus; *adj.* metatarsal

meter (m)

an SI/MKS unit of length; a distance equal to 1,650,763.73 wavelengths of the radiation emitted corresponding to the Krypton-86 atom transition between levels $2p_{10}$ and $5d_5$ in vacuum

meter-angle

the amount of eye convergence when the visual axis of each eye is centered on an object one meter distant from the cornea of each eye

Meter-Kilogram-Second system (MKS)

that measurement subset of the metric system which uses the meter, kilogram, and second as its basic units; *syn.* MKS system; *see also Centimeter-Gram-Second system*

method

a technique, orderly sequence of steps, or set of operations used to perform some given task

method of adjustment

a psychophysical methodology in which the subject actively varies some aspect of a stimulus until the variable stimulus appears either to match or be just noticably different from a fixed reference stimulus, as specified for the test

method of constant stimuli

a psychophysical methodology in which stimuli are presented to the subject who is to make judgments about how they differ from a standard stimulus, whether greater or lesser along some dimension

method of equal-appearing intervals

a psychophysical methodology in which the subject adjusts a set of stimuli

until the elements of the set appear equidistant from each other along some dimension in an attempt to establish interval-level data; *syn.* method of equal-sense distances

method of equal-sense distances

see method of equal-appearing intervals

method of limits

a psychophysical methodology in which some dimension of a stimulus is changed in small increments in an ascending/descending manner until the subject either ceases responding or changes his response

method of loci

a visualization type of mnemonic in which a sequence of locations is used to remember a sequence of events

method of magnitude estimation

a psychophysical methodology in which a subject assigns relative quantitative values to stimuli based on their intensity compared to a reference value

method of paired comparisons

a psychophysical methodology in which all possible pairs of stimuli are presented to a subject for comparison along one or more dimensions; *syn.* paired comparisons

method of rank order

an ordinal-level psychophysical methodology in which stimuli are presented to a subject for ranking along a specified dimension

method of ratio estimation

a psychophysical methodology in which a subject is instructed to adjust or rate a stimulus along some dimension such that it is a specified ratio of a reference stimulus; *syn.* method of ratio production

method of ratio production

see method of ratio estimation

method study

see methods study

methodology

the standard technique used to accomplish different tasks

methods analysis
> see *methods study*

methods design
> the process of developing improved work methods to improve job performance; see *methods engineering*

methods engineering
> the analysis, design, and implementation of improved work methods and systems where human effort is used; see *methods design*, *methods study*

methods study
> a systematic examination of the techniques, factors, and resources involved in the component parts of one or more operations; also method study; *syn.* methods analysis; see *methods engineering*, *methods design*
>> Comment: with the intent of improving techniques and productivity, while reducing costs

Methods Time Measurement (MTM)
> a predetermined motion time system
>> Comment: exists in several versions

metric system
> the measurement system predominating in most of the world and using units such as meter, liter, kilogram, second, etc.; see also *MKS system*, *CGS system*, *English system*

mho ($\mho$)
> a unit of conductance; equal to the reciprocal of the ohm

Michigan Anthropometric Processor (MAP)
> a software program developed by the University of Michigan for analog-to-digital acquisition and real-time checking of anthropometric data

micro- (μ)
> (prefix) one-millionth or 10^{-6} of the basic unit; small

Micro-Matic Methods and Measurement (4M)
> a computerized predetermined motion time system

microbe
> a microorganism, especially referring to any bacterium

microchronometer
> a large-faced electric clock with marked time units in decimal minutes with rapidly-moving hands used for noting the time in micromotion studies; see also *wink counter*

microelement
> an element of work which occurs in an interval of time too short to allow it to be adequately observed with the unassisted capacity of the human eye/perceptual system; *opp.* macroelement

microgravity
> any environment in which objects of significant size and mass appear to remain suspended indefinitely, usually due to hypogravitational conditions on the order of 10^{-6} g; *syn.* zero gravity

microgravity growth factor
> that proportional increase in body height, primarily within the torso due to release from intervertebral disk compression, which an individual experiences when exposed to microgravity conditions
>> Comment: generally about 3%

micromotion analysis
> see *micromotion study*

micromotion study
> the use of normal or high-speed photographic or videographic frame rates for the frame-by-frame analysis of events or processes which occur too rapidly for adequate real-time observation by the eye; *syn.* micromotion analysis; see *memomotion study*

micron
> a micrometer; equal to 10^{-6} m
>> Comment: use of this unit is not recommended

microorganism
> any living organism too small to be seen or studied with the naked eye

microphone
> a transducer which is sensitive to sound waves, and converts the mechanical

energy to electrical energy

microreciprocal degree (mired)
see microreciprocal kelvin
Comment: an older term

microreciprocal kelvin (mirek)
a unit of reciprocal color temperature, equal to $10^{-6}/T_k$

microsleep
a very brief sleep period, usually on the order of a few seconds to minutes, in an individual who is not permitted to go to sleep but is too fatigued to remain awake

microspectrophotometer
a device for measuring the wavelength and intensity of light absorbed as it passes through a transparent substance

microwave
radiation within that portion of the electromagnetic spectrum having wavelengths of approximately 3 mm to about 30 cm, between radio waves and infrared

microwave cataract
a partial or complete opacity of the eye lens due to microwave radiation exposure

microwave dosimetry
the study or measurement of the amount of microwave energy to which a system is exposed

microwave hearing effect
an auditory sensation apparent as a clicking or buzzing sound in humans exposed to pulsed microwave energy

microwave nonthermal effects
the presumed non-heating effects from exposure to low power microwave energy; also nonthermal microwave effects

microwave thermal effects
an alteration in biological systems due to the heat produced by absorbed microwave energy; also thermal microwave effects

micturition
the process of urine secretion; *syn.* urination

mid-
located at approximately the axial center of some entity

midaxillary line
an imaginary or marked vertical line passing through the antero-posterior center of the axilla and down the side of the trunk

midaxillary line at umbilicus level skinfold
the thickness of a skinfold at the umbilicus level in the midaxillary line
Comment: measured with the individual standing comfortably erect and arms hanging naturally at the sides

midaxillary line at xiphoid level skinfold
the thickness of a horizontal skinfold at the xiphoid level in the midaxillary line
Comment: measured with the individual standing comfortably erect, the arms slightly abducted, and the shoulder extended

midaxillary plane
a vertical plane extending through the centers of the armpits which divides the body into anterior and posterior segments

middle ear
that portion of the ear within the temporal bone in which the auditory ossicles are located, between the tympanic membrane and the oval and round windows

middle ultraviolet
that portion of the ultraviolet radiation spectrum from about 200 to 300 nm

midline
an imaginary line or plane which divides a structure into two approximately symmetrical parts

midpatella
a point on the patella which is midway between the superior and inferior margins of the patella

midpatella height
the vertical distance from the floor or

other reference surface to midpatella; *syn.* knee cap height; *see also patella top height, patella bottom height, knee height (standing)*

> *Comment:* measured with the individual standing erect, but with relaxed leg musculature, and his weight evenly distributed between both feet

midsagittal

pertaining to the midsagittal plane or a point on it; also mid-sagittal

midsagittal plane

an imaginary plane which divides the body or other (approximately) symmetrical structure into right and left sections; also mid-sagittal plane; *syn.* medial plane, median plane

midshoulder

a point half way between the neck-shoulder junction and acromion

midshoulder height, sitting

the vertical distance from the seat upper surface to midshoulder

> *Comment:* measured with the individual sitting erect, with the head and back against a wall

midthigh

a position midway between the inguinal crease and superior aspect of the patella along the midline of the leg as described in the thigh length measure; also mid-thigh

midthigh circumference

see thigh circumference, midthigh

mil [1]

one thousandth of an inch

mil [2]

one thousandth of a radian

mileage death rate

see motor vehicle incidence rate

Military Standard (MIL-STD)

a mandatory standard issued by the U.S. Department of Defense for use by contractors or others in manufacturing items for DOD use; *see also Department of Defense Standard*

milli- (m)

(prefix) one thousandth or 10^{-3} of the basic unit

millicurie-of-intensity-hour

see sievert

milligram-hour (mg-h)

a unit of radiation dose, equivalent to 1 milligram of radium for one hour

Minamata disease

a disease due to methyl mercury consumption/ingestion which results in narrowing of the visual field, ataxia, sweating, or other symptoms

miner's cramp

see heat cramp

miner's helmet

a safety helmet with an attached lamp

mineral

a naturally-occurring inorganic substance

mini-Gym

see MK-I, II

minimal angle of resolution

see minimum resolution angle

minimal passageway

that minimal height and width of a corridor which allows an individual clothed for specified working conditions to pass without conflict with boundaries or other persons

minimal perceptible erythema

see erythemal threshold

minimal separable acuity

see resolution threshold; also minimum separable acuity

minimal weight

the least amount a person can weigh without endangering lean body mass and essential fat storage

minimum

the lowest active, existing, or permissible value; *adj.* minimal; *opp.* maximum

minimum cost life

see economic life

minimum dose

the smallest quantity of an agent which will produce a physiological effect

minimum lethal dose

the average of the smallest dose of

some agent which kills and the largest dose which fails to kill living organisms under controlled conditions; *see* **median lethal dose**

>   *Comment:* an older term no longer used

minimum resolution angle

the smallest angular or linear separation at which an individual can resolve two visual objects as separate under a specified set of conditions; *syn.* angle of resolution, minimum angle of resolution, resolution angle, minimum/minimal separable acuity

minimum separable acuity

see **resolution acuity**; also minimal separable acuity

minometer

an instrument for measuring stray radiation from radioactive sources

minor axis

the shorter axis defining an ellipse; *opp.* major axis

minor defect

a defect which may affect appearance, slightly reduce functionality, or other characteristics, but which causes no serious malfunction; *opp.* major defect

minor injury

an occupational or other injury in which no significant amount of time from work is lost and no major medical costs are incurred; *opp.* major injury

Minor's sweat test

an examination to measure possible damage to the sympathetic nervous system by determining which dermatomes of the body do not perspire; *syn.* sweat test

minute respiratory volume

the total volume of air moved into and out of the respiratory system per minute

mired

see **microreciprocal degree**

mirek

see **microreciprocal kelvin**

mirror

(n) a highly specularly reflecting surface

mirror

(v) create a mirror image in computer modeling or graphics

mirror image

a structure which would correspond at least in part to the reflection of another part of an original object about some plane

mirror stereoscope

a laboratory device used to present separate images of a scene to each of the eyes by a system of mirrors; *syn.* Wheatstone stereoscope

mission

that designated activity at a particular location which a system is intended to accomplish

mission reliability

the probability that a given product or system will complete a specified mission

mist

the presence of finely-divided liquid droplets in air

mitten

a type of fitted handwear for covering the hand which has a slot for the thumb, but does contain separate finger slots

MK-I, II

a small commercial exercise device flown in earth orbit on Skylab for exercising arm and back muscles; *syn.* mini-Gym

MKS system

see **Meter-Kilogram-Second system**

mnemonic

any formal technique for aiding in memory storage or recall

mobile

having the freedom or ability to physically move about from one location to another through relatively independent means; *n.* mobility

mobility aid

any physical device which enhances one's mobility, especially with regard to the handicapped

mobility analysis

a determination of which employees

have the skills, training, experience, or other capability to move to other jobs if it becomes necessary

mockup

a full-scale, representative physical layout of a workstation, equipment, or situation used for training or as a design tool; also mock-up; *syn.* service test model

modal

see mode

modal time

that element time which occurs with the highest frequency during a time study

modality

any sense, such as vision or hearing

mode [1]

that value or set of values which occurs most frequently in a data set; *adj.* modal

mode [2]

an operating state

model

(v) generate a model, or simulate a system

model

(n) a system for providing quantitative estimates of results under specified conditions

     Comment: may be physical, graphical, computer/electronic, or theoretical/abstract

moderate work

that level of work activity which has a gross metabolic cost of 180–280 calories per square meter of skin surface per hour

modified Cooper-Harper scale

see Cooper-Harper Scale, modified

Modified Rhyme Test

a multiple choice test in which an individual is to select the word he believes he heard spoken from a selection of rhyming alternatives

modular

see module

modular design

consisting of modules

modular workstation

a workstation which may be assembled from modular components in a variety of different configurations; *see also cluster workstation*

modulation

the variation in value of some parameter characterizing a periodic oscillation

module

a standard unit which may serve as a building block for larger structures; *adj.* modular

modulus

the numerical value assigned to a standard stimulus, against which other stimuli are judged and assigned relative values

moisture vapor transmission rate

the mass of water vapor passing through a specified area of one or more fabrics per unit time

molecule

the smallest divisible electrically-neutral portion of a substance more complex than an atom which is held together by chemical bonds and retains the properties of that substance

moment [1]

a statistic measure, represented by the sum of the deviations from the mean, raised to some power, and divided by the number of terms used in accumulating the sum; *see* **arithmetic mean, variance, skewness, kurtosis**

moment [2]

the tendency of a force to generate rotation in a body or torsion about an origin; *see* **torque, torsion**

moment arm

that component of the vector representing the distance from a point of rotation which is perpendicular to the line of action of a force creating a torque

moment concept

the idea that lifting stress is also a function of the bending moments at the spine, not just of the weight lifted

moment of force
see torque

moment of inertia
the tendency of an object to retain its current rotational motion about an axis; *syn.* rotational inertia

$$I = \sum m_i r_i^2 = \int r^2 dm$$

where:
m = mass element
r = distance from the axis of rotation

momentary hold
the maintenance of some position for a brief period of time; *opp.* sustained hold

momentum
the tendency of an object to continue in its current motion unless acted on by an outside force; *see linear momentum, angular momentum, Newton's first law of motion*

monaural
hearing with or presented to only one ear; pertaining to sounds transduced by a single channel, regardless of whether presented to one or both ears

monitor [1]
(v) observe, listen to, keep track, or exercise surveillance of ongoing progress, events, or situations by any appropriate means; *ger.* monitoring

monitor [2]
(n) an individual or instrument who/which performs a monitoring function

monitor [3]
(n) the physical device housing the electronics and some controls supporting the display for a computer or video system

monochromasia
total color blindness, in which all the red, green, and blue cones are missing or non-functional; also monochromasy
   Comment: the individual sees only shades of gray, lightness

monochromat
an individual having monochromasia

monochrome
pertaining to a screen display or hardcopy having a single color image against a background

monocular
pertaining to only one eye, or vision using one eye; *syn.* uniocular

monocular visual field
that part of the visual environment which can be seen by a single eye at any given instant with the head and eye stationary; *see also binocular visual field*

monoplegia
the paralysis of a single limb, or a single muscle group

monotone
see monotonic

monotonic
pertaining to a function in which the dependent variable either continuously increases or continuously decreases in magnitude with an increase in the independent variable throughout the range of values under consideration, such that each point for either function uniquely defines one point for the other; *syn.* monotone; (n) monotonicity; *opp.* non-monotonic, *see monotonic increasing, monotonic-decreasing*

monotonic-decreasing
pertaining to a function in which the dependent variable continuously decreases in magnitude with an increase in the independent variable; *see monotonic; opp.* monotonic-increasing

monotonic-increasing
pertaining to a function in which the dependent variable continuously increases in magnitude with an increase in the independent variable; *see monotonic, opp.* monotonic-decreasing

monotony
the psychological state created by the lack of variety due to the repeated performance of a non-challenging task or a long-duration task; *adj.* monotonous

Monte Carlo Method
a probabilistic technique for obtaining solutions to problems by statistical sampling methods; *syn.* Monte Carlo Technique

Monte Carlo Technique
see *Monte Carlo Method*

moonlight
work a second job

morale
a measure of the level of confidence and enthusiasm of an individual or group

morbidity
the ratio of the number of individuals having some disease to the total population

morbidity rate
the number of cases of a specific disease occurring in a population within a specified time interval; *syn.* incidence rate

morning person
(sl) an individual who typically wakes up easily in the morning, ready for the day, and has trouble staying up late at night; *opp.* evening person

morphology
the study of the structure and form of biological organisms

mortal
subject to death; *n.* mortality

mortality rate
the number of deaths occurring per 1000 population in a specified time period; *syn.* death rate

motion aftereffect
any illusion of continuing motion which begins on cessation or change of a particular motion

motion analysis
the acquisition, processing, organization, and use of data obtained from human physical activity, whether of certain specific joints, body segments, or the body as a whole

motion and time study
see *time and motion study*

motion cycle
the entire set of physical activities required to perform a given work cycle one time

motion economy
see *motion efficiency*

motion efficiency
the concept that body motions in performing a task should be reduced to the minimal, simplest, least fatiguing possible set

motion efficiency principles
a set of some common sense or empirically determined concepts dealing with human movements for the industrial/manufacturing workplace to simplify and improve the effectiveness of manual work and minimize fatigue; *syn.* principles of motion economy, principles for motion improvement, characteristics of easy movement; see *workplace design, display-control layout*
Comment: general principles:
(a) use natural, rhythmic, easy movements
(b) establish habitual movements
(c) use both hands simultaneously in parallel motions, not sequential
(d) minimize movements
(e) involve the fewest body segments possible in performing the work
(f) distribute actions among the various muscles of the body
(g) use ballistic movements rather than slower, controlled movements
(h) use momentum to aid performance
(i) minimize momentum when muscular effort must be used to overcome it
(j) use continuous, curved movements, not straight lines involving rapid changes in direction

motion sickness
a condition in which the signs or symptoms of nausea, vomiting, and/or physiological effects produced by either real or perceived motion of the body or its surroundings

motion study
see *motion analysis*

Motion Time Analysis (MTA)
a predetermined motion time system

motions inventory
the nature and quality of possible motions within the capabilities of an individual under specified circumstances

motivation
a psychophysiological construct which is involved in the initiation, direction, and sustenance of behavior by an individual or group toward accomplishing some goal; *adj.* motivational

motoneuron
see motor neuron

motor
pertaining to the activation of muscles by efferent neurons or nerves

motor activity
any pattern of muscular activity concerned with locomotion or the moving of a limb or body part

motor end plate
see end plate

motor fitness
a measure of an individual's physical suitability for a particular task

motor homunculus
a representation of the human body on the surface of the motor cortex, whose distribution is proportional to the density of innervation in various parts of the body; *opp.* sensory homunculus

motor learning
any form of learning involving the coordinated activities of muscles

motor nerve
an efferent nerve which provides motor innervation to a muscle

motor neuron
an efferent neuron which sends/carries information toward a neuromuscular junction; *syn.* motoneuron

motor point
a location on the skin at which electrical stimulation will cause contraction of the underlying muscle

motor skill
the ability to move some or all parts of the body in a coordinated fashion toward the performance of some task

motor unit
the combination of a motor neuron, its axon, the neuronal terminal branches, and the muscle fibers they innervate

motor vehicle accident incidence rate
see motor vehicle incidence rate

motor vehicle incidence rate
a factor for rating the number of deaths from motor vehicular accidents by miles of vehicular travel; *syn.* motor vehicle accident incidence rate

$$MVIR = \frac{number\ of\ vehicle\ accidents}{total\ vehicular\ miles\ traveled} \times 100,000\ miles$$

mountain sickness
see altitude sickness

mounting height
that vertical height above the floor, table, ground, or other surface at which an illumination source is located

mouse
a computer input device having one or more buttons and capable of two-dimensional rolling motion which can drive a cursor on the display and perform a variety of selection options or commands

mouse keys
an interactive feature for handicapped individuals which will allow them to use certain keys on a computer keyboard to control a cursor normally operated by a mouse

mouth
that portion of the head which is bounded anteriorly by the lips, posteriorly by the fauces, laterally by the internal tissues of the cheeks, superiorly by the palate, and inferiorly by the tongue and lower jaw; *syn.* oral cavity

mouth breadth
see lip breadth

mouth breadth, smiling
see lip breadth, smiling

mouth stick

a rod for allowing a quadriplegic or other handicapped individual to operate various forms of equipment by holding the device in his mouth and using pressure to operate the equipment; *syn.* typing stick

move [1]

(v) execute one or more isotonic muscular contractions, resulting in a change in position of one or more parts of the body; *n.* movement

move [2]

(v) transfer (cut and paste) a segment of text, graphics, or other material in a computer system from one location to another

move [3] (M)

(n) a physical basic work element involving motion of the hand carrying one or more objects

movement disorder

any pathological condition which results in an abnormal deviation from an intended movement, an inability to execute a desired movement, or an undesired involuntary movement

moving average

an arithmetic mean based on a fixed number of samples over time, in which as each new sample is added, the oldest sample is dropped

multi-factor plan

an incentive plan in which employee awards are based on more than one factor

multilevel sampling

the selection of a primary, large or high-level unit, followed by secondary, tertiary, et cetera, units, each selected from within the next higher level unit; *syn.* multistate sampling

multilimb coordination

the ability to meaningfully integrate the movements of two or more limbs to fulfill some purpose such as manipulating a control, an object, or locomotion

multimedia

pertaining to the presentation of information which stimulates more than one sensory modality

multiple activity chart

see multiple activity process chart

multiple activity operation chart

see multiple activity process chart

multiple activity process chart

a process chart showing the chronological activities involving a work system, with each component of the system allocated a separate vertical column to show relative or coordinated activities; *syn.* multiple activity chart, multiple activity operation chart

multiple correlation

the degree of relationship between a criterion variable and two or more predictor variables

multiple correlation coefficient (R)

a numerical value representing the correlation between a set of two or more predictor variables and one criterion variable; *syn.* coefficient of multiple correlation, multiple R

multiple machine work

a work assignment which has a worker attending to two or more machines

multiple R

(sl); *see multiple correlation coefficient*

multiple regression

the analysis or use of the combined and individual contributions from more than one predictor variable for predicting the value of a single criterion variable

multiple sclerosis (MS)

a disease resulting in demyelination within the CNS and the corresponding movement, speech, and other difficulties

multiple watch timing

see accumulative timing

multisensory

combining or related to more than one sensory modality; *syn.* heteromodal

multistate sampling
see *multilevel sampling*

multitasking [1]
the processing of more than one dataset or application at a time, usually with the operator working directly only on one application; also multi-tasking

multitasking [2]
the assignment of a worker to more than one task or job

multivariate [1]
having more than one dependent variable

multivariate [2]
pertaining to more than one variable

multivariate analysis
any statistical analysis involving more than one independent variable and/or more than one dependent variable

multivariate analysis of variance
(MANOVA)
an analysis of variance involving two or more of both independent and dependent variables

Munsell chroma (C)
saturation in the Munsell color system

Munsell color system
a color ordering system for surfaces which divides colors into perceptually uniform segments for ordering and specifying with regard to hue, chroma (saturation), and value (lightness)

Munsell hue
hue

Munsell value
a measure of lightness in the Munsell color system, on a scale ranging from 1 (black) to 10 (white); *syn.* Munsell value scale

muscle [1]
a structure composed of a mass of muscle tissue, usually enclosed by some type of sheath, and forming a distinct unit; *adj.* muscular

muscle [2]
see *muscle tissue*

muscle action
any muscle activity which results in a change in length or an increase in tension in the muscle

muscle capacity
see *muscular endurance*

muscle fatigue
see *muscular fatigue*

muscle fiber
a muscle cell; *see also* **intrafusal fiber, extrafusal fiber**

muscle group
a collection of individual skeletal muscles which have similar innervation and perform a similar/common/related function

muscle hemoglobin
see *myoglobin*

muscle spindle
see *neuromuscular spindle*

muscle testing
any procedure intended to measure the performance of a restricted number of muscles on some graded basis

muscle tissue
an irritable, contractile, extensible elastic tissue composed of long tubular or spindle-shaped cells; *see* **skeletal muscle, smooth muscle, cardiac muscle**

muscle tone
a state of continuous mild muscle contraction; *syn.* muscle tonus

muscle tonus
see *muscle tone*

muscular endurance
the maximum time under stated conditions which a muscle or muscle group can maintain a given measure of external force; *syn.* muscle capacity; *see also* **endurance, strength**

muscular fatigue
the buildup of lactic acid in muscle tissue due to prolonged heavy exertion; *syn.* muscle fatigue

muscular strength
see *strength*

musculoskeletal system
the integrated system of muscles, bones, and joints in the body

musculospiral nerve
see *radial nerve*

mustache
that long term accumulation of hair growth which originates on the face generally above the upper lip, medial to the lip margins, and beneath the nasal septum base

mutagen
any agent capable of producing a genetic mutation

myelin
a lipid-rich material, consisting primarily of membrane material from neural support cells, which insulates the axonal membrane

myelin sheath
the collective concentric wrapping of the membranes of many neural support cells around an axon at intervals along its length; *see also* **node of Ranvier, saltatory conduction**
Comment: each support cell forms one internode; permits saltatory conduction

myoclonus
an isolated rapid involuntary contraction of one or more muscles

myoelectric limb
a limb prosthesis which senses muscle electrical activity in the proximal remaining portion of the limb or the trunk region and uses those signals to drive one or more motors to operate the prosthesis

myofibril
the basic unit of contractile structure in skeletal muscle cells

myoglobin
a protein in muscle which may function as an oxygen carrier; *syn.* muscle hemoglobin

myography
see *electromyography*

myoneural junction
see *neuromuscular junction*

myopia
a refraction error in which parallel light rays from a distant object are focused anterior to the retina under relaxed accommodation; *syn.* nearsightedness

myosin
a globular muscle protein involved in contraction; *see also* **actin, actomyosin**

myositis
an inflammation of muscle tissue, often due to overuse

N

N589

a classical acoustics modeling software package

nail

the elastic protein tissue covering the dorsal portion of the terminal phalanges of the hand and foot; *see also fingernail, toenail*

nail body

the exposed portion of the nail

nail fold

the rounded skin at the lateral and proximal portions of a nail

nail groove

the depressed region between the nail and the nail fold

nail matrix

that structure beneath the skin from which nail tissue is formed

nail root

that portion of the nail which lies beneath the skin between the lunula and the nail matrix

nano- (n)

(prefix) one-billionth or 10^{-9} of the basic unit

nap [1]

a brief period of sleep

nap [2]

the short, small fibers on a fabric surface

nape

the back of the neck; *syn.* nucha

napestrap

a strap-like device extending from a piece of headgear over the nape of the neck to assist in headgear retention

napier

see néper

narcolepsy

a disorder in which an individual experiences numerous severe occasions of sleepiness during the day

narcosis

a profound stuporous or lethargic state

naris

the passage at either the anterior or posterior nasal cavity; *pl.* nares; *see internal naris, external naris*

narrow band

pertaining to a frequency band consisting of a few hertz on either side of a center frequency; *opp.* wide band

narrow band analysis

a type of frequency analysis in which sound intensity level measurements are restricted to a few hertz on either side of a center frequency

nasal [1]

see nose

nasal [2]

pertaining to the type of sound produced when the velum is lowered to allow air passage through the nasal cavity

nasal bone

the bone forming part of the bridge of the nose and extending in an anterior-inferior direction to form the base for the protruding portion of the nose

nasal breadth

see nose breadth

nasal cavity

the region between the external nares and the nasopharynx

nasal field

the medial portion of the eye's field of view; *opp.* temporal field

nasal height
see *nose height*

nasal reflex
the induction of sneezing due to stimulation of the nasal mucous membranes

nasal root
the junction of the nasal bone with the frontal bone

nasal root breadth
the minimum horizontal linear distance across the base of the nose between the eyes
Comment: measured with the facial muscles relaxed

nasal root depression
the concave region where the bridge of the nose meets the forehead between the eyes

nasal root height
see *sellion height*

nasal septum
the collective tissues separating the right nostril from the left

nasalize
produce a sound with the nasal portion of the vocal tract open; *n.* nasalization

nasion
the horizontal and vertical midpoint of the nasofrontal suture on the skull

nasolacrimal duct
the tubular structure interconnecting the medial portion of the eye to the nasal cavity for drainage of tears

nasopharynx
the uppermost cavity of the pharynx, lying behind the internal nasal cavity and the soft palate

National Aeronautics and Space Administration Standard (NASA-STD)
a document containing standards published by NASA for use in the U.S. space program and related aerospace or medical work

National Health Interview Survey
a survey conducted by the National Center for Health Statistics via interviews of household samples of the U.S. civilian, non-institutionalized population on various health and health status issues; *syn.* Health Interview Survey

Natural Color System (NCS)
a color ordering system which describes colors in terms of the perceived relative amounts of basic colors and specified by blackness, chromaticness, and hue

natural environment
that environment relatively unaffected by man; *opp.* induced environment

natural fiber
any fiber having a plant or animal origin

natural frequency
the free oscillation frequency of a system without external input

natural language
a computer language in which the rules approximate those of the user's normally written language

natural selection
that process through which workers who cannot maintain the pace or quality of work required in a given job will move to less demanding jobs

natural wet bulb temperature
see *wet bulb temperature*

nausea
a feeling of discomfort in the stomach region, accompanied by a food aversion and a possible tendency to vomit

navel
see *umbilicus*

navicular bone (foot)
a bone in the posterior portion of the foot; *syn.* scaphoid bone

navicular bone (wrist)
the largest wrist bone, located in the proximal row of bones on the thumb side; *syn.* scaphoid bone

navigation
those activities involved in directing the movement of a vehicle toward its intended destination

navigation aid

any instrument or other object which assists in navigation; *syn.* navigational aid

near field

that acoustic radiation near a sound source; *opp.* far field

near infrared

that portion of the infrared radiation spectrum just beyond the visual range, from about 750 nm to 1400 nm; also near IR; *syn.* short wavelength infrared; *opp.* far infrared

near ultraviolet

pertaining to that portion of the ultraviolet radiation spectrum having wavelengths ranging from about 300 to 380 nm; also near UV; *syn.* long wavelength ultraviolet; *opp.* far ultraviolet

near vision

the ability to see the close physical environment

near vision chart

any of a number of cards with letters, words, or paragraphs for determining the smallest font size which can be easily read under given conditions

nearsightedness

see myopia

neck

the region of the body comprised of those tissues which connect the trunk with the head, including the cervical vertebrae

neck breadth

the horizontal linear distance from one side of the neck to the other at the vertical midpoint between otobasion inferior and the shoulder

Comment: measured with the individual standing erect, the facial and neck musculature relaxed, and without flesh compression

neck – bustpoint length

the surface distance from the neck–shoulder junction to the tip of the bra

Comment: measured with the individual standing erect, the facial,

neck, and torso musculature relaxed

neck – cervicale length

the surface distance from cervicale to the point at which the neck–shoulder junction becomes the vertical portion of the neck

Comment: measured with the individual standing erect, the neck and scalp musculature relaxed

neck circumference, maximum

the maximum surface distance around the neck, including the thyroid cartilage in the male

Comment: measured without flesh compression, with the individual standing erect, looking straight ahead, and the neck musculature relaxed

neck circumference, minimum

the minimum surface distance around the neck inferior to the laryngeal prominence

Comment: measured without flesh compression, with the individual standing erect, looking straight ahead, and the neck musculature relaxed

neck depth

the horizontal linear distance from the most anterior protrusion of the neck to the nape of the neck

Comment: measured with the individual standing erect, the facial and neck musculature relaxed, and without flesh compression

neck – shoulder junction

the level of the lateral point at which the shoulder and neck meet and the angle of the surface arc is 45° above horizontal; *syn.* shoulder–neck junction

neck – waist length

the surface distance from the superior point of the neck–shoulder junction over the front midline of the body to the midsagittal waist level

neckrest

any padded structure which provides

support to the neck, especially when sitting

needs analysis
the breakdown of identified needs into their component parts to determine the causes/reasons for the needs

needs assessment
the determination/identification of what knowledge, skills, abilities, or other characteristics are required for a task, job, or operation

negative acceleration
see deceleration, negative g

negative afterimage
an image seen on a bright background following the removal of a stimulus, and which is approximately the complementary color of the original stimulus; *syn.* afterimage; *opp.* positive afterimage

negative feedback
a signal which tends to decrease the output of a system; *opp.* positive feedback

negative g
an acceleration acting along the body longitudinal axis in a superior direction; *syn.* headward g; *opp.* positive g

negative pressure respirator
a type of respirator which functions by the wearer inhaling to draw air through the filter; *opp.* positive pressure respirator

negative reinforcement
that which causes a weakening or a decrease in the frequency or size of a response as a result of contingent reinforcement; *opp.* positive reinforcement

negative skew
having a distribution curve with the mean less than the mode; *opp.* positive skew

negative transfer
a condition in which previous experience causes interference with the learning of a new task, usually due to conflicting stimuli or response requirements; *syn.* negative transfer of learning, negative transfer of training; *opp.* positive transfer; *see also proactive inhibition*

negative transfer of learning
see negative transfer

negative work
that dynamic work done by a person using external forces and eccentric muscle contractions; *opp.* positive work

negligence
having exhibited an act of omission or commission without reasonable caution or care or without regard for the circumstances

nem
a nutrition unit, based on the caloric content of one gram of standard composition breast milk; equal to about 0.6 Calorie

neonate
a newborn infant; *adj.* neonatal

neoplasm
a new growth of abnormal cells and/or tissues; *syn.* tumor

néper (Np)
a unit of absorption/attenuation for sound waves, based on the natural logarithm of the ratio of two quantities; also napier

nerve
a collection of one or more axons bound together by connective tissue and having a defined origin and termination

nerve cell
see neuron

nerve deafness
a hearing impairment due to some abnormality in the auditory nerve

nervous system
a system comprised of neural and various supporting tissues which is capable of taking input, integrating that input, and providing motor output

nested
located within some larger structure

net metabolic cost
that metabolic activity incurred only from a particular activity, with the basal

metabolic rate subtracted from the gross metabolic cost

network
a linking of computers, people, activities, or other entities for communication or other purposes

neuroanatomy
the study of nervous system structure

neurocirculatory
pertaining to all or part of both the nervous and circulatory systems or the interaction between them

neuromuscular
pertaining to all or part of both the motor aspects of the nervous system, the muscular system, or the interaction between them

neuromuscular junction
that point of interface between the motor neuron and muscle tissue at which the synapse occurs; *syn.* myoneural junction

neuromuscular spindle
a capsular proprioceptive sensory structure located within skeletal muscles which contains several intrafusal fibers and is responsive to stretch for providing nervous system feedback to prevent damage by overstretching

neuromuscular stimulation
the stimulation of nervous and/or muscle tissue(s) via electrical, chemical, or other means

neuron
the basic functional unit of the nervous system; a nerve cell, generally consisting of a cell body, dendrites, and an axon; *adj.* neural

neuropathy
any nervous system pathology, whether due to injury or disease; *adj.* neuropathic

neurophysiology
the basic physiology of neurons and the nervous system in general, from simple metabolism to the generation and conduction of impulses

neurotendinous spindle
see Golgi tendon organ

neurotransmitter
a chemical which is released from one neuron at a chemical synapse and for which a receptor is located nearby on the same or another neuron; *syn.* transmitter

neutral body posture
that posture which the body tends to assume when relaxed with the eyes closed or covered in a microgravity environment; the arms lie in front of the body with the elbows, the neck, the hips, and knees all somewhat flexed

> *Comment:* given sufficient volume to assume this position

neutral body posture height
see neutral body posture stature

neutral body posture stature
the maximum perpendicular linear distance from a plane the most distal part of the feet to a plane at the highest point on the head when the subject is in the neutral body posture; *syn.* neutral body posture height

neutral density filter
an optical filter which reduces the intensity of light without appreciably changing the relative spectral distribution; *syn.* neutral filter, gray filter

neutralizer
a muscle which functions to prevent some undesired action of another muscle

neutron
an atomic particle having no electrical charge and a mass approximately equal to that of a proton and electron combined

new candle
see candela

newton (N)
the SI unit of force; equal to that force resulting in an acceleration of one meter per second per second to a 1 kg mass

Newton's first law of motion
every mass maintains its current state of motion unless acted on by one or more non-equilibrating forces; *syn.* law of inertia

Newton's laws of motion
three physical laws which govern the basic interactions of physical objects and forces; *see Newton's first law of motion, Newton's second law of motion, Newton's third law of motion*

Newton's second law of motion
the force required to impart a given acceleration to an object is proportional to the mass of that object

$$\vec{F} = m\vec{a}$$

Newton's third law of motion
for every action, there is an equal and opposing reaction

newton-meter (N-m)
a unit of torque in the SI/MKS system, equal to a 1 N force acting perpendicularly at 1 meter from a point of rotation

night blindness
a reduced capability for night or low light vision, regardless of the cause; *syn.* nyctalopia

night shift
see third shift

night vision
see scotopic vision

night vision goggles (NVG)
a light image intensifying device for enabling an individual to see terrain, objects, or other items of interest at very low light levels

nip point
the nearest point of intersection or near contact of two oppositely rotating circular surfaces or a rotating circular surface and a planar surface

nipple [1]
a projection from near the center of the breast, usually of different hue than normal flesh, and which provides the terminal milk duct outlets

nipple [2]
any structure resembling or serving a function similar to the human nipple

nit (nt)
a unit of luminance, equal to 1 candela/m^2

nitrogen (N_2)
a colorless, odorless, and chemically relatively inert gas
Comment: makes up about 78–79% of earth's atmosphere

nitrogen dioxide (NO_2)
a reddish-brown gas which is the combustion product of nitrogen and oxygen, usually occurring only under electrical discharge

nitrogen narcosis
a condition, due to breathing of nitrogen gas under high pressures, whose symptoms range from joviality and lack of concern, to drowsiness and weakness, to unconsciousness and death, depending on the pressure and duration of exposure; *syn.* rapture of the deep/depths; *see also inert gas narcosis*

no brain rule
(sl) a task design guideline that the workplace should prevent the worker from getting hurt even if he doesn't think before acting

nocturnal
pertaining to a species or an individual who prefers to be active at night

nodal point (of the eye)
an imaginary point in the eyeball at which light rays from any point in the visual field will intersect the visual axis

node [1]
a junction; *adj.* nodal

node [2]
a point, line, or surface in a standing wave at which the wave has essentially zero amplitude

node of Ranvier
a gap in the myelin sheath of a nerve fiber in which the axonal membrane is exposed
Comment: enables the neural impulse to jump from node to node, providing for faster transmission

noise [1]
any undesired component of a measurement which accompanies a signal

and interferes with signal clarity, usually of a wideband nature; *adj.* noisy

noise ²

any unwanted or undesirable sound; *adj.* noisy

noise cancellation

an active noise reduction technique using a device which monitors an incoming signal with noise and produces an opposing signal to effectively neutralize the noise prior to passing the signal to the observer

noise control

the process of achieving a more nearly acceptable environment through the use of any noise reducing techniques

noise criterion curve (NCC)

any of several sets of criteria for providing a single number rating the acceptability of continuous environmental noise, based on curves of noise intensity or sound pressure level vs. frequency

Comment: ratings are given for the NC curve which is not exceeded; each curve is named for the dB level where the curve crosses the 2 KHz point

noise exposure

the cumulative amount of acoustic stimulation which reaches the ear of an individual over some specified period of time

noise level

the intensity of unwanted sound

Comment: usually specified in dB, relative to some specified reference

noise meter

see *sound level meter*

noise pollution

(sl) an excessive amount of noise in the environment

noise rating number

the perceived noise level of specified acoustic conditions that is tolerable

noise reduction

a decrease in the noise sound pressure level at a specific point using one or more devices or structures

noise reduction coefficient (NRC)

the average sound absorption coefficient for a material over the logarithm of frequency in the range from 256 to 2048 Hz

noise reduction rating (NRR)

the degree of attenuation in decibels of noise intensity under continous noise conditions for hearing protective devices

noise suppressor

an electronic circuit which is capable of automatically inhibiting the amplifier of a radio receiver to eliminate background noise when no signal is being received; *syn.* squelch

noise-induced

caused by noise

noise-induced hearing loss

a loss of hearing acuity due to high noise levels, which typically begins at about 4 KHz and spreads to either side of that frequency; *see also noise-induced permanent threshold shift, permanent threshold shift, noise-induced temporary threshold shift, temporary threshold shift*

noise-induced permanent threshold shift (NIPTS)

a permanent hearing loss due to extremely high noise exposure levels

noise-induced temporary threshold shift (NITTS)

a temporary hearing loss due to high noise exposure levels

noisy shoulder

(sl) emitting a grating noise on elevation or depression of the shoulder

Comment: often due to a snapping tendon over the scapula

Nomex®

a fire-resistant material

nominal bandwidth

the range between the specified upper and lower cutoff frequencies of a system

nominal group technique (NGT)

a method for generating innovative product ideas in which the individuals within the group communicate verbally with each other only at specified periods of time, using their individual creativity the remainder of the time

nominal scale

a basic measurement scale in which items are categorized or classified using only labeling methods

nomological validity

an aspect of construct validity concerned with the fit between theoretical postulates and empirical data

non-accidental injury

any injury which cannot be traced to a specific accident

non-adaptive response

a reaction to a situation which does not support continued survival

non-auditory noise effect

any physiological or psychological effect of noise other than via the auditory system

non-ballistic movement

see controlled movement

non-blackbody

any surface which reflects at least some of the radiation impinging on it

non-causal association

a statistical association in which no cause-and-effect relationship is apparent between two variables

non-certified color additive

any of a category of substances which are approved by the FDA for cosmetic use without special safety testing

non-combustible

pertaining to a substance which is essentially incapable of burning or supporting a fire

non-cyclic element

a segment or step of a process or operation which doesn't occur within each cycle

non-destructive testing

any study of the properties of a material or structure not requiring its damage or destruction; also nondestructive testing

non-deterministic

pertaining to any event or condition which cannot be reliably predicted given certain prior events and currently known laws; *syn.* random; *opp.* deterministic

non-disabling injury

any injury not resulting in death, permanent disability, or temporary total disability

non-disruptive

pertaining to an activity which doesn't interfere with any other ongoing activity; *opp.* disruptive

non-flammable

pertaining to any material or substance which does not readily burn; also nonflammable; *opp.* flammable, inflammable

non-interactive window

an active window which is not receptive to user input; *ant.* interactive window

non-interlaced display

a visual display in which the entire display is rewritten in one step; *opp.* interlaced display

non-invasive

pertaining to those clinical or experimental procedures which do not require breaking the skin, insertion into any body cavity except the mouth, and which do not cause extreme discomfort; *opp.* invasive

non-ionic detergent

any detergent with molecules which don't ionize in water

non-ionizing radiation

that electromagnetic radiation, generally consisting of those visible light wavelengths or longer, which does not directly cause ion formation in materials

non-linear correlation

a correlation which does not follow the linear relationship $Y = a + bX$

non-linear damping

damping due to a force that is not

proportional to velocity

non-linear equation
see *non-linear function*

non-linear function
any function which can't be expressed in the form $y = a_1x_1 + a_2x_2 + ... + a_nx_n$; *syn.* non-linear equation

non-linear regression
any type of regression involving a function or curve which has other than a directly or inversely proportional relationship between variables; *opp.* linear regression

non-loading secondary task
a secondary task which may be attended to when the operator's primary task doesn't require attention; *opp.* loading secondary task

non-meritorious
see *frivolous*

non-monetary incentive
any incentive plan not involving monetary compensation, such as improved working conditions and social benefits

non-monotonic
pertaining to a function which contains cyclic or both increasing and decreasing aspects within the region of interest; also non-monotone; *opp.* monotonic

non-parametric statistics
any statistical analysis which makes no assumption about the population distribution; *opp.* parametric statistics

non-rapid eye movement sleep
any phase of sleep in which rapid eye movements are not present; also non-REM sleep

non-recoverable light loss factor
any light loss factor due to equipment or other conditions which cannot be remedied through normal maintenance, specifically including temperature, lamp position/tilt, equipment operation, luminaire surface deterioration, line voltage, and ballast characteristics; *syn.* unrecoverable light loss factor; *opp.* recoverable light loss factor

non-repetitive
pertaining to an operation, process, or job which is frequently changed or altered in some way

non-standard
differing from established specifications, conditions, or requirements

non-stationary
pertaining to a condition or function where the mean, spectral density, and probability distribution vary with time; *opp.* stationary

non-stationary time series
a stochastic time series whose characteristics change with an integral increase in the time axis; *opp.* stationary time series

non-woven fabric
any type of cloth produced by a semi-random arrangement of fibers, whether synthetic or natural, held together by adhesives or needling

noradrenalin
see *norepinephrine*

norepinephrine
a catecholamine which serves both as a hormone and a neurotransmitter; *syn.* noradrenalin, arterenol

normal [1]
that which conforms to some standard; typical or commonplace; healthy

normal [2]
pertaining to or having a Gaussian (normal) distribution

normal [3]
perpendicular to a vector, plane, or other entity

normal [4]
the moving average of temperature which is recognized as a standard for a given location

normal curve
see *normal distribution*

normal distribution
a symmetrical probability distribution function having the peak at a point which corresponds to the mean, mode, and the median; *syn.* Gaussian distri-

bution, Gaussian curve, normal curve; *see also probability density, probability distribution function*

$$f(X) = \frac{1}{\sqrt{2\pi\sigma^2}} e^{\frac{-(X-\mu)^2}{2\sigma^2}}$$

normal effort
that amount of effort required or expended in manual work by an average operator with average skill and attention to the task

normal element time
a statistically-determined element time based on the expected or required performance by an average qualified worker working at a normal pace; *syn.* leveled elemental time, normal elemental time

normal line of sight
that line of sight which is assumed by an individual in a relaxed posture and is typically directed approximately 10° to 15° below the horizontal plane

normal operator
an operator who is adapted to his position and attains normal performance when using prescribed methods and working at a normal pace

normal pace
the manual productivity level required or achieved by a normal effort

normal performance
that output expected from an average qualified operator working with prescribed methods at an average pace

normal probability distribution
see normal distribution

normal saline
a solution of 0.9 g of sodium chloride in 100 ml of water, which is isotonic with body fluids; *syn.* isotonic sodium chloride, physiological saline

normal time
that temporal period required for a qualified worker to perform some task or operation while working at a normal pace without personal, fatigue, or other allowances; *syn.* base time, leveled time

normal working area
the approximately planar region of a work surface which is bounded distally by the arc swept out by a worker's extended fingertips and proximally by the body while pivoting the shoulder laterally (lateral rotation) in the normal working position; *see also normal working area (one-handed), normal working area (two-handed), normal working volume, normal working position*

normal working area, one-handed
the normal working area for only the right or only the left arm

normal working area, two-handed
the normal working area for that overlapping area between the two individual one-handed working areas

normal working position
see normal working posture

normal working posture
the typical posture assumed by a worker for a given task, generally taken as a standing or sitting position, with the upper arm hanging in a relatively stationary position close to the body and the elbow flexed at about 90°; *syn.* normal working position

normal working volume
the three-dimensional region bounded proximally by a worker's body and distally by the arc swept by the fingertips of one or both hands with a range of elbow flexions and/or body rotations about its vertical axis; *see normal working area, reach envelope*

normalize
carry out a transformation on a variable to obtain a linear function, a normal distribution, or a desired range, or to have the sum of the parts equal 1.0

normative
pertaining to or the establishment of a norm or standard for evaluation

normotonic
having normal muscle tone

normoxic

having a normal oxygen level

nose

the collection of bone, cartilage, and other tissues which protrude from the general anterior facial surface and form the external nares; *adj.* nasal

nose breadth

the maximum horizontal linear distance across the nose, at whatever level it occurs; *syn.* nasal breadth

Comment: measured with the facial muscles relaxed and without compressing tissue

nose clip

any spring device which pinches off the nostrils to prevent entry of water or air

nose height

the linear vertical distance from subnasale to sellion; *syn.* subnasale – sellion length, nasal height

nose height – breadth index

the percentage value of the ratio between the nose height and the nasal breadth

nose length

the linear distance from sellion to pronasale

Comment: measured parallel to the ridge of the nose

nose protrusion

the linear horizontal distance from subnasale to pronasale

Comment: measured with the individual standing or sitting erect, with the facial muscles relaxed

nostril

see external naris

nox

a unit for measuring low levels of illumination; equals 10^{-3} lux

noxious

pertaining to that which is harmful or poisonous

noy

a unit for representing the perceived noise level relative to the perceived

noise level of random noise of bandwidth 1000 ± 90 Hz at a sound pressure level of 40 dB, referenced to 2×10^{-4} microbar; *pl.* noys

np chart

a graph or display, tracking over time, the number of non-conforming units in samples when the number of items in each sample is constant; *see p chart*

nub

an intentional knot or tangle in a fabric which gives it an irregular texture

nucha

see nape; adj. nuchal

nuchale

the lowest point in the midsagittal plane of the occiput that can be palpated among the muscles in the posterior-superior part of the neck

nuclear fission

see fission

nuclear fusion

the joining of a hydrogen nucleus with another hydrogen or heavier nucleus in a thermonuclear reaction to form heavier nuclei, with the release of energy; *syn.* fusion; *see also fission*

Nuclear Magnetic Resonance (NMR)

see magnetic resonance imaging

nucleic acid

a polymer of purine and pyrimidine bases, each chemically combined with a five carbon sugar and phosphoric acid; *see ribonucleic acid, deoxyribonucleic acid*

nucleon

a proton or neutron

nucleus pulposus

the viscous fluid in the center of an intervertebral disk

nuclide

an atomic species having the same number of neutrons, the same atomic proton number, or the same atomic mass number

nude [1]

having a minimal amount of clothing (e.g., underwear), in which many anthropometric measurements are taken

nude [2]

without clothing; *syn.* naked

nude body dimensions

anthropometric measures which have been taken with a nude subject

nuisance impact

see accidental impact

null

having a quantity of zero; a non-existent entity

null gravity

see microgravity; also null *g*

null hypothesis

a statement proposing that there is no statistically significant difference with respect to one or more given variables between two or more groups

numb

having an impaired ability or no ability to feel tactile sensations

number facility

the ability to perform basic arithmetic processing correctly within a reasonable time limit

Comment: e.g., add, subtract, multiply, and divide — individually or in combination

numbering system

any plan for the assignment of numeric values to items, cases, or events as a means for classification

numerical analysis

the use of mathematical approxima-

tion techniques to solve problems

numerical control

a method for precisely controlling the motions of a mechanical device, usually some type of machine tool, via a mathematical description of the object being manufactured

numerical display

any electrical display involving numbers, as on a panel or instrumentation; *syn.* digital display

nurture

the use of environmental factors to aid growth and development

nutrient

a substance which nourishes

nutrition

the requirements and processes of the living body involved with activity, growth, maintenance, and repair; *adj.* nutritional

nutritional anthropometry

see kinanthropometry

nyctalopia

see night blindness

nylon

any of a set of long-chained amide polymers used in fabrics

nystagmogram

a recording or display of nystagmus

nystagmus

an involuntary, back and forth, rotating, or up and down rhythmic oscillation of the eye

O-scale

a system for estimating the degree of obesity using transformed, geometrically-adjusted skinfold values with body weight

obesity

that condition resulting from a prolonged condition in which caloric energy intake exceeds energy output, the excess being converted to fat and deposited within the body

object [1]

any physical entity which can be viewed or manipulated

object [2]

any structure which can be displayed or manipulated by a computer system

objective

pertaining to a measure or aspect which can be observed or evaluated by more than one person independently; *ant.* subjective

objective basic element

any of a set of work elements which involve an observable element

objective rating

a type of performance rating which has an objective, as opposed to subjective, basis

objective tree

a qualitative form of relevance tree which may be used simply to place variables in perspective

observation

a data point measured and recorded by viewing or sensing an event or process

observation board

a clipboard or similar tool used to support the timing device and hold any forms in gathering time and motion data

observation form

any generic or specially-designed form for recording the different work elements in a particular time study

observe

view to acquire data for documentation or study; *n.* observation

observed rating

that rating applied to a worker's pace by a time and motion study individual relative to that individual's judgment of what the standard pace should be

observed time

that time required for the accomplishment of a defined task or task element, as measured by some timing device; *syn.* actual time, raw time

observer (O)

an individual who makes the observations in a study

observer error

any error due to intra- or inter-observer unreliability or differences in judgement

Occam's razor

a rule that, given two theories which explain a phenomenon, the simpler is preferred

occasional element

a job element which occurs at irregular intervals, less than once in a given work cycle or operation

occipital bone

a curved, flat bone forming a portion of the posterior and inferior skull

occipital condyle

one of a pair bilaterally-distributed condyles at the base of the skull which articulate with the atlas bone

occipital lobe

a pyramid-shaped structure at the posterior portion of the cerebrum whose primary function is visual processing

occipital pole
the posterior tip of the occipital lobe of the brain

occiput
the posterior portion of the head

occluded gases
those gases forced into a closed space or tunnel with blowers

occupancy
the number of people permitted to occupy a building or region within a building

occupation
that trade, profession, or other activity which occupies one's time for compensation; *adj.* occupational

occupational acne
an occupational skin disorder involving acne resulting from regular exposure to acne-causing material(s) such as tar, wax, and chlorinated hydrocarbons, and which disappears on removal from those material(s); *see also occupational dermatosis*

occupational biomechanics
the study of the volitional acts of the individual in loading the musculoskeletal system in the working environment

occupational dermatosis
any of a class of occupational skin disorders involving one or more regions of the skin, such as contact dermatitis, eczema, or rash; *see also occupational acne*

occupational disease
any functional or organic disease derived from the worker's exposure to the operations or materials involved in the working environment; *syn.* occupational illness

occupational ecology
the study of the worker, the working environment, and the interaction between the two

occupational ergonomics
the study and/or practice of human factors in the workplace

occupational health
a subset of occupational medicine dealing with promoting the maintenance of worker mental and physical well-being, including means of disease prevention

occupational illness
see occupational disease

occupational injury
any physical trauma resulting from, or related to, one's employment; *syn.* work injury

occupational medicine
the field within medicine which is concerned with occupational health while examining the causes and treatment of diseases found in industrial environments; *syn.* industrial medicine

occupational neurosis
any neuropsychological disorder, not caused directly by an individual's occupation, but which is characterized by symptoms such as pain or fatigue involving those parts of the body normally used in his occupation

occupational noise
that noise found in the workplace

occupational nystagmus
an ocular nystagmus resulting from prolonged exposure to poor lighting conditions or retinal fatigue

occupational paralysis
a muscular weakness or atrophy due to nerve compression resulting from the working environment

occupational physiology
see work physiology

occupational psychiatry
a specialty within psychiatry concerned in business and industry with: (a) the promotion of mental health; (b) diagnosis and treatment of mental illness; and (c) the dealing with the psychological aspects of personnel problems

such as hiring, absenteeism, vocational adjustment, and retirement

occupational safety
the study and/or implementation of principles intended to recognize hazards and prevent accidents in work-related situations

occupational skin disease
see occupational skin disorder

occupational skin disorder
any occupational disease involving the skin; *see occupational dermatosis, occupational acne; syn.* occupational skin disease

occupational strain
the reaction of one or more parts of the body to occupational stressors; *syn.* work strain

occupational stress
an internal condition resulting from any forces exerted on the individual as a result of performing some task in the work environment; *syn.* work stress

occupational stressor
any stressor present in the workplace; *syn.* work stressor

occupational therapist (OT)
one who is licensed or otherwise qualified to practice occupational therapy

occupational therapy
the training in or use of certain occupational skills for therapeutic or rehabilitation purposes

occurrence
an incident

octave
a range of frequencies, usually specified by the center frequency, whose highest frequency is twice that of the lowest; *syn.* octave band

octave band
see octave

ocular
pertaining to the eye

ocular dominance
a condition in which one eye is subconsciously relied on more than the other; *syn.* eye dominance

oculogram
a surface electrical recording of activity adjacent to the eye which indicates eye movement patterns

oculogravic illusion
an illusion indicating a tilting of the visual field produced when a change in gravity vertical occurs, as in a centrifuge or other linear acceleration

oculogyral illusion
a visual illusion involving a sense of rotation in the opposite direction produced when an abrupt change in rotational velocity occurs

oculomotor
pertaining to eye movements

oculomotor nerve
the third cranial nerve, which provides motor input to the intrinsic and some extrinsic eye muscles

odor
a property of any substance which results in a sensation of smell; *adj.* odorous

odorant
any relatively volatile substance which is added to an odorless or offensive material to give the latter a distinctive odor for safety, attractant, or other purposes; *see also malodorant*

odoriferous
having an odor

odorimetry
the study or measurement of the effects of odors on the olfactory sensory structures

odorize
add an odorant to another substance

off time
that period within a given day when an individual is not scheduled to be at work

off-line
pertaining to a terminal or other hardware not ready for access to a computer or network; *ant.* on-line

offgassing
the release of adsorbed or occluded

substances from a solid or liquid material, often by exposure to heat; *syn.* outgassing

office

any location in which management, supervision, and administrative support personnel are housed and their respective functions are performed

office automation

the use or implementation of computers or electromechanical devices for communications or manipulating, storing, or sending documents

office layout

the arrangement of desks, filing cabinets, photocopiers, other associated equipment, and the personnel who occupy an office

ogive

a cumulative distribution curve, generally resembling an "S" shape, depending on the distribution

ohm (Ω)

a unit of electrical resistance; equal to the electrical resistance between two points of a conductor when a constant potential of 1 volt is applied between the two points and produces a current of one ampere

oil acne

a form of occupational acne due to direct exposure to oils or oil-soaked clothing

olecranon

the proximal end of the ulna which forms the elbow prominence; *syn.* olecranon process

olecranon fossa

a depression in the posterior distal end of the humerus, into which the olecranon process of the ulna fits when the elbow is extended

olecranon height

the vertical distance from the floor or other reference surface to the underside of the elbow

 Comment: measured with the individual standing erect, the elbow

flexed 90°, and the upper arm vertical

olecranon process

see olecranon

olfaction

the sense of smell; *adj.* olfactory

olfactometer

any device for measuring the sensitivity of smell

olfactory nerve

the first cranial nerve, which conveys sensory information regarding smell to the brain

omphalion

the vertical and lateral midpoint of the navel

omphalion height

the linear vertical distance from the floor or other reference surface to omphalion; *syn.* umbilicus height, waist height, omphalion waist height

 Comment: measured with the individual standing erect, with his body weight equally distributed on both feet

omphalion waist height

see omphalion height

on-demand

supplied as a result of a user-initiated response

on-line

pertaining to a fully connected, powered, and ready for operation terminal or other hardware access to a computer or network; *ant.* off-line

on-off control

any simple control mechanism which has only two possible discrete outcomes, either full on or full off, with no intermediate state possible

on-off switch

a type of on-off control which consists of a manual, remote, or automatic switch

on-the-job training (OJT)

the training of an employee by doing the tasks or job he will be expected to perform when training is completed, rather than by classroom or other training techniques

one-handed normal working area
see *normal working area (one-handed)*

one-hole test
a psychomotor skill test in which an individual is required to grasp, move, and position a small cylindrical object in a hole with close tolerances

one-point discrimination
the ability to localize a point on the body surface where pressure is being applied

one-tailed test
a test of statistical significance in which a directional hypothesis is used, stating that a sample value will be exclusively either less than or greater than some value; see also *two-tailed test*

one-third octave band
a division of an octave into three parts, with boundaries differing by a factor of the cube root of 2 (approximately 1.26); *syn.* third octave band, third octave, 1/3 octave band

ontogeny
the study of the origin and development of an individual organism, from the zygote to adult; *opp.* phylogeny

opacity
a property of a substance or material which prevents light passage through it; *adj.* opaque

opalescence
the clouded, iridescent appearance of a translucent substance or material when illuminated by more than one frequency of visible light

open shop
a facility in which employment is available to both labor union members and non-union workers; *opp.* closed shop

open stope
pertaining to an underground workplace which is either unsupported or supported only by occasional timbers or rock pillars

open timbering
a technique for supporting the soil or rock in a shaft or tunnel in which vertical supports are located some distance apart with overhead horizontal struts between them

open window
a display window which is perceptually and functionally available to the user; *opp.* closed window

open-loop system
any system in which its own output provides insignificant or no input back to the system, with all or the remainder of the input coming from another source; also open loop system; *opp.* closed-loop system

opening shock load
see *parachute opening shock*

operant conditioning
a form of learning/training in which an organism provides a certain response to obtain a reward, which then reinforces the occurrence of that response in the future; *syn.* instrumental conditioning

operate
manipulate, support, and operationally maintain a system or piece of equipment; *n.* operation

operating system
that software which provides the basic operational building blocks for using a computer system

operation
the act of performing any planned job or task by one or more humans with or without machines/equipment in which value is added to a product or information is input, processed, or output

operation analysis
a systematic review and study of the purpose, procedures, time and motions required, tools and equipment used, materials used, standards, workplace design, and working conditions for any operation; also operations analysis

operation analysis chart
a form which lists all relevant variables involved in an operation

operation breakdown
see job breakdown

operation card
see standard practice sheet

operation chart
see two-handed process chart

operation process chart
an abbreviated flow process chart consisting of a graphic/symbolic description providing a top-level view of the sequence for an entire operation, specifying such information as the actions and inspections involved, materials used and points of introduction, etc.; *syn.* outline process chart

operational
ready for immediate use, or in the process of being used

operational containment
an active process for preventing an interface between entities which should be kept separate; *see also physical containment*

operational effectiveness
a measure of satisfaction of the work accomplished or the rate at which work is being done within a given total system-environment

operational maintenance
any minor inspection, cleaning, servicing, adjustment, or parts replacement in equipment which can normally be performed by an operator without any specialized training or high-level technical skills

operational readiness
a state or condition in which a system is not functioning due to scheduling or other reasons, but will perform its intended function when called upon to do so

operational research
see operations research

operational suitability
a measure of the ease of use or usability of a manufactured product

operations
the sum of all activities of an organization

operations analysis
see operation analysis

operations research
the application of scientific, statistical, and/or modeling methodology toward obtaining information for management to make objective, quantitatively-based decisions using specified criteria regarding the men, machines, materials, and money under their control

operative temperature
a measure of heat stress

operator
an individual or robot whose functions may include manipulating, supporting, and operational maintenance of a system or piece of equipment

operator error
see human error

operator input
that information or data presented to/received by an operator via instructions, displays, observing equipment/system operation, or the general working environment

operator instruction sheet
any form of written instructions for providing the operator with details for performing a given task or job; *see also standard practice sheet*

operator output
any physical or verbal action taken by an operator

operator overload
a condition in which an operator is expected to do more than he is capable of performing effectively within the given workplace, environment, or other constraints

operator performance (P)
any measure of the work output of an operator

operator process chart
an operation process chart describing the activities of a single worker with-

out differentiation between the two hands; *syn.* man process chart

operator productivity
see productivity

operator training
instruction which is intended to enable or enhance an individual's performance on a job or task

operator utilization
the ratio of actual working time to total clock time

operator workload assessment
the use of any relevant physiological, cognitive, or other measure to determine operator workload

ophthalmic
pertaining to the eye

opinion
a view or belief based on a judgment about what is believed to be true regarding some issue, object, or event, but without absolute certainty or knowledge

opisthocranion
that point on the occipital bone in the mid-sagittal plane which marks the posterior extremity of the largest skull diameter measure

opponent color
one of a set of pairs of opposing colors; *see opponent process theory*

opponent colors system
a color ordering system in which specified color pairs are considered to be at the ends of a single dimension: red vs. green, blue vs. yellow, and white vs. black

opponent process theory
a theory that there are receptors in the eye for red or green, for blue or yellow, and for white or black; *syn.* Hering's opponent process theory

optic
pertaining to the eye

optic chiasm
the point at which some of the neural fibers from the retina cross to the opposite side of the brain; also optic chiasma

optic nerve
the third cranial nerve, which carries visual information to the brain

optical
see optics

optical axis
an imaginary straight line extending along a horizontal plane of the eye through the midpoint of the cornea, the pupil, and the retina; *syn.* geometrical axis, pupillary axis; *see also visual axis*

> *Comment:* the optical axis is separated from the visual axis by about 4°

optical character reader
a device having the capability to scan a single or limited types of standardized text

optical character recognition (OCR)
the study or use of photoelectric methods to identify printed or handwritten characters

optical density (OD)
a measure of the opacity of a translucent medium, represented by the logarithm of the reciprocal of the transmittance by a transparent material

$$OD = \log\left(\frac{1}{T}\right) = \log\left(\frac{I_o}{I_t}\right)$$

where:
T = transmittance
I_0 = incident intensity
I_t = transmitted intensity

optical element
any structure within an optical device involved in shaping or directing light passage through that device

optical glass
a glass which meets certain standards in being free from imperfections which would adversely affect its light transmission, e.g., bubbles, seeds

optics
the study of the generation, transmission, refraction, reflection, and detection of electromagnetic radiation be-

tween X-rays and radio waves; *adj.*
optical

optimal

the most desirable

optimal menu hierarchy

that hierarchy of menu structures which
yields either the lowest average access
times or fewest number of steps in
getting to a specified point or in the
most common uses

optimistic time

the shortest possible time in which a
given operation, task, or other activity
could be completed

optimum location principle

the concept in designing a man-machine interface that each display and
control should be placed in the best site
according to one or more criteria for its
intended use

optimum replacement interval

see economic life

optokinetic nystagmus (OKN)

that nystagmus in a normal individual
caused by a succession of objects moving across the visual field

> *Comment:* the movement is relative
> — may be achieved with the indi-
> vidual stationary and moving ob-
> jects or a moving individual pass-
> ing a number of stationary objects

optometer

an instrument for determining the visual capacities of the eyes

optometry

the study and/or measurement of the
human eye's capabilities; *adj.* opto-
metric

oral [1]

pertaining to the mouth

oral [2]

spoken

oral cavity

see mouth

oral comprehension

see oral verbal comprehension

oral expression

see oral verbal expression

oral fissure

that approximately elliptical opening
formed by the separation of the facial
lips

oral verbal comprehension

the ability to understand spoken language; *syn.* oral comprehension

oral verbal expression

the ability to use spoken language to
communicate with others; *syn.* oral
expression

orbit

the eye socket; *adj.* orbital

order [1]

see order of magnitude

order [2]

a written or verbal direction or command from someone in authority

order [3]

having some systematic structure or
pattern; lack of chaos

order [4]

a request for a specific number and
type of goods or services; *syn.* purchase order

order entry

the process of inputting the information pertaining to a purchase order into
a computer for processing

order of magnitude

an integer value representing an exponent of some number or expression;
syn. order

ordered metric scale

a variant of the four basic measurement scales in which the rank orders of
both the subjects and the intervals are
known, but the interval magnitudes
are not known

ordinal scale

a basic measurement scale in which
items can be classified by rank, using
some magnitude measure, but with no
specification about the absolute magnitudes or magnitudes of differences
between items

ordinance

a municipal statute or regulation

ordinate [1]

the vertical or dependent axis on a two-dimensional graph, typically labeled the y axis; *adj.* ordinal; *opp.* abscissa

ordinate [2]

a particular value on a graph, represented by the perpendicular distance from the abscissa

ordnance

any and all types of military weaponry and the ammunition to support them

organ

a bodily structure composed of differentiated, specialized tissues for the performance of some function; *see also* **cell, tissue**

organ of Corti

a tissue consisting in part of the sensory transducing hair cells within the cochlea; *syn.* spiral organ of Corti

organic [1]

pertaining to chemical compounds composed of carbon chains, which may include hydrogen, oxygen, nitrogen, and other atoms

organic [2]

see **organ**

organization

that structure of people, concepts, or other entities which exist or are created to carry out or assist in one or more specific objectives

organization chart

a graphic representation of the interrelationships within an organization, which may indicate lines of authority and areas of responsibility; sl. org chart

organizational climate

those properties of the working environment which may have effects on employee productivity

organizational psychology

that field of study and practice involving the structure and function of organizations

organo-

having a carbon base

orientation

the process of providing a new employee with some background information on the organization, its policies, and its procedures

orienting reflex

see **orienting response**

orienting response

a mild psychophysiological response involving a sudden shift of attention to process information associated with some sudden event; *syn.* orientation reflex; *see also* **startle response**

origin [1]

the proximal point of attachment to bone of a muscle tendon; *opp.* insertion

origin [2]

the null reference point for a coordinate system, at which all axes meet and are usually assigned values of zero

originality

the ability to produce new, unusual, or clever thoughts on a given topic

oropharynx

the middle region of the pharynx, from the level of the soft palate to the level of the hyoid bone, and from the posterior pharyngeal wall to the fauces

orthoaxis

the true anatomical axis of rotation for a limb

orthocenter

the instantaneous anatomical center of rotation for a joint

orthogonal [1]

being perpendicular or at right angles

orthogonal [2]

completely independent or separable

orthograde

standing or walking in an upright position

orthonormal

pertaining to orthogonal vectors or functions whose lengths or products are 1, respectively

orthopedics

that medical field concerned with the

treatment or correction of deformities or diseases of the structures involved in human movement; also orthopaedics

orthoptics
the study of visual responses and reactions, and any corrections required to effect normal binocular vision; *adj.* orthoptic

orthosis
any device for aiding the strength and/or dexterity of a weakened, damaged, or atrophied body part or for preventing bone or other deformity; *pl.* orthoses; *adj.* orthotic

orthostatic
pertaining to an upright stance

orthostatic hypotension
a fall in blood pressure on standing; *see also* **orthostatic intolerance**

orthostatic intolerance
the inability of the cardiovascular system to supply adequate blood pressure and flow to the brain on standing; *opp.* orthostatic tolerance; *see also* **orthostatic hypotension**

orthostatic tolerance
the capacity of the cardiovascular reflexes to maintain arterial pressure to enable an individual to tolerate an erect stationary posture; *opp.* orthostatic intolerance

orthotic
see **orthosis**

os calcis
see **calcaneus**

OSA color system
a color ordering system which attempts to relate the distance between points on a color diagram to the perceptual difference between any two colors, and is based on specifications of lightness (L), yellowishness (j), and greenness (g)

oscillate
execute repeated reversals in velocity, generally between a range of values; *n.* oscillation

oscillating conveyor
a trough-like mechanism which moves

loose materials along by vibrating the bed; *syn.* vibrating conveyor

oscillator
any device which produces oscillations

oscillogram
a hardcopy record of electrical waveforms, typically biological in origin

oscillograph
an instrument used for obtaining an oscillogram

oscilloscope
a cathode-ray-tube instrument used for real-time display, analysis, and/or storage of electrical waveforms

OSHA 200 log
see **OSHA Form 200**

OSHA Form 200
a form for recording occupational injuries, illnesses, and deaths for submission to the U.S. Department of Labor

OSHA General Industry Standard (GIS)
Part 1910 of Chapter XVII of Title 29 of the Code of Federal Regulations dealing with OSHA occupational safety and health standards; *syn.* General Industry Standard

OSHA reportable accident
any accident resulting in a fatality, hospitalization, lost workdays, loss of consciousness, medical treatment, or job termination/transfer for one or more workers; *syn.* reportable accident

osseous labyrinth
an interconnected series of cavities within the temporal bone which comprise the inner ear, including the vestibule, semicircular canals, and cochlea

osseous semicircular canals
see **semicircular canals**

ossicle
see **auditory ossicle**

osteochondritis
an inflammation of both bone and cartilaginous tissues

osteochondritis dissecans
a joint condition in which a fragment of articular cartilage and the underlying bone have partially or completely sepa-

rated; *syn.* osteochondrosis dissecans

osteochondrosis dissecans
see osteochondritis dissecans

osteodystrophy
a reduced capacity to support the body weight due to osteoporosis

osteometry
the study of the dimensions and proportions of skeletal structure

osteonecrosis
the death of regions of bone tissue due to the loss of blood vessels which supply the bone, usually at joints; *syn.* avascular necrosis

osteoporosis
a bone disorder characterized by an overall loss of mineral mass through resorption, resulting in larger marrow spaces, decreased cortical bone, and decreased structural strength; *adj.* osteoporotic

    Comment: accompanied by an increased risk of fracture

Ostwald color system
a color ordering system which specifies colors by hue, a white-to-black dimension, and a depth-vividness dimension

otitis externa
any inflammation of the external auditory canal

otitis interna
an inflammation of the inner ear

otitis media
an inflammation of the middle ear

otobasion
the region of attachment of the anterior portion of the auricle to the skin of the face; *see otobasion superior, otobasion inferior*

otobasion inferior
the lowest point of attachment of the auricle to the skin of the face

otobasion superior
the highest point of attachment of the auricle to the skin of the face

otolith
a calcium-based particle within the utricle and saccule of the inner ear used for sensing head position

ounce (oz)
a unit of mass in the English system of measurement

ouncedal
a unit of force equal to that which imparts an acceleration of 1 ft/sec^2 to a 1 oz mass

out of phase
pertaining to waveforms having the same frequency but which are not at the same point in their respective cycles at the same time; *opp.* in phase

outcome
that projected or actual result due to the implementation of certain decisions

outdoor air pollution
that pollution found outside living, working, or other structures, generally consisting of airborne hazardous or offensive substances such as dust, ozone, hydrocarbons, and smoke

outdoor work
see outside work

outer ear
see external ear

outerwear
any type of clothing typically worn outside or over other clothing

outgassing
see offgassing

outlier
a value within a data set which is located at such an extreme point from the remainder of the distribution that its presence can't be reasonably attributed to any known cause; *syn.* wild value

outline process chart
see operation process chart

outpatient
an individual who enters a hospital or clinic for treatment and/or diagnosis, but who does not occupy a bed overnight

output [1]
any data, information, display, or other results provided by a computer system

output [2]

the total production of product, services, energy, or other quantity from some individual or unit within a specified period of time

output device

any means through which output from a computer is provided; *syn.* output equipment

output equipment

see output device

Comment: an older term

output standard

that number of products or services which is expected within a specified period of time from a worker or unit using a specific method

outside air

that air drawn in from a source external to a ventilation system

outside work [1]

see external work

outside work [2]

that work done outside of some building or structure where the worker is exposed to the atmospheric elements; *syn.* outdoor work

outside-in display

a display using the outside world as a frame of reference, such that the display reflects the way an external operator would see an object react; *opp.* inside-out display

outsource

hire an outside firm to perform some specific function, especially with respect to performing computerized information services

oval window

that junction between the middle ear and the inner ear at which the base of the stapes and its ligaments join the temporal bone, and through which sound is conducted to the cochlea

ovary

that female organ which releases certain hormones and the eggs for reproduction; *adj.* ovarian

overachiever [1]

an individual whose performance is better than would be expected from known characteristics or previous performance

overachiever [2]

an individual who feels a need to excel and does more work than is required

overall cycle time

see work cycle time, standard time

overall sound pressure level

the summation of all acoustic energy in the frequency bands from 22.4 Hz to 11.2 KHz

overall study

the measurement and recording of the cycle time to verify a time study standard

overarching weight

that force from the overhead ground or rock in a tunnel or shaft

overcast

pertaining to a sky with complete cloud cover, and no visible disk of the sun

overhaul

the process of disassembling, inspecting, refinishing, replacing part(s), adjusting, reassembling, and/or testing as required to return hardware or equipment to service

overhead [1]

pertaining to the region above the head

overhead [2]

that portion of the cost of doing business which cannot be allocated to a particular operation, project, system, or product

overhead work

that work performed with the arms raised above the head and/or shoulders

overlap

partially obscure another display entity

overlay

completely or partially obscure one display entity by another

overlearning
  that learning or practice beyond the point of a single correct response

overload
  experiencing a load greater than that which an entity can safely handle

overload principle [1]
  the concept that a system will fail to function properly or not function at all when saturated by the task requirements

overload principle [2]
  a rule that the strength, endurance, and hypertrophy of a muscle will increase only as a result of the muscle working against loads greater than those normally encountered for a period of time

override
  take over the operation of a control system which has been controlling one or more systems by a human or another control system

overshoot
  a motor response which goes beyond the intended target point or value

overstrike
  a computer operating mode in which the keystrokes entered by the user replace those on the display, with the originals being lost; *syn.* typeover; *opp.* insert

overt behavior
  any objective behaviorial act; *opp.* covert behavior

overt lifting task
  the lifting and / or manipulation of significant loads; *opp.* covert lifting task

overtime
  those hours worked by an hourly worker beyond the standard workday or workweek for that worker; *syn.* premium time

overtone
  a constituent of a complex tone whose frequency is an integral (greater than unity) multiple of the fundamental frequency; *see* **harmonic**

overuse syndrome
  see repetitive motion injury

overutilization hypoxia
  a tissue oxygen deficiency due to a greater demand for oxygen than can be supplied; *syn.* oxygen deficit; *see also* **oxygen debt**
  > *Comment:* usually encountered during prolonged heavy muscular activity

oxidizing agent
  any substance which contains an excess of oxygen or certain other substances which are readily available for chemical combination with other substances and release energy on doing so

oximeter
  an instrument for measuring the oxygen content in a given environment

oximetry
  the measurement of the amount of oxygen in a given environment

oxygen (O_2)
  a chemical element; a tasteless, odorless gas at normal temperature and pressure which serves as the primary oxidizing agent for biological systems

oxygen capacity
  the maximal amount of oxygen which can be absorbed by a specified amount of blood when equilibrated with an excessive amount of oxygen

oxygen consumption rate ($\dot{V}_{O_2}$)
  the amount of oxygen used by the body within a specified unit time

oxygen debt
  that additional amount of oxygen above normal resting requirements which is required during the recovery phase following strenuous exercise to reconvert the anaerobic by-products of metabolism generated during overutilization hypoxia

oxygen deficient atmosphere
  an atmosphere which contains less than the approximately 20 – 21% of oxygen found in normal air or less than the partial pressure of oxygen required to support normal bodily respiration

oxygen deficit
see overutilization hypoxia

oxygen mask
any mask covering at least the nose and mouth and used for oxygen inhalation

oxygen safe tolerance curve
see oxygen tolerance curve

oxygen saturation
the value of the ratio of actual oxygen content of hemoglobin to its oxygen capacity

oxygen tolerance curve
a graphic representation of the amount of time an average individual can be expected to breathe pure oxygen at different water depths or at higher barometric pressures without harmful effects; *syn.* oxygen safe tolerance curve

oxygen toxicity
see chronic oxygen toxicity, acute oxygen toxicity

oxyhemoglobin (HbO_2)
hemoglobin combined with oxygen

ozone (O_3)
an unstable, toxic gas molecule consisting of three oxygen atoms, and having a pungent odor

P

p chart
a graph or display, tracking over time, the fraction of non-conforming units within samples; see *np chart*

P-300
an apparently endogenous brain event-related potential which often occurs about 300 msec following an unexpected stimulus
Comment: has a large onset time variance, however

pace
regulate the rate at which some task or function is performed; see *external pacing, internal pacing*

paced work
any repetitive job or task which is to be carried out at a specified pace; see *restricted work, unrestricted work*

pacemaker
any device which has a direct influence on the rate at which some process s carried out or the duration of the cycle of a process; see *circadian pacemaker, cardiac pacemaker*

Pacinian corpuscle
a layered sensory receptor for kinesthetic information

package
place in a container in some specified pattern or sequence

packaging inefficiency
that amount of volume which cannot be used in packaging due to factors such drawer frame volume, drawer slides, shelves, or object structure; *syn.* packing inefficiency

packer
one whose job is to place items into containers for shipment, sale, storage, or other function

packing house
a food processing plant in which animals are slaughtered, processed, and placed in cold storage

packing inefficiency
see *packaging inefficiency*

pad [1]
any small fatty tissue mass, such as on the terminal phalanges for the fingers, thumb, or toes

pad [2]
one or more layers of soft materials which act to protect a part of the body from impacts or hard surfaces

page [1]
(v) move through a displayed file one page at a time; see also *scroll*

page [2]
(n) that amount of text, graphics, or other material which comprises a printed page in hardcopy

page [3]
(n) that segment of a file which may be displayed on a screen at one time

pain
an unpleasant sensation which causes stress and/or suffering

pain threshold
the minimum level at which one perceives pain; also threshold of pain

paired comparisons
see *method of paired comparisons*

palatal
articulated with the tongue on or near the hard palate

palate
the upper boundary of the mouth; see *hard palate, soft palate; adj.* palatal, palatine

palatine bone

a skull bone forming part of the hard palate, the nasal cavity, and lower orbit

palatine velum

the posterior portion of the soft palate which partially separates the mouth from the pharynx; *syn.* velum

pallet

a low-profile, dual-sided platform, often with slots between the sides for storing, stacking, and/or transporting packaged materials or equipment

palletize

place on a pallett

pallette

a list of possible selections, especially on a computer display

palm

the generally soft, frontal aspect of the hand in the region of the metcarpals, over which the phalanges may flex; *adj.* palmar

palm length

the linear distance from the base of the hand to the furrow where the middle finger joins the palm

Comment: measured with the hand and fingers extended

palmar arch

an arterial loop within the palm

palmar reflex

a tendency for the fingers to flex when the palm is stimulated

palpate

examine or locate by touch and/or pressure; *adj.* palpable

palpebral fissure

the approximately elliptical region between the upper and lower eyelids when separated

pan [1]

(v) scan horizontally across the surrounding visual environment such that the center of the field of view changes

pan [2]

(v) shift the center of the field of view across a modeled image on a display

Panel Layout And Integrated Design (PLAID)

see *PLAID*

pant

breathe deeply and rapidly after prolonged heavy physical exertion

pants

an article of clothing consisting of a single oval or tubular structure to cover the lower torso, which bifurcates into two smaller tubular-shaped structures for covering at least part of the legs, often having some type of fastener at the waist for holding in place

Panum's area

a small region within the retina in which images from the two eyes are fused, even though they may not fall precisely on corresponding points; *syn.* Panum's fusional area

paperwork flow chart

see *forms process chart*

parabola

a concave or convex planar curve having an equation of the form below; *adj.* parabolic

$$y = ax^2 + bx + c$$

parachute

(n) a device, consisting of a canopy, tethers, and an attachment mechanism for a payload, which uses the canopy to retard and/or direct motion through air

parachute opening shock

the combination of forces and torques acting on the body or other attached mass as the parachute canopy fills with air and decelerates the system; *syn.* opening shock load

paracusia

any hearing impairment; also paracusis

paradigm

a model or schematic representation

paralexia

a reading disability in which words or syllables are transposed or substituted

parallax
a difference in the apparent relative positions of objects when viewed from different points

parallax error
that observational error which may occur when reading an instrument with a dial and pointer meter due to the observer's eye not being aligned perpendicularly to the dial face

parallel [1]
pertaining to more than one operation or process ongoing at the same time

parallel [2]
having extent or traveling in the same direction, but separated by some constant distance

parallelism
the state of being parallel

paralysis
the loss of muscle function and/or sensation

parameter [1]
an arbitrarily-defined constant value under a given set of circumstances and from which other values or functions may be defined

parameter [2]
a descriptive property of a population; *opp.* statistic

parametric modelling
varying the size and/or orientation of one or more entities in a computer model without altering the basic geometry

parametric statistics
those statistical analyses assuming a known, typically a normally-distributed, population; *opp.* non-parametric statistics

paraplegia
a condition in which the two lower limbs are affected by paralysis or are no longer usable; *adj., n.* paraplegic

parasite
any organism which resides in or on a host, and from which the organism obtains nourishment and shelter for much or all of its life

parasympathetic
pertaining to that division of the autonomic nervous system originating from the brain stem and sacral sections of the spinal cord, has its ganglia generally placed near the innervated organs, and which generally opposes the sympathetic division by performing restorative, digestive functions; *syn.* craniosacral; *opp.* sympathetic

paratonia
the resistance to passive movements of a limb

parent [1]
a document, substance, or nuclide from which another document, substance, or nuclide is obtained or derived

parent [2]
a mother or father

parent menu
the initial menu displayed on entry into an application

paresis
a weakness or incomplete form of paralysis in a limb; *adj.* paretic

paresthesia
a form of decompression sickness resulting in an itching, tingling, crawling, or burning sensation associated with the skin
    Comment: believed due to evolved subdermal gas bubbles

paretic gait
a type of gait comprised of short steps, usually with the feet dragging and the legs laterally separated

Pareto analysis
the arrangement of data by priority and using that information to solve problems

Pareto diagram
a graphical plot of a Pareto distribution, whether in curve or histogram form

Pareto distribution
a numerical listing or graphical plot

comparing value and percentages of items representing that value

Pareto's Law
the concept that a majority (about 80%) of a given problem or other situation can often be explained by a minority (about 20%) of the cases; *syn.* rule of 80-20

parietal
pertaining to that which is situated on or forms a dividing structure, such as a wall

parietal bone
a large, flat skull bone forming much of the lateral and superior part of the cranial cavity

parietal lobe
the posterior-superior portion of the cerebrum

paroxysm
a sudden appearance, reappearance, or intensification of symptoms

part 572 dummy
see Hybrid III

part family
a set of parts having some specifiable similarities

part learning
a learning situation in which the entire set of material to be learned is divided into segments where each segment is learned separately, with eventual learning of all segments; *see also whole learning*

part-task simulation
that type of simulation which provides an individual or group the ability to learn only selected portions of a total task

part-task trainer (PTT)
a training device which provides an individual or group with the ability to learn only portions of the total task

part-time employment
that working arrangement involving an employee having a set or variable number of hours less than the standard work week, or less than about 35 hours per week; *syn.* part-time work

part-time work
see part-time employment

partial disability
any disability other than death or total disability resulting in some loss of use of a bodily member or function

partial pressure
that portion of the total pressure exerted by one of the gases in a gas mixture

partial tone
a simple tone which is a component of a complex tone

particle
any small part of matter, ranging in size from sub-atomic to several millimeters; *adj.* particulate

particulate
any particle dispersed within solid, liquid, or gaseous matter

partition
any dividing structure which has relatively small thickness compared to height and length such as a wall or cardboard which serves to isolate individuals or objects from each other for sound attenuation, protection, or other purposes

partly-cloudy
pertaining to a sky which has about 30–70% cloud cover

pascal (Pa)
the SI/MKS unit of pressure; equals 1 N/m^2

Pascal's Law
a rule that a confined fluid transmits any externally-imposed pressure equally in all directions

passageway
a corridor which is of sufficient height and width to permit movement between two points

passenger
a rider in a vehicle who normally has no active part in operating or controlling the vehicle

passive
resulting from external causes and

without volitional effort by the entity being acted upon; *opp.* active

passive isolation

that energy attenuation through the use of a system or mechanism not requiring energy to operate and acting near or within another system which is generating some undesired energy output; *opp.* active isolation

passive movement

the movement of some limb or more of the body by another person or device without conscious active participation by the individual being moved; *opp.* active movement

passive restraint

any type of restraint in which an individual must make a conscious effort to remain in place; *opp.* active restraint

passive safety measure

any means of implementing safety precautions which does not require any action on the part of the individual for operation or utilization of the mechanism or device; *opp.* active safety measure

paste

place a section of previously cut text or data into a document or computer file; *opp.* cut

pastel

pertaining to an unsaturated color or color with low saturation; *syn.* weak color

patella

a sesamoid bone anterior to the knee joint; *adj.* patellar; *syn.* knee cap

patella bottom height

the vertical distance from the floor or other reference surface to the lowest point on the inferior border of the patella; *syn.* knee cap height; *see also midpatella height, patella top height, knee height (standing)*

Comment: measured with the individual standing erect, his weight equally distributed, and the leg/hip muscles not tensed

patella height

see midpatella height, patella top height, patella bottom height

patella top height

the vertical distance from the floor or other reference surface to the highest point on the superior border of the patella; *syn.* knee cap height; *see also midpatella height, patella bottom height, knee height (standing)*

Comment: measured with the individual standing erect, his weight equally distributed, and the leg/hip muscles not tensed

patent [1]

open or accessible

patent [2]

a grant by some governing body having the appropriate authority to permit the exclusive right to manufacture or sell an invention for a specified period of time

patent ambiguity

that uncertainty due to wording which can be interpreted in more than one way

pathogen

any agent capable of causing disease; *adj.* pathogenic

pathogenesis

the origin and development of a disease in the body; *adj.* pathogenic

pathological reflex

any reflex which differs in strength, type, or reaction time from the norm

pathology

the study of disease processes or the disease itself; *adj.* pathological

patten

a structural support which provides a high sole on one shoe which reduces the weight-bearing requirement on the other leg

pattern [1]

a form or mold used in the manufacturing of an item

pattern [2]

an integration of many separate elements which is perceived as a larger unit

pattern coding
any (set of) perceptual indicator(s) which may be used to differentiate areas of interest to an observer or reduce operator search time

pattern recognition
an automated, electronic process for identifying a scanned or other input image consisting of a predetermined set of alphanumeric characters, symbols, or other shapes

pause
a temporary cessation of an ongoing activity

pay differential
that difference in pay to workers for working under certain conditions, such as shiftwork

payload
that mass/volume comprising the cargo which a delivery vehicle carries to a certain location and which is in addition to that necessary for vehicle operation

payout time
that temporal period required to recover an original investment

peak [1]
a maximum positive or a maximum negative value in a waveform

peak [2]
a maximum value or highest point

peak height velocity
that point during physical maturation at which stature increases at its fastest rate

peak load
the maximum rate or capacity which a system or component is either designed for or is able to perform or support

peak performance
a behavior which exceeds an individual's expected or predicted level of ability

peak weight velocity
that point during physical maturation

at which body weight increases at its fastest rate

peak-to-peak amplitude (p-p)
the algebraic difference between the maximum and minimum quantities within a cycle of a waveform; also peak-peak amplitude; *syn.* peak-to-peak value

peak-to-peak value
see *peak-to-peak amplitude*

Pearson product-moment correlation coefficient (r)
a numerical value, ranging from -1.0 to $+1.0$, which indicates the degree of linear correlation between two normally-distributed variables

$$r_{xy} = \frac{N \Sigma XY - \Sigma X \Sigma Y}{\sqrt{N \Sigma X^2 - (\Sigma X)^2} \sqrt{N \Sigma Y^2 - (\Sigma Y)^2}}$$

(*computational*)

$$r_{xy} = \frac{\Sigma xy}{N \, sd_x \, sd_y} \; (theoretical)$$

where:

X, Y	=	raw score values of X and Y variables, respectively
x, y	=	deviation scores from the mean for X and Y, respectively
sd_x, sd_y	=	respective standard devia tions
N	=	number of cases in the sample

Comment: 0.0 represents no correlation, with correlation increasing as the value approaches ± 1.0

pectoral muscle
a large skeletal muscle overlying the ribs in the region between the sternum and the shoulder; *syn.* pectoralis major
Comment: the underlying pectoralis minor muscle is not normally of significance in human factors

pectoral skinfold
the thickness of a skinfold taken over the pectoral muscle beginning at the

anterior axillary fold and with its length directed toward the nipple; *syn.* chest skinfold

> *Comment:* measured with the individual standing erect, the neck, shoulder and torso muscles relaxed

pedal
(n) a control, normally operated by the foot, which involves some type of motion for operation; *see rotary pedal, translational pedal, reciprocating pedal*

pedestrian
any person on or near a roadway and not in a vehicle

pelvic bone
see coxal bone

pelvic breadth
see bi-iliocristale breadth

pelvic girdle
the right and left coxal bones joined to form the complete bilateral hip structure

pelvimetry
the measurement of the inlet and outlet size(s) of the birth canal

pelvis
the combination of the pelvic girdle, sacrum, and coccyx bones

penale height
the vertical distance from the floor or other reference surface to the upper edge of the junction of the penis with the abdomen

> *Comment:* measured with the individual standing erect, his weight equally distributed between both feet, and the penis flaccid

penis
the sensitive male erectile structure involved in urination and copulation

penumbra
that region of an ionization track within a tissue or other material beyond the umbra, and which is due to recoiling electrons

perceived noise level (PNL)
a calculated value intended for use as a single measure of environmental noise; *syn.* perceived noisiness level

perceived noisiness level
see perceived noise level

percentile
a value or score representing the percentage of people at or below a certain measurement on a given dimension within a specified cumulative distribution; *see also centile*

perception
the process of becoming aware of and interpreting external objects, events, and relationships based on experience following the receipt of sensory information

perception deafness
auditory nerve or cochlear deafness; also perceptive deafness

perceptive deafness
see perception deafness

perceptual adaptation
a semipermanent change in perception or perceptual-motor coordination which serves to effectively reduce or eliminate an apparent discrepancy between or within sensory modalities or the errors introduced by this discrepancy

perceptual load
a measure of the amount of information an individual is expected to acquire and process within a given period of time

perceptual overload
a condition in which more information is being received by the senses than can be processed or understood

perceptual skill
the relative proficiency in detecting and interpreting information received from sensory inputs

perceptual speed
the rate at which one can accurately compare presented objects, either simultaneously or sequentially

perceptual work
any activity which requires the senses

for information gathering and involves any necessary integration of that information

perceptual-motor performance
a measure of the ability to conduct any activity which involves a combination of the individual's sensory, cognitive, and motor functions; also perceptual motor performance

perceptual-motor task
any activity involving an overt movement to a non-verbal stimulus

percutaneous absorption
the absorption of materials through the skin or other exposed body surfaces

performance [1]
see *human performance*

performance [2]
a measure of the output or capability of some system

performance aid
any job aid, training, or other device or system which is intended for or capable of improving human performance

performance analysis
an examination of performance indices, measures, and standards for their relevance, appropriateness, and correctness

performance curve
a graphical curve illustrating quantitative accomplishments plotted against some reference variable such as time

performance decrement
a decline in human or machine output quality or quantity for whatever reason

performance factor
see *performance shaping factor*

performance index
the value of the ratio of some standard work output to actual operator work output; *syn.* performance ratio; *see also* *operator productivity, performance rating*

performance measure
any objective or subjective instrument developed to evaluate personnel or equipment effectiveness

performance measurement
the process of accumulating or assessing worker, group, or equipment accomplishments with respect to a performance standard

performance operating characteristic
a performance tradeoff curve or function indicating how performance on one task is affected by concurrent performance on a second task

performance rating [1]
a quantitative value representative of worker performance with respect to what is considered normal performance; *syn.* pace rating; *see also* ***operator productivity, performance index***

performance rating [2]
the process of using any system or method for evaluating or setting employee performance standards; *see also* *leveling*

performance ratio
see *performance index*

performance sampling
the use of any technique for observing one or more workers during the execution of a given task as part of a performance evaluation

performance shaping factor
any aspect of the individual and environment which predispose humans toward a certain performance level; *syn.* performance factor

performance standard
any criterion or benchmark measure of performance, against which actual performance may be compared

performance technologist
one who is qualified by experience, education, and training to practice performance technology

performance technology
that acquired knowledge which provides for the conceptualization, design,

development, implementation, and analysis of systems intended to optimize performance

perfusional change
a change in the blood flow rate

perilymph
a clear fluid within the osseous labyrinth of the inner ear, and within which the membranous labyrinth is located

perineum
the region at the base of the torso in the pubic crotch near the midsagittal plane, extending anterior-posterior from the pubic arch to the coccyx, and laterally between the ischial tuberosities

period
that time required to complete one cycle of a repetitive event; the reciprocal of the frequency of an event

periodic data
that form of deterministic data which has a clearly established repeating cycle

peripheral nervous system (PNS)
that portion of the nervous system outside the brain and spinal cord; *see also autonomic nervous system, central nervous system*

peripheral neuropathy
any disease involving the peripheral nervous system

peripheral vision
the ability to see or sense objects, motion, or light due to stimulation of the portions of the retina away from the fovea centralis while the gaze is directed straight ahead at a fixed point; *see also foveal vision*

peripheral visual field
that portion of the visual field which lies beyond that which impinges on the foveal region of the retina; *opp.* central visual field

periscope liberty
the permission to observe outside a submerged submarine through the periscope

permanent disability
an irreversible disability of any non-

fatal degree; *syn.* permanent impairment

permanent impairment
see permanent disability

permanent menu
a menu which is constantly and inseparably part of a display

permanent night shift
a non-rotating shift in which workers remain assigned to the third shift indefinitely

permanent threshold shift (PTS)
a irreversible reduction in sensitivity to stimuli in one or more sensory modes due to any condition, *opp.* temporary threshold shift; *see also noise-induced permanent threshold shift*

permeability index (i_m)
a unit for describing the efficiency of fabrics in transferring moisture and associated latent heat away from the body; numerically equal to the ratio between actual evaporative heat transfer of a clothing system to the ideal evaporative heat transfer of a wet-bulb system

permissible exposure limit (PEL)
any of a set of OSHA-established maximum levels of exposure to certain substances; *syn.* maximum allowable concentration; *see time weighted average, ceiling, short-term exposure limit*

perseveration
a movement disorder characterized by repeated motions after the task which required the motions has been completed

person-type flow process chart
see worker-type flow process chart

personal allowance
the provision of time during the workday for the employee to take care of personal needs; *syn.* personal time, personal needs allowance

personal flotation device
any device intended to keep an individual afloat in water

personal leave
that period of time which an individual may be gone from work without penalty and which is separate from vacation time, holidays, and sick leave

personal needs allowance
see *personal allowance*

personal protection equipment
see *personal protective equipment*

personal protective equipment (PPE)
any outer layers of clothing, material, or device which is worn on the body and intended to protect an individual from exposure to or contact with one or more harmful substances, forces, or energies; *syn.* personal protection equipment

personal space
a region surrounding an individual, the size of which he prefers to regulate; *syn.* personal zone

personal time
see *personal allowance*

personal zone
see *personal space*

personnel audit
an aspect of organizational analysis involving a comprehensive examination of all personnel functions, including selection, training, retention, and needs

personnel selection
the process which includes some or all of the following steps: (a) performing job/task analyses, (b) selecting, administering, and validating one or more tests to give to candidates, (c) evaluating candidate performance on the test(s), and (d) deciding which candidates to hire

perspective
the compensation of relative positions in space for a model displayed on a CRT

perspiration
see *sweat (n)*

perspire
see *sweat (v)*; *n.* perspiration

phalanx
any of the bones in the fingers or toes; *pl.* phalanges

phantom limb
the kinesthetic sensations which amputees occasionally experience and cause them to believe they still have the amputated member

pharynx
a tube-like structure which extends from the posterior nasal cavity to the esophagus and larynx

phase [1]
a temporal or physical subset of a project which is in existence for some portion of the total time

phase [2]
that fraction of a complete period though which a periodic entity has passed, relative to some origin or reference point
> *Comment:* expressed in angular terms, with a complete period being 360° or 2π radians

phase angle
the difference in phase between two periodic waveforms

phase shift
that change in time, either advancing or delaying, of a periodic waveform from a known or standard position to a new position, usually represented by an angle in sinusoidal waveforms

phenomenal zero
that physical level or intensity of some stimulus along a given dimension at which an individual judges it to be non-existent or have no value

philtrum
the vertical depression which extends from the base of the nasal septum to the superior margin of the upper lip

philtrum length
the vertical linear distance from the superior margin of the upper lip to the base of the nasal septum
> *Comment:* measured with the facial muscles relaxed

phon
a unit for the loudness level of a sound
Comment: absolute threshold = 0
phons

phonation
the production of speech sounds using
the vocal cords to interrupt air flow
from the lungs

phoneme
the smallest unit or sound of speech
which can have meaning or result in a
change of meaning in a word

phonetic alphabet
a standardized word list used to enhance voice communication over radio
or telephone

phonetically balanced (PB)
a type of test used in measuring speech
intelligibility, in which a monosyllabic
word list is used as a standard

phosphor
any substance emitting light when exposed to radiation

phosphor mercury lamp
see mercury-fluorescent lamp

phosphorescence
that light emission following radiation
absorption which may continue for
some time after the stimulating radiation ceases

phot (ph)
a unit of illumination with the centimeter as the unit of length rather than the
meter; equal to a one lumen flux on a
one centimeter area

photochromatic interval
a range of luminance between the absolute rod threshold and the cone color
identification threshold within which
no perception of color can be made

photochromic effect
exhibiting a color change on light exposure

photogrammetry
the use of photographs to track motion
or determine center of mass/gravity of
body; *adj.* photogrammetric

photokeratitis
an inflammation of the cornea due to

UV light exposure

photokeratoconjunctivitis
an inflammation of the corneal conjunctiva from exposure to UV light

photometer
an instrument for measuring aspects of
radiant energy in the visible portion of
the electromagnetic spectrum, such as
luminance, illuminance, luminous intensity, and luminous flux; *see physical photometer, visual photometer*

photometric brightness
see luminance
Comment: an outdated term

photometric unit
a unit of measurement for radiant energy in relation to its effect on visual
receptors

photometry
the study, measurement, and/or calculation of one or more quantifiable
aspects of light; frequent *syn.* light
measurement

photon [1]
a quantum unit of electromagnetic energy

photon [2]
see troland
Comment: an old, outdated unit

photonics
the study and/or technology of generating, containing, transmitting, processing, detecting, and using light and
other forms of electromagnetic energy
whose basic unit of measure can be
treated as a photon

photopic
pertaining to relatively high levels of
illumination; *opp.* scotopic

photopic adaptation
see light adaptation

photopic luminous efficiency function
*see photopic spectral luminous effi-
ciency function*

**photopic spectral luminous efficiency
function** ($V(\lambda)$)
a mathematical representation or graph
describing the relative sensitivity of
the retina to wavelengths of light un-

der moderate to high illumination; *opp.* scotopic spectral luminous efficiency function

photopic vision
that aspect of visual perception due to stimulation of normally functioning retinal cones under moderate to high illumination levels resulting in the sensation of color; *syn.* color vision; *opp.* scotopic vision

photoreceptor
any sensory mechanism or physical device which is capable of detecting radiation within and/or near the visible region of the electromagnetic spectrum

physiatrics
the practice of physical or rehabilitation medicine

physiatrist
one who practices physical medicine

physiatry
see physiatrics

physical age
see chronological age

physical anthropology
a sub-discipline of anthropology which specializes in the study of human biological variation and evolution

physical basic element
one of a set of basic elements of work characterized by the ASME as involving some form of physical activity

physical constant
a physical quantity which does not vary in numerical value

physical containment
a passive isolation, once established, involving solid structure between two substances which are to be kept separate; *see also operational containment*

physical deconditioning
the loss of muscular, cardiovascular, or other body tone due to lack of physical activity, as might occur in extended periods of inactivity, bedrest or microgravity

physical disability
any disability involving a structural or motor aspect

physical effort
the use of biomechanics, physiology, and body structures in carrying out some function

physical handicap
a physical disability which requires the use of one or more physical devices to enable the individual to have a significant degree of independence and function in the normal environment

physical health
a state in which one's bodily structure, strength, and joint motion capabilities are within normal limits; *see also health*

physical medicine
the use of electrotherapy and other physical techniques to effect the rehabilitation of injured patients; *syn.* physical medicine and rehabilitation

physical photometer
an instrument which uses light sensitive transducers to measure luminance; *see also visual photometer*

physical therapist
one who practices physical therapy

physical therapy (PT)
the use of physical stimuli and/or exercise to treat injury or disease

physical work capacity (PWC)
the maximum rate of oxygen consumption or rate of work output for an individual performing a very strenuous physical task, corresponding to a specified heart rate; *syn.* physical working capacity
> *Comment:* the measured heart rate is often included with abbreviation as a subscript, generally 150, 170, or 180 beats per minute

physical working capacity
see physical work capacity

physical workload
any measure of the physical labor or effort involved in some activity; *opp.* mental workload

physiological response
  any of the body's reactions to an internal or external stimulus

physiological saline
  see **normal saline**

physiological work measurement
  the application of work physiology techniques to determine the physiological task load/severity on the worker; *syn.* work measurement

physiologically inert
  having no functional or chemical effect on the body

physiology
  the study of the biological, biochemical, and biophysical functions of living organisms; *adj.* physiological

pi (π)
  a physical constant, equal to about 3.14159

pia mater
  a membrane lying adjacent to and closely following the contours of neural tissue in the brain and spinal cord

pico- (p)
  (prefix) 10^{-12} of the basic unit

picture element
  see **pixel**

pie chart
  a circular graphic which may be subdivided radially and used for representing proportions of a whole, by analogy with a pie

piece rate
  an incentive plan which provides a constant monetary compensation per production unit completed; *syn.* rate, piecework; *see also* **wage rate**

piezoelectric effect
  a property exhibited by some dielectric crystals in which the application of a mechanical force results in polarization of electric charge; or inversely, the application of a voltage between some faces produces a mechanical deformation

piezoelectric touchscreen
  a display with an overlying glass pane separated from the display by a set of pressure-sensitive crystals which are capable of locating the touch point

pilot
  (n) an individual who pilots

pilot
  (v) maneuver or control a vehicle which is not normally limited to motion on a relatively hard, fixed surface

pilot study
  a preliminary, small research effort undertaken prior to a full research experiment
  Comment: may be to gather preliminary data, test procedures, or some other function

pinch
  bring two structures together and apply pressure to an object between them

pinch grasp
  a position in which the thumb pad and the anterior-lateral side of the index finger are together with the intent to generate forces adequate to hold some relatively thin object between them

pinch point
  any location where it is possible to be caught between: (a) moving parts of a machine, (b) moving and stationary parts of a machine, or (c) material and parts of a machine

pineal body
  see **pineal gland**

pineal gland
  a small, cone-shaped structure, attached to the midline of the upper brainstem, which is the source of the hormone melatonin, and is believed to be involved in physical development and biological rhythms; *syn.* pineal body

pink noise
  a type of noise with equal energy in a given constant percentage bandwidth, having the effect of reducing the power in higher frequency components compared to white noise

pinkeye
(sl) *see* *conjunctivitis*

pinna
see *auricle*

pistol grip
that structure on a tool or other device which resembles the grasping structure on a pistol, is used for holding the tool/device during carrying or operation, and may have a trigger-like mechanism for operation

pitch [1]
the movement or orientation of an object about its lateral (side-to-side) axis

pitch [2]
the perception of the quality of a tone, involving primarily frequency, which permits ordering on a scale from low to high

pitch discrimination
the ability to distinguish frequency differences in pure sounds; *syn.* tonal discrimination

pivot joint
a joint in which motion is limited to rotation about an axis perpendicular to the contact surface

pivot point
an approximation of the center of rotation for various types of hinge joints

pixel
the smallest resolution point on a CRT display or vidicon-type detector
Comment: a contraction of picture element

placebo
a null treatment given to a control group

placeholding cursor
a cursor which indicates the location of last entry or at which the next text or other data entry will occur

PLAID
an interactive, three-dimensional computer modeling program used for designing aerospace crew stations and integrating crew anthropometry, lighting, and other aspects; *syn.* Panel Layout And Integrated Design

plan (Pn, PL)
a therblig which consists of the cognitive/mental process of determining what action must be taken next

planar wave
see *plane wave*

Planckian locus
see *blackbody locus*

plane of fixation
see *fixation plane*

plane of work
see *work surface*

plane wave
a waveform composed of parallel planes perpendicular to the direction of propagation in which all points in a given plane are at the same phase; *syn.* planar wave

planimeter
a device used for determining the area of a planar surface by determining and working with the areal boundary

planned maintenance
see *routine maintenance*

plant layout
(n) the arrangement of the physical facilities, machinery, and equipment within a given production/manufacturing location

plantar
pertaining to the sole of the foot

plantar flexion
a rotation about the ankle joint which results in a relative downward motion of the anterior portion of the foot; also plantarflexion; *opp.* dorsiflexion

plantar flexor
any muscle which depresses the anterior foot about the ankle joint; also plantarflexor; *opp.* dorsiflexor

plasma [1]
a highly or completely ionized gas

plasma [2]
that fluid portion of blood without the cells

platelet
a small, enclosed blood-carried structure used in the clotting process

platykurtic
pertaining to a normal distribution that is more flattened than peaked; *opp.* leptokurtic

play for position
see pre-position

plethysmograph
an instrument used to detect changes in the volume of some body part

pleura
the membrane surrounding the lung and lining the thoracic cavity; *syn.* pleural sac

pleural sac
see pleura

plosive
pertaining to a sound produced by turbulent air flow when the oral tract is opened to release pressure

pneumatic
pertaining to or using air

pneumoconiosis
any disease of the lung caused by the inhalation of mineral dusts which produce an increase in fibrous connective tissue and a chronic hardening of the lung tissues

pneumotachograph
an instrument used to record the breathing/pulmonary ventilation rate

pneumothorax
an accumulation of air or gas in the pleura

point [1]
(v) move a cursor on a display to a certain item using a direct manipulation computer input device

point [2]
(n) a measure of type size; 1 point is 1/72 in.

point [3]
(n) an output measure in the Bedaux system consisting of that production by one standard operator in one minute's time

point and click
a method of operation with a graphical user interface in which a mouse or other computer input device is used to drag a pointing cursor, often an arrow, to a certain display location and execute a command by clicking when the cursor overlies a certain block or region; also point-and-click

point biserial r (r_{pb})
the correlation coefficient between a continuous variable and a dichotomous variable

point of fixation
see fixation point

point of inflection
see inflection point

point of observation
the midpoint of an imaginary line connecting the pupil centers of the two eyes

point of operation
the zone in which the primary functional parts of a machine or tools perform their function(s)

point of subjective equality (PSE)
the value of some variable or stimulus in psychophysical work at which the observer cannot tell a difference between a reference value and the variable value — i.e., the two appear equal

point source
an energy source whose dimensions, relative to its distance from the point of observation, are insignificant for practical purposes in any calculations; *opp.* extended source
> *Comment:* for visual work, anything less than than about 10′ of arc is considered a point source

pointer
a highlighting technique which uses some directional device or indicator to locate that portion of a display toward which attention is desired or required

pointing cursor
a position-indicating cursor which shows the user's relative location among display structures

pointmark
a specific point of interest on a body landmark

Poisson distribution

a discrete distribution having a probability distribution function of

$$p(n) \simeq \frac{m^n e^{-m}}{n!}$$

where:

p(n)	=	probability of n in N
m	=	Np = mean = variance
n	=	the number of occurrences desired
N	=	the number of possible occurrences

polarize

orient light waves in a specific plane

policy allowance

a non-bonus time allowance added to the standard time to maintain or provide a satisfactory earning level under unusual circumstances

polygraph

a multi-channel chart recorder for recording several physiological measures simultaneously; frequent *syn.* oscillograph

polymer

a compound formed by the combination of two or more molecules, usually of the same basic substance with a resulting material of high molecular weight

pop-up menu

a menu which is displayed overlaying other screen entities, generally in the center of the screen, following an appropriate keystroke, mouse click, or other form of operator input and which remains displayed until a selection is made or the menu is closed or exited

popliteal

pertaining to the tendons, fossa, and other tissues posterior to the knee

popliteal crease

the junction of the biceps femoris and gastrocnemius muscles or their tendons posterior to the knee when the knee is flexed about 90°; *syn.* popliteal line

popliteal crease clearance, horizontal

the horizontal linear distance between the most anterior part of the seat pan and the popliteal crease

Comment: measured with the individual sitting erect, the knee flexed about 90°, and the feet flat on the floor or other reference surface

popliteal fossa

the natural depression in the posterior portion of the leg at the knee; *syn.* popliteal region

popliteal height

the vertical distance from the footrest surface to the biceps femoris tendon at the underside of the thigh immediately behind the popliteal crease; also popliteal height, sitting

Comment: measured with the individual sitting upright with the knee flexed 90°

popliteal line

see popliteal crease

popliteal region

see *popliteal fossa*

popliteus

a flat skeletal muscle underlying the distal portion of the popliteal fossa

population

all possible members of a group having a distinct set of characteristics

population stereotype

a common behavioral expectation in a perceptual-motor situation

portable lighting

any type of lighting designed to be easily transported manually from one location to another

position (P)

a therblig which consists of placing an object in a predetermined location for future operations

Position Analysis Questionnaire (PAQ)

a structured questionnaire used in an attempt to quantify job descriptions based on a set of job elements

position control

a type of control mechanism in which a displacement of an input device leads to a proportional displacement of an output device; *syn.* zero-order control; *see also* **rate control**

position index (P)

a factor for the relative average luminance of a source located within the visual field which produces a sensation at the comfort-discomfort boundary

positioning movement

re-locating one or more body segments relative to other object(s) or a reference point

positive acceleration

see **acceleration**

positive afterimage

that color image of an object which continues to be seen after ceasing to look at the object; *syn.* afterimage; *opp.* negative afterimage

positive click

the incorporation of momentary stops with an audible click or tactile sensation in such motions as turning a dial or pushing a button to provide sensorimotor feedback to the operator

positive feedback

a signal which tends to enhance or prolong the output of a system; *opp.* negative feedback

positive *g*

an acceleration vector directed inferior along the body longitudinal axis, usually referring to a magnitude greater than that of normal earth gravity, as in an aircraft maneuver which results in blood pooling in the lower extremities; *opp.* negative *g*

positive gravity

see **gravity, positive g**

positive pressure air mask

see **positive pressure respirator**

positive pressure respirator

a type of respirator which functions by having pure air at higher than ambient pressure directed into the mask for inhalation and to prevent entry of contaminants; *syn.* positive pressure air mask; *opp.* negative pressure respirator

positive reinforcement

the cause of strengthening or increasing the frequency of a response as a result of contingent reinforcement; *opp.* negative reinforcement

positive skew

having a distribution curve with the mean greater than the mode; *opp.* negative skew

positive transfer

a condition in which an individual's previous experience aids learning of a new task; *opp.* negative transfer; *see also* **transfer of training**

positive work

the application of a force through some vertical distance in opposing gravity; *opp.* negative work

post-rotary nystagmus

that nystagmus caused by deceleration of the vestibular system as head rotation is stopped

posterior

referring to a location behind or toward the back of the body

posterior cricoarytenoid

a skeletal muscle in the larynx which causes the glottis to open on contraction; *opp.* lateral cricoarytenoid

posterior neck length

the surface distance from nuchale to cervicale

Comment: measured with the individual standing erect, the head, neck, and shoulder muscles relaxed

postflight earblock

the pressure differential generated due to the absorption of oxygen by tissues within the middle ear after breathing pure oxygen on a high-altitude flight

postprandial

following a meal

postural movement

any bodily movement resulting from the neuromuscular coordination of the postural muscles and nervous system to maintain or attempt to maintain postural stability

postural muscles

those muscles normally involved in maintaining an erect posture, usually the slow muscles of the legs, back, and abdomen

postural sway

any non-volitional pendular movements of the head, trunk, or entire body which occur while standing or sitting at rest; *syn.* postural tremor

posture [1]

a set of anatomical coordinate frames at various body points whose origins are defined quantitatively relative to a comprehensive three-dimensional coordinate system; *adj.* postural

posture [2]

a qualitative description of the general position of the body (i.e., standing, sitting); *adj.* postural

potential energy (PE)

stored or available energy; *opp.* kinetic energy

potential for improving performance

the value of the ratio of outputs of a master performer to an average performer

$$PIP = \frac{Master\ performer\ output}{average\ performer\ output}$$

potentiometer

a continuously variable electrical resistor

pound

see pound force, pound mass

pound force (lbf)

a unit of force in the English system

pound mass (lbm)

a unit of mass in the English system

power [1]

the statistical probability of rejecting a hypothesis when it is not true; equal to $1 - \beta$, where β = the probability of making a type II error

power [2]

the rate at which work is done or energy transferred with respect to time; defined in several ways, depending on the situation:

$$P = \frac{dW}{dt} = Fv = \tau\omega$$

where:

$\frac{dW}{dt}$	=	instantaneous work output
F	=	force
v	=	velocity
τ	=	torque
ω	=	angular velocity

power density spectrum

see power spectrum; also power-density spectrum

power function

any relationship in which one variable is described by a constant multiplied by another variable raised to some exponent, e.g., $A = kB^n$; *see also inverse power function*

power grip

a type of grip in which the fingers and palm are partially flexed around an object, with an opposing thumb and thumb crotch

power law

see Stevens' power law

power spectral density

the mean square value of that portion of the energy within a narrow frequency band of a power spectrum; *syn.* autospectral density, spectrum level, power density

power spectrum

a plot of power spectral density versus frequency; *pl.* power spectra; *syn.* frequency spectrum, power density spectrum

practice

the repetition of some activity in an attempt to become more proficient in that activity

practice effect

that improvement in performance observed over time due to learning prior to that point at which an individual achieves stable/asymptotic performance

pre-employment examination
see *pre-employment physical examination, pre-employment test*

pre-employment physical examination
a physical examination of a job applicant prior to employment; *syn.* pre-employment examination

pre-employment screening
the use of information from pre-employment examinations and/or background checks to verify that an individual passes certain criteria prior to employment

pre-employment test
any physical skill/ability or mental function test given to job applicants prior to employment; *syn.* pre-employment examination

pre-position (PP)
a therblig which consists of placing an object in a specified location so that it may be grasped for another operation when needed; *syn.* play for position

pre-program
place into memory a software or firmware set of instructions and data into a device prior to its undertaking the activity using that instruction set

pre-programmed movement
see *ballistic movement*

pre-prototype
a model constructed from commercially-available materials or components to demonstrate form, fit, and/or general function

preamplifier
an electrical device for amplifying very low amplitude electrical signals prior to input to a standard amplifier; sl. preamp

preauricular point
that location anterior to the opening of the external auditory canal representing the most posterior aspect of the

zygomatic arch

precaution
any action taken to reduce the probability of an accident

precision
the extent to which a measurement protocol yields the same results when applied repeatedly to the same situation

precision grip
that type of grip utilizing only the more distal phalanges of the hand, and which is intended for optimum control rather than strength

predetermined motion time system (PMTS)
any of a number of work measurement techniques which use some combination of the following practices and synthesizing the times required to perform a job or task: (a) a methodology for determining the basic human motions (and possibly strength requirements) involved in performing a job or task, (b) obtaining the times required for those motions, (c) determining what the performance level for that job or task ought to be, (d) the organization and storage of this information in a database or other format for prediction of future similar jobs or tasks

predetermined time
see *predetermined motion time system*

predetermined time system
see *predetermined motion time system*

predicted four-hour sweat rate (P4SR)
a measure of heat stress based on empirically-determined values using physically-fit, acclimatized males for estimating the maximum workload via the amount of water perspired with consideration for the clothing worn, the metabolic rate, and the environmental conditions; *syn.* heat stress index
Comment: an older measure

predictive display
a display which shows an operator,

through the use of extrapolation from current data, an estimate of the position of a point or object at a later time; *syn.* predictor display

predictive maintenance
that type of maintenance performed when one or more sensors or other indicators present information that a system or piece of equipment is about to fail

predictive validity
having a high correlation between applicant test results and later performance on the job

predictor display
see predictive display

predictor variable
a variable used for predicting the observed result in a correlation or regression study; *opp.* criterion variable
Comment: analogous to the independent variable in experimental work

preferred limb
a preference in the use of one limb over the other

preferred line of sight
that deviation from the horizontal plane at which an individual aligns his line of sight for the task at hand
Comment: typically about 10° to 15° below the horizontal plane for VDT tasks

preferred speech interference level (PSIL)
the average sound pressure levels of noise (in dB) in three octave bands with 500 Hz, 1 KHz, and 2 Khz center frequencies; also preferred-octave speech interference level; *see also speech interference level*

prehensile
adapted for grasping; *n.* prehension

preloading
having an individual carry out a fatiguing task prior to engaging in a task which measures performance

premium bonus
an incentive plan in which the worker's award is based on the amount of time saved from the alloted time for a task

premium time
see overtime

prenatal
prior to birth

prerequisites of biomechanical work tolerance
a set of recommendations covering postural, engineering, and movement considerations for improving performance, reducing stress, and possibly preventing occupational injuries; *see also workplace design, motion efficiency principles*
Comment: proposed by Tichauer

presbycusis
the loss of hearing through aging, usually beginning with deterioration at the higher frequencies; also presbyacusis

presbyopia
a refractive disability due to the loss of lens elasticity with age, resulting in a focal point posterior to the retina and the inability to see near objects clearly or read small print

prescriptive zone
that environmental temperature above which the body cannot maintain thermal equilibrium for the working conditions and in which precautions for heat stress should be implemented

presentation human factors
that field dealing with the study or use of oral or written presentations and those variables involved in the communication of information

pressure joystick
see isometric joystick

pressure vessel
any container designed for holding fluids at pressures signficantly greater than a standard atmosphere

prevalence rate
the number of existing cases of some disease or injury within a defined population at a given time

preventive maintenance
that form of maintenance performed

with the intent of avoiding a major breakdown

preview control
having the use of predictive displays for a teleoperator

primary color
any of the colors red, green, or blue

primary flight display (PFD)
a display from any of the primary flight instruments

primary flight instrument
any of the following types of instruments in an aerospace vehicle cockpit: attitude indicator, airspeed indicator, vertical speed indicator, altimeter, turn coordinator, and heading indicator

primary line of sight
see **line of sight**

primary motor vehicle controls
those controls involved in moving, directing, and stopping a motor vehicle, including the throttle, steering mechanism, and brake pedal

primary positioning movement
the first movement an operator makes in positioning a control device; syn. gross adjustment, slew

primary radiation
that ionizing radiation coming from an original source, whether natural or artificial; see also **secondary radiation**

primary standard
an original standard, typically found only in national or international laboratories, and from which other standards are derived

primary task
that task to which an individual should pay the greatest attention and which is of the most importance or highest criticality; see also **secondary task**

primary viewing area
that portion of a display, screen, or window on which are presented the data and/or text generated by the operator or by a computer in response to a query or computation

prime mover
any muscle which produces or main-

tains a specific motion or posture; syn. agonist

primitive solid
in computer modeling, a basic three-dimensional form which may be used to build more complex forms

principle of equivalence
a rule which states that forces imposed by acceleration on a body, whether due to motion or gravity, are equivalent

principles for motion improvement
see **motion efficiency principles**

principles of motion economy
see **motion efficiency principles**

print
put text, graphics, or other form of computer output into a form for display, especially for hardcopy

proactive inhibition
that reduced efficiency in retention of a set of information caused by previous learning; see also **negative transfer**, **retroactive inhibition**

probabilistic process
a mathematically-described phenomenon for which the instantaneous amplitude values can't be specified uniquely at any given instant of time

probability [1] (p)
the value of the ratio of the number of ways one or more specified events can occur to the total number of events which may occur
    Comment: expressed as a number between 0 and 1

probability [2]
the likelihood of observing a particular result or event, especially within a specified time or a given set of circumstances

probability density
the ordinal value for a point corresponding to a certain deviation measure on a probability distribution function

probability density function (PDF)
see **probability distribution function**

probability distribution function (PDF)
a non-monotonic graphical, math-

ematical, or tabular representation of a function whose peak is the mode and whose ordinal values represent the probability density of that function; *syn.* probability density function; *see also* **cumulative distribution function**
> Comment: represents the bell-shaped curve for the normal and t distributions, varying curve shapes for other distributions

probability theory
a mathematical system for estimating the likelihood of one or more certain events occurring out of a larger set of events

problem analysis
the identification and study of the cause(s) of a problem

problem sensitivity
the ability to notice when something is faulty or likely to become so

procedural analysis
the use of flowcharts, decision tables, etc. to develop and/or represent the sequence in which a task is to be performed

procedure
any instruction set or sequence of actions used to accomplish a given task

procedure flow chart
see **form process chart**

process [1]
any predetermined or planned series of continuous or repetitive steps or operations, usually involving the movement of people, materials, forms, or other entities from one location to another, with the intent to bring about some desired product or result

process [2]
a prominent projection from a bone

process chart
a graphic/symbolic, diagrammatic, and textual description of the events occurring in some activity; *syn.* flowchart, product analysis chart, production process chart; *see also* **man process chart, multiple activity process chart, operation process chart, operator pro-**

cess chart, flow process chart, form process chart, simultaneous motion chart

process chart symbol
one of a standard set of five graphical objects for representing actions or events to be used on a variety of process charts: operation, transportation, inspection, delay, and storage

process control
the manual or automatic direction of processing or operating conditions to effect a desired output by maintaining certain standard or specified operating tolerance conditions

process control chart
a chronologically-organized graphical/symbolic tabular presentation of a process, including pertinent data from each sub-process

process control engineering
that engineering field involving developing and implementing the techniques used in process control

process costing
a cost determination in which manufacturing costs are spread over the units produced by department

process engineer
an engineer practicing process engineering

process engineering
the selection and designation of the processes, operational sequences, and tools to be used in making a product

process layout
a type of plant layout in which machinery which performs similar functions is placed in the same area

process planning
a procedure for determining the operations required to produce a certain product or attain a desired goal

process sheet
a sequential listing, sketch, or diagram of a process

process time
the time expected, allowed, or required to complete a process

product
    any single entity resulting from an integrated effort consisting of one or more steps
        Comment: may be a physical item or a service

product analysis chart
    see process chart

product layout
    a type of plant layout in which all machinery and tools concerned with the manufacturing of a particular product or product line are located together

product liability
    a concept that the manufacturer(s), supplier(s), and others involved are legally responsible for any damage or injuries caused by or resulting from use of their product

product line
    a class of products manufactured or sold by a single company

production capacity
    the maximum potential output of a facility for a given set of conditions, including time, personnel, and cost

production economics
    the analysis or practice of attempting to optimize input and output costs to make a profit

production flow analysis
    the study of the routing of a part, component, or system through the various machines and workplaces and the operations it undergoes in a manufacturing or integration facility

production item
    a finished product intended for full and complete use, with the production line operational

production process chart
    see process chart

production standard
    any performance or quality standard established for a manufacturing or service facility, such as standard time or number of rejects

production study
    an extensive, continuous analysis of the components of production-related activities
        Comment: normally to check a standard or determine the variables and their effects on output

production time
    the total time required for facility preparation, manufacturing, and testing of a product

productive labor
    see direct labor

productive time
    that time during which useful work is performed; *opp.* idle time, non-productive time

productivity [1]
    the value of the ratio of actual output to standard output; *syn.* operator productivity

productivity [2]
    any measure of the rate of output relative to the personnel and financial cost supporting that output

productivity improvement
    having an individual or group produce more goods or services within a given time, compared to some previous measure

productivity index
    the value of the ratio of objective production output to employee hours and other resources used

proficiency
    the level of an individual's acquired knowledge or skill in a particular field or task

profile analysis
    the study of groupings of persons or objects; *syn.* cluster analysis, pattern analysis

profit sharing
    an incentive plan through which a company pays its employees based on company profits in addition to the employees' regular pay

program
    an organized effort involving several groups or people toward accomplishing some goal

Program Evaluation and Review Technique (PERT)

a sophisticated management technique for defining and interrelating the various tasks which must be performed to complete a job on time, then tracking the work as the job progresses

Program for European Traffic with Highest Efficiency and Unprecedented Safety (PROMETHEUS)

a project involving the European Community with automotive and electronics manufacturing companies to develop future highway-vehicle systems

programmable function key

a function key whose action may be altered within an application or between applications; *opp.* fixed function key

progress chart

any graphical representation of the status of the work underway; *see also Gantt chart*

progressive resistance exercises (PRE)

a system for increasing the amount of loads lifted, both within a session and across sessions; *syn.* DeLorme exercises

projected anthropometric measurement [1]

an estimate of the future value of an anthropometric measure from a current measure, either of the same individual or the population as a whole

projected anthropometric measurement [2]

an estimate of an unknown anthropometric measure on an individual from two or more other known anthropometric measures, whether from the same or other individuals

prompt

a visually-displayed message or other cue which either requests some action of an operator or user or indicates that the system is ready for input

pronasale

the most anterior point on the nose; the tip of the nose

pronasale to back of head

the horizontal linear distance from in-

ion to the tip of the nose; *equiv.* pronasale to wall

Comment: measured with the individual standing or sitting erect, and looking straight ahead

pronasale to top of head

the vertical distance from the tip of the nose to the level of the top of the head

Comment: measured with the individual standing or sitting erect and looking straight ahead

pronasale to wall

the horizontal linear distance from a wall or other vertical surface to the tip of the nose; *equiv.* pronasale to back of head

Comment: measured with the individual standing or sitting erect, his back against the wall and looking straight ahead

pronate

rotate the hand, wrist, and forearm counterclockwise as viewed along the arm axis from the shoulder; *n.* pronation; *opp.* supinate

pronator

any muscle which causes a pronating motion; *opp.* supinator

prone [1]

pertaining to a posture having the frontal portion of the body downward, with the torso parallel to the reference surface, and generally with the hips and knees extended

prone [2]

having a tendency to behave in a certain way

proof pressure

that pressure to which pressure vessels are subjected to meet acceptance criteria for a given use, usually the maximum operating pressure multiplied by the proof factor

propellant

the pressurization gas in a spray can

proportional control

any gradable type of activity between all and none which exercises a controlling function

proprioception

the sense of posture or the physical position/movement of the limbs in relation to one's environment; *adj.* proprioceptive, proprioceptic

proprioceptor

any mechanoreceptor sensitive to position and movement of the body or its parts, including joints, muscles, and tendons

prosthesis

an artificial replacement for an organ or limb whose appearance may or may not resemble the original structure, and which may have some of the functionality attributed to the original structure; *pl.* prostheses; *adj., syn.* prosthetic

prosthetics

the study of the design, manufacture, and use of prostheses

protanomaly

a color vision deficiency in the ability to discriminate the red content of colors due to weak red cones; *adj.* protanomalous

protanope

one having protanopia or protanomaly

protanopia

a color blindness involving an inability to discriminate the red content of colors due to the absence of red cones; *syn.* red-blindness; *adj.* protanopic

Comment: often some effect on green discrimination as well

protective clothing

any article or set of clothing designed and worn with the intent to protect a worker from a hazardous environment

protective cream

any substance designed to protect skin areas during exposure to harmful materials or conditions

protein

any of a set of complex organic molecules consisting of specific sequences of amino acids

proton

a nucleon having a single positive charge and a mass of just less than one

prototype

a model or preliminary version of a product which is produced prior to fabrication of the production item and is representative of the final system for testing and evaluation

protuberance

any local region on the body or body tissue which projects above the background

proxemics

the study of the nature and effect of the preferred separation distance by individuals in interpersonal situations as a function of culture, psychology, and environmental factors

proximal

referring to a portion of the body or a body segment which is closer to the central longitudinal axis than another part; *opp.* distal

proximity measure

an indication of the distance between some detector and a surface

proximity operations

activity by one entity within a specified distance or volume of another entity

prudent

attentive, careful, and sensible in one's conduct

psychogenic deafness

a lack of hearing due mostly to psychological factors, and in which the individual doesn't realize that he can hear better than he reports; *syn.* functional deafness

psychological refractory period

that phenomenon observed when an individual is attending to two or more stimuli such that the reaction time to each stimulus increases when the inter-stimulus interval between the stimuli decreases; *syn.* refractory period

psychological shock
a sudden disturbance of mental equilibrium; *syn.* shock

psychology
the study of human behavior and its perceptual/cognitive bases; *adj.* psychological

psychometric function
a mathematical or graphical function showing the relationship between a set of stimuli varying quantitatively along a given dimension and the relative frequency with which an observer will give a certain category of response regarding some property of the stimulus

psychometrics
the measurement of psychological processes using experimental design and statistical techniques

psychomotor
pertaining to both motor and mental processes or activity

psychomotor performance
a measure of the achievement level displayed by an individual in executing a psychomotor task

psychomotor skill
any acquired muscular action in response to sensory stimuli and/or mental processes

psychomotor task
any task involving coordination of sensory/cognitive processes and some related motor activity

psychophysical characteristic
any measurable mental or physical quality of an individual, such as reaction time, strength, sensory acuity, or dexterity

psychophysical measurement
the process of obtaining data about any of an individual's psychophysical characteristics, or the resulting data from such a process

psychophysical method
any of a set of standardized procedures for presenting stimuli ranging from fully quantifiable physical stimuli to presently unquantifiable stimuli based on opinions or emotional feelings for an individual's response and applying numerical data to that response

psychophysical quantity
the human perceived value corresponding to some presented stimulus

psychophysical scale
any range of values which describes a function of human sensitivity or other capability, and which has some type of dimensional unit associated with it

psychophysics
that area of experimental psychology which attempts to quantify relationships between stimuli and their psychological or psychobiological responses; *adj.* psychophysical

psychosocial
pertaining to any combination or the interaction of psychological and sociological variables, conditions, or effects

psychosocial evaluation
the measurement or study of socially conditioned behaviors or reactions in the social environment

psychosocial factor
any social influence related to or affecting the psychological factors of human behavior

psychosomatic
pertaining to mind and body interrelationships

psychosomatic medicine
that medical field dealing with health and disease involving both the mental/emotional and physical components and their interactions

psychosomatic reaction
a bodily response resulting from a stimulus which evokes emotion

psychrometer
a frame containing both a dry-bulb thermometer and a wet-bulb thermometer; *syn.* wet and dry bulb thermometer; *adj.* psychrometric; *see also wet bulb temperature, dry bulb temperature*

Comment: one of the techniques for

determining relative humidity

psychrometric calculator
any simple device for determining the dew point or relative humidity values from dry and wet bulb temperature readings and the barometric pressure

psychrometric formula
an empirically-determined formula for determining water vapor pressure based the on barometric pressure and psychrometer readings

psychrometric table
a table of values for determining water vapor pressure, relative humidity, and dew point from psychrometer readings

psychrometric wet bulb temperature
see *wet bulb temperature*

pternion
the fleshy tip of the most anteriorly-projecting toe with all the foot phalanges fully extended

ptosis
the slippage or drooping of a structure below its normal position

pubic crotch height
the vertical distance from the floor to most superior portion of the crotch in the midsagittal plane; see also *symphyseal height*
> Comment: measured with the individual standing erect and his weight balanced equally on both feet

pubic crotch length
the surface distance from the anterior waist midpoint through the crotch and over the maximum protrusion of the buttock to the posterior waist level above the buttock

pubic symphysis
the fibrocartilaginous joint in the midline between the two coxal bones

Pulfrich effect
a binocular visual phenomenon occurring when differing amounts of light are admitted to the two eyes and a pendulum swung in a plane perpendicular to the line of sight appears to move in an elliptical path having depth;

syn. Pulfrich pendulum effect

pull-down menu
a vertical menu which is displayed from the top of the display following the press of an appropriate key or selection button and which disappears on a selection being made, the press of another key, or the release of the selection button

pulmonary
pertaining to the lungs

pulmonary hyperinflation syndrome
an overdistension and rupture of the lung by expanding gases during a decrease in environmental air pressure; syn. burst lung

pulmonary ventilation ($\dot{V}_E$)
the volume of gases which move into and out of the lungs per unit time

$$\dot{V}_E = \Sigma TV = f_r \times \overline{TV}$$

where:
TV = tidal volume
f_r = respiratory frequency
$\overline{TV}$ = mean tidal volume

pulse [1]
a brief, large deviation from a signal baseline

pulse [2]
a detectable peripheral measure of a heart beat

pulse code modulation (PCM)
a type of waveform encoding for communications in which an analog signal is electronically sampled at a certain rate, quantized to a specified level, then binary encoded

pulse duration
that period of time from the onset of a pulse to the signal return to baseline or to within some percentage of the pulse height from the baseline

pulse noise
see *impulse noise*

pulse pressure
the difference between the systolic and diastolic blood pressure readings

pulse rate
  see *heart rate*

pulse shape
  the waveform exhibited by a pulse

punitive damages
  that compensation awarded to a victim
  by a court as a means of punishing an
  entity for wrongdoing or negligence;
  see also **compensatory damages**

pupil
  the approximately circular opening in
  the iris, through which light passes en
  route to the retina; *syn.* pupillary aperture

pupillary aperture
  see *pupil*

pupillary axis
  see *optical axis*

pupillary muscles
  those smooth circular and radial
  muscles of the iris which determine
  pupil diameter, the pupillary sphincter and pupillary dilator muscles, respectively

pupillometry
  the measurement of the eye pupil size

purchase
  (v) buy goods, materials, services as
  well as maintain the necessary record-
  keeping to support that activity

purchasing
  the organization or department responsible for buying goods, materials, and
  services

pure tone
  a sound wave which varies as a simple
  sinusoidal function of time and has a
  unique pitch

purity ¹
  see *excitation purity*

purity ²
  a measure or indication of the lack of
  contaminants in a signal or substance

Purkinje effect
  a decreased sensitivity of the human
  eye to light of longer wavelengths as
  illumination decreases proportionately
  across the spectrum; *syn.* Purkinje shift,
  Purkinje phenomenon

Purkinje phenomenon
  see *Purkinje effect*

Purkinje shift
  see *Purkinje effect*

purple boundary
  that straight line interconnecting the
  termini of the spectrum locus in a chromaticity diagram; *syn.* purple line

pursuit tracking
  a task in which the subject is required
  to maintain the position of some object
  on or within certain limits of a moving
  target

pursuitmeter
  any equipment designed to involve
  tracking a moving point or region for
  measuring eye-hand coordination

pushbutton
  a small control device which operates
  using short-travel, in-and-out linear
  movement, usually intended for operation by a finger

put away
  any activity which involves removing
  a work item, tools, or other materials
  from the workplace

put away time
  that time required to perform a put
  away function

pyknic
  a Kretschmer somatotype having the
  characteristics of rounded contours,
  large body cavities, and a large amount
  of body fat

pyramid
  one of two large, bilateral efferent neural fiber tracts on the posterior brain
  stem which innervates many of the
  skeletal muscles; *adj.* pyramidal

pyramidal system
  that motor system comprised largely
  of neurons originating in the cerebral
  cortex, with their axons passing along
  the posterior brainstem without an intervening synapse to the spinal cord,
  and which is involved in rapid voluntary body movements; *syn.* cortico-
  spinal system; *opp.* extrapyramidal system

Q

Q

see quality factor, quartile, semi-interquartile range

Q factor

see quality factor

quadriceps femoris muscle

a muscle group on the anterior thigh consisting of four muscles which flex the hip and extend the knee: rectus femoris, vastus lateralis, vastus medialis, and vastus intermedius

quadriplegia

a condition in which all four limbs are affected by paralysis; *adj., n.* quadriplegic

qualified handicapped individual

a handicapped individual who is capable of performing a specific job with reasonable accommodation

qualified operator

a worker having the adequate physical/mental attributes, training, education, and experience who has demonstrated that he is capable of safely performing a given task involving the operation of machinery or equipment with acceptable quality and quantity

qualified worker

a worker having the adequate physical/mental attributes, training, education, and experience who has demonstrated that he is capable of performing a given task using basic tools with acceptable quality and quantity

qualitative

pertaining to a description with essentially nominal or non-numerical features; a more subjective explanation or categorization; *opp.* quantitative

quality

a condition in which a product satisfies a set of requirements involving such aspects as strength, durability, function, appearance, and user-satisfaction

quality analysis

an examination or study of product quality goals

quality assurance (QA)

that set of activities involving initial product design, product improvement, quality control, etc. with the intent of establishing and maintaining a certain level of quality

quality circles (QC)

a small group of people involved in a similar type of work who voluntarily meet weekly on paid time in an attempt to identify, analyze, and solve some of the problems in their work environment; *syn.* quality control circles, employee participation team

quality control (QC)

that set of activities, including inspection, analysis, and testing, which is involved in gathering data and demonstrating that a product meets or has a certain probability of meeting a set of specifications

quality control chart

a chart used for recording data regarding product quality

quality control circles

see quality circles

quality factor [1] (Q, QF)

the number intended to represent the effectiveness of various types of ionizing radiation based on relative biological effectiveness or linear energy transfer; *syn.* Q factor

X-ray, beta	1
protons, fast neutrons	10
alpha particles	20

quality factor [2]

a measure for expressing the rate of attentuation (sharpness) of time-varying energy as a function of frequency; *syn.* Q, Q factor

quality of lighting

see lighting quality

quality of working life

see habitability

quantitative

having a description with numerical values, especially in relation to other numerical values; *opp.* qualitative

quantity of light

see light quantity

quartile (Q_i)

the dividing point between two adjacent quarters of a distribution

questioning technique

a method for analyzing and attempting to improve work processes, generally by asking questions like: (a) what is the purpose for some activity, (b) why is it done at a particular location, (c) why is a particular sequence followed, (d) why does a particular person perform that job, and (e) is the method being used to accomplish the task the best possible

questionnaire

a written set of questions intended to obtain an individual's responses on his attitudes, issues, et cetera

queue

(v) place a discrete unit in position for sequential flow in a single processing channel; *ger.* queuing

queue

(n) a location from which units may be selected for processing when processing time or other requisite conditions permit; *syn.* waiting line

queuing theory

a quantitative rule describing the patterns pertaining to arrivals, service times, and the sequence in which arrivals are handled; *syn.* waiting line theory

quickening

a display technique involving the use of time derivatives for aiding an operator in tracking or control operations involving motion

QWERTY keyboard

a keyboard with a letter distribution pattern of QWERTY on the left side of the top row of alphabetic characters; *see also Dvorak keyboard*

R meter

an instrument for measuring and displaying the intensity of ionizing radiation in roentgens

R value

see *thermal resistance value*

race [1]

a breakdown of the human species by certain genetically-determined characteristics such as skin color, bodily proportions, hair type, and stature; *adj.* racial

race [2]

a form of competition where the purpose is to complete a prescribed path or function with a minimal time or ahead of others in the competition

rad

see *radiation absorbed dose*

radappertize

radiate food with ionizing radiation to sterilize it; *n.* radappertization

radar

a system using emitted and reflected electromagnetic energy, principally radio frequency and microwaves, to detect and track objects or weather
  Comment: an acronym derived from RAdio Detection And Ranging

radar display

the graphic presentation of a radar scan on a radarscope

radarscope

the CRT or other device used to project the scan of a radar beam for operator viewing

radial [1]

pertaining to the radius bone

radial [2]

on a line directed outward from the center of circle

radial deviation

a movement of the wrist such that the longitudinal axis of the hand is directed toward the lateral/radial/thumb side of the forearm; *opp.* ulnar deviation

radial keratotomy

a surgical procedure in which radial incisions are made in the cornea to improve myopic vision

radial nerve

a spinal nerve innervating the upper arm, forearm, and the dorsal-thumb side of the hand; *syn.* musculospiral nerve

radiale

the uppermost point on the lateral margin of the proximal end of the radius bone

radiale height

the vertical distance from the floor or other reference surface to radiale
  Comment: measured with the individual standing erect and the arms hanging naturally at the sides

radiale – stylion length

the linear distance from radiale to stylion parallel to the long axis of the freely hanging lower arm; *syn.* forearm length, lower arm length
  Comment: measured with the individual standing, the arm hanging naturally at the side, and the palm facing the thigh

radian

a planar angular measure in which the arc length of the subtended angle at the center of a circle equals the radius (approximately 57.3 degrees)

radiance (L)

the value of the ratio of the radiant flux

to the solid angle and the perpendicular surface projection

radiant emittance

see *radiant exitance*
   Comment: an older term

radiant exitance (M)

the radiant flux density leaving a specified surface; *syn.* radiant emittance

radiant flux

the rate of flow of radiant energy per unit time

radiant flux density

that amount of radiant power which flows onto or through a unit area

radiant heating

that heating which occurs solely by radiation

radiation

the emission and propagation of any form of energy; see *electromagnetic radiation, sound radiation*

radiation absorbed dose (rad)

a unit of measure for the amount of any type of ionizing radiation absorbed; equals 100 ergs per gram of material; see also *gray*
   Comment: replaced the roentgen as the dosage unit

radiation accident

any accident exposing humans to excessive radiation or involving the spread of radioactive materials beyond their intended containment

radiation dosimetry

a measurement of the amount of ionizing radiation exposure by individuals, materials, or equipment at a specific location

radiation effect

any of the documented effects from high levels of ionizing radiation exposure, such as hair loss, cancer, nausea, cataracts, or death; see also *radiation sickness*

radiation fluence

see *fluence*

radiation length

the mean distance required to reduce

the energy of charged particles by 1/e in passing through a material

radiation monitoring

a form of environmental monitoring in which periodic or continuous measurments are taken to determine the radiation levels present in a specific environment

radiation protection

any measure to reduce the exposure of humans and/or equipment to radiation, whether through legislation, regulations, policies, or physical measures

radiation safety

the study and/or implementation of equipment and procedures to prevent excessive radiation exposure to personnel or radiation release to the environment

radiation sickness

any form of illness resulting from exposure to large amounts of radiation, whether from radiation therapy, a nuclear explosion, or an accident, and usually consisting of symptoms including nausea and vomiting; *syn.* radiation syndrome; see also *radiation effect*

radiation syndrome

see *radiation sickness*

radiation therapy

the use of radiation to treat disease

radiator

any device which radiates energy

radical

see *free radical*

radio frequency (rf)

the portion of the electromagnetic spectrum normally used for communications, from about 10 KHz to 100 GHz

Radio Automobile Communications System

a Japanese government-private industry project to develop future street/highway vehicle systems

radioactive contamination

contamination with radioactive material(s)

radioactive decay
the emission of a particle or photon from an unstable nuclide as the nuclide is transformed into one or more different nuclides

radioactive decontamination
the removal of contaminating radioactive materials from a given location

radioactive half-life
that time required for one half the original amount of unstable nuclei in a radioactive sample to decay; *syn.* half-life

radioactive waste
any form of radioactive material which is no longer useful or cannot be processed for use

radioactivity
that activity in an unstable nuclide which results in radioactive decay; *adj.* radioactive

radiography
the art, technique, and/or process of making a film record from X-rays; *syn.* roentgenography

radioisotope
an unstable nucleus which exhibits radioactive decay; *syn.* radionuclide

radioluminescence
the emission of visible light from ionizing radiation or the decay of radioactive isotopes causing the excitation of crystals or phosphors

radiometer
an instrument used for measuring radiant electromagnetic power within or near the visible spectrum; *syn.* radiometric photometer

radiometric photometer
see radiometer

radiometric unit
a unit of measurement for radiant electromagnetic energy in terms of energy or power, without regard to biological effects

radiometry
the study or measurement of radiant energy; *adj.* radiometric

radionuclide
see radioisotope

radius [1]
the linear distance from the center of a circle to its edge; *adj.* radial

radius [2]
the lateral forearm bone

radius of curvature
the radius of a circle whose arc matches a curve or surface at a given point

railing
a rail and its supports which form a barrier

rainbow passage
a statement to be read by the wearer of a respirator as part of fit test to generate a wide variety of facial muscle movements

ramp
a lightly sloped surface which serves as a mobility aid for handicapped individuals or movement of goods to a different height or level, usually adjacent to a set of steps or stairs

ramus
the posterior, vertical portion of the mandible; *pl.* rami; *syn.* ascending ramus

Rand formula
an agreement under which employees pay dues to a union without being required to join the union

random [1]
determined only by chance

random [2]
see non-deterministic

random error
a non-systematic error, which may only be predicted statistically

random noise
an oscillation whose instantaneous magnitude cannot be specified other than probabilistically within some range for any given time instant

random observation method
see activity sampling

random process
a collection of all possible sample func-

tions which might be produced by random phenomena; *syn.* stochastic process, random signal

random sample

a sample in which each member of the population has an equal probability of selection; *syn.* simple random sample

random signal

see random process

random variable

a real function whose value is determined randomly; *syn.* chance variable, stochastic variable

randomize

select or assign randomly; *n.* randomization

randomized design

an experimental design in which the subjects are assigned randomly to groups representing different conditions or levels of the independent variable

range [1]

the upper and lower limits of a set of values or the difference between those values

range [2]

the distance required to stop an ionized particle; *syn.* ion range

range of motion (ROM)

the spatial extent through which a combination of joints, limbs, or links can be normally moved; *see also* **joint range of motion**

rank

(v) place a set of scores or other numeric variables in rank order

rank [1]

(n) the location of a score in a set of rank-ordered scores

rank [2]

(n) the maximum number of linearly independent rows in a matrix

rank correlation

see rank-order correlation

rank order

a sequence in which objects are organized sequentially by their quantitative score on some descriptor, variable, or parameter

rank-difference correlation

see Spearman rank-order correlation

rank-order correlation

any non-parametric correlation test for significance between at least ordinal-level paired observations in a random sample; *see Kendall rank-order correlation, Spearman rank-order correlation*

rapid eye movement (REM)

any short, quick movement of the eyes, especially that occurring during the rapid eye movement phase of sleep

rapid eye movement sleep

that phase of sleep during which the eyeballs can be observed to move rapidly and the EEG resembles an awake, alert state; *syn.* REM sleep

Comment: typically thought to be involved with dreaming

rapture of the deep/depths

(sl); *see nitrogen narcosis*

Comment: an older term

rarefaction

the momentary reduction in pressure during the trough in a sound wave

raster display

a video- or CRT-type display in which the screen is written in a standard, preprogrammed sequence

rate [1]

(n) the frequency at which a certain event or circumstance occurs within a specified or commonly understood time period

rate [2]

(n) the quantity of output produced, expressed as either per unit time or percent of capacity/normal

rate [3]

(n) *see speed*

rate [4]

(n) *see wage rate, piece rate*

rate [5]

(v) judge the relative or absolute amount of some quality of an entity or process, using some scale or other basis

rate change [1]

any alteration in a production time or output standard

rate change [2]

any alteration in worker compensation, whether time-based or output-based

rate control

the ability to adjust a controlling device as a function of changes in the velocity of a continuously-moving object or pattern; *syn.* first-order control, velocity control

rate cutting

an arbitrary reduction in the incentive pay rate or scale

rate determination

the study and analysis to decide what a standard time or incentive pay rate should be

rate of closure

the rapidity with which different pieces of information can be integrated into a meaningful pattern; *syn.* speed of closure

> *Comment:* the pattern is initially unknown

Rate of Perceived Exertion

see Rating of Perceived Effort

rate setting

the establishment of standard time values or any monetary pay scale for a given operation, based on a rate determination

rated activity sampling

a more detailed activity sampling in which a rating is determined for each work element to establish the work content in addition to that time occupied by delays and other activities

rated average element time

see normal element time

Rated Perceived Exertion

see Rating of Perceived Effort

Rated Perceived Exertion Scale

see Rating of Perceived Effort Scale

rating [1]

that class or level at which an operator is qualified

rating [2]

that assessment of a worker's pace or output relative to the standard pace or output

rating factor

that level of skill and effort displayed by an operator during the period of study, based on 100% as normal skill and effort

Rating of Perceived Effort (RPE)

an individual's estimate of how hard a task is, in terms of very, very light to very, very hard; *syn.* Rate of Perceived Exertion, Rated Perceived Exertion, Rating of Perceived Exertion; *see also Rating of Perceived Effort scale*

Rating of Perceived Effort scale

a 15-point scale ranging from 6 to 20 which is used by an individual to report the amount of effort/exertion he is putting forth; *syn.* Borg Scale, Rating of Perceived Exertion scale; *see Rating of Perceived Effort*

Rating of Perceived Exertion

see Rating of Perceived Effort

Rating of Perceived Exertion scale

see Rating of Perceived Effort scale

rating scale

any rank ordering scale for recording worker performance

ratio

a mathematical relationship between two numerical variables or values in which they are expressed as a fraction (as A/B) and may be evaluated as a quotient

ratio scale

a basic measurement scale meeting the criteria of an equal-interval scale and in which a known valid zero exists such that the ratio of numerical measures can be interpreted as a ratio of their magnitudes

ratio-delay study

a study in which a large number of

instantaneous work samples are taken randomly

Raub scale

see *computer anxiety scale*

raw material

an unprocessed material or material which has undergone some preliminary processing elsewhere which is used as input to a processing operation

raw time

see *observed time*

ray tracing

the calculation of a light ray path through an optical system

Rayleigh disk

a thin, lightweight, circular disk made of sound-reflective material which is mounted at 45° to the incident sound and used to determine sound wave particle velocity from the torque induced on the disk; also Rayleigh disc

Rayleigh ratio

the ratio of the intensity of incident light to the intensity of scattered light at a specified distance

Raynaud's phenomenon

an occupational disease involving a constriction of the blood vessels in one or more fingers or toes, resulting in numbness, a tingling sensation, or a reduction of prehensile ability; *syn.* dead hand, white finger, vibration white finger

Comment: generally due to excessive vibration in the use of heavy vibrating tools

re-entry

a return to the work force or other aspect of society following a period of absence

reach [1]

a straight-line anthropometric measure taken along the longitudinal axis of a body extremity

reach [2]

the capability of achieving a certain point in space through any self-directed orientation of limbs

reach [3] (R)

a work element involving the use of any combination of flexion, extension, or other joint movements to attain a point in space for some operation

reach envelope

the volume or solid bounded proximally by the body, clothing surface, or proximal reach and by the distal boundary in any direction which an individual or robotic device can attain by any combination of postures, rotations, and flexion and/or extension movements under specified conditions; *syn.* span of reach

reach from wall, maximum

the distance from a wall to the tip of the middle finger

Comment: measured with the individual's back and the contralateral shoulder pressed against the wall, the shoulder of the arm being measured held as far forward as possible, with the arm and hand extended horizontally

reactance platform

see *force plate*

reaction time (RT)

that interval of time required for a response initiation to occur following the presentation onset of a stimulus; *see also response time, choice reaction time, cognitive reaction time, simple reaction time*

reaction time delay

a time factor used in the modeling of man-machine systems to simulate the human reaction to an event

reactionless tool

any tool which compensates internally for forces or torques induced as a result of its actions such that there is essentially no external force or torque applied to the user

reactions inventory

see *response inventory*

read-out

a meter reading or other form for displaying visual information to an individual; *see also* **call-out**

readability

a quality of text or numbers which allows groups of related alphanumeric characters to be easily discriminable and recognized as words or number sequences; *syn.* legibility

readiness potential

a change in cortical potential in the motor cortex just prior to a planned, volitional movement

readiness time

that period of time necessary for a system to be prepared from a specified, inactive state to a state in which it is stabilized and can perform its intended function

reading point

see **breakpoint**

reading radius

the distance from the front of the eyeball to the object or display to be read

reading speed

the number of words of text read per minute by an individual

real ear attenuation at threshold (REAT)

a methodology for determining the attenuation provided by a hearing protection device fitted in a prescribed manner on a group of normal wearers

real time [1]

having essentially no perceptible delay between the occurrence of an event and the knowledge of the event at another location; also real-time, realtime

real time [2]

having a control system which delivers the necessary inputs to the system being controlled at or prior to the times they are required

realistic job preview (RJP)

a pre-employment description or new-employee orientation in which accurate and precise information about the job is related to the individual

realizable

pertaining to a component or system which is physically manufacturable,

not merely theoretical

realization rate

that proportion or percentage of actual annual hours worked by all employees in a workplace or company compared to the planned or expected normal annual hours to be worked

rear projection

pertaining to a display which is presented to one side of a translucent screen for viewing from the other side

reasonable accommodation

that modification to the workplace or other environment which enables a qualified handicapped individual to work in a given situation and does not impose an undue hardship on an employer

recall

(n) a statement by a manufacturer or distributor of some item that it may have one or more defects and should be returned for replacement or repair

recall

(v) access information stored in memory and output that information

receiver operating characteristic (ROC)

a graphical presentation of detector performance in signal detection theory, including the combined effects of sensitivity and response bias on operator performance; also receiver-operating-characteristic; *see* **signal detection theory**

receiving

that organization or those activities involved in the receipt and distribution within the plant of raw materials, equipment, and supplies from external sources

recency error

an error due to recent events which bias a rating or other measure; *syn.* recency of events error

recency of events error

see **recency error**

receptive field

a region of the skin, retina, or other

structure having an extended sensory apparatus within which appropriate stimulation affects the response of a given sensory neuron

receptor
any structure intended for receiving chemical and/or physical sensations from the environment, particularly a neural structure

recessed
pertaining to any device or object which is embedded, either entirely or in part, within another structure such that the device or object is still visible

reciprocal color temperature
a chromaticity measure which more nearly provides equal perceptible divisions of color temperature

reciprocal megakelvin (MK⁻¹)
see microreciprocal kelvin

reciprocating pedal
a foot-operated device which operates as a member of a pair of pedals by moving in opposing directions about a common shaft with a rotational capability limited by the pedal radial motion

reclining
pertaining to a posture involving an intermediate position between sitting and lying in which the posterior aspect of the body is against some reference surface, with the torso approximately between 45° and horizontal, and some degree of flexion of the hips and knees

recognition
the process in which an observer interprets or computer processor matches with data in memory the information available from a stimulus or object to arrive at a conclusion about the stimulus or object

recognition time
that temporal interval required for a particular stimulus to be recognized

recognized hazard
any hazard declared by OSHA to be such, as well as any hazard known or

suspected to exist within a workplace by management

Recommended Daily Allowance (RDA)
that amount of a specified mineral, vitamin, or other substance which is recommended for normal health

recompression therapy
the treatment for decompression sickness using a hyperbaric chamber to increase the ambient pressure

recoverable light loss factor
any light loss factor due to conditions which can be remedied through activities such as maintenance, normal servicing, or cleaning, specifically including lamp burnout, lumen depreciation, luminaire dirt depreciation, and room surface dirt depreciation; *opp.* non-recoverable light loss factor

recruit
search for possible new employees

rectangular coordinate system
an n-dimensional coordinate system composed of n perpendicular axes, where n is any positive integer value; *syn.* Cartesian coordinate system

rectilinear
pertaining to a straight line

rectilinear motion
that type of motion in which every point on a body moves in a straight line; *syn.* translational motion, linear motion

rectilinear teleoperator
a teleoperator having the capability for rectilinear motion, and which is usually restricted to that motion

rectum
a segment of the large intestine just proximal to the anal canal; *adj.* rectal

rectus abdominis muscle
a voluntary skeletal muscle which has a vertical extent and is located over the abdomen just lateral to the midline of the body

recumbent anthropometry
the taking or study of anthropometric measurements taken from the very

young, disabled, or persons in a lying, supine, or reclining posture, with the variations necessary to obtain or use the anthropometric measures

recumbent length

the length of the body from the top of the head to the bottom of the heels

Comment: measured with the individual lying on a flat, hard surface (e.g., a recumbent-length table), the head positioned such that the line of sight is vertical, the hips and knees fully extended, and the longitudinal axis of the feet vertical

recumbent-length table

a hard, flat surface, which is wider than the shoulders and longer than the body stature, and which has a fixed headboard and a sliding/removable baseboard for measuring lengths while the subject is lying down; also recumbent-length board

recycled air

that portion of the atmospheric gases retained in a ventilation system for recirculation

red

a primary color, corresponding to that hue apparent to the normal eye when stimulated with electromagnetic radiation approximately between wavelengths of 650 nm to 725 nm

red blindness

see protanomaly, protanopia

red blood cell (RBC)

see erythrocyte; syn. red blood corpuscle

red blood corpuscle

see erythrocyte

red marrow

that type of marrow which produces certain types of blood cells, including erythrocytes

red muscle

a skeletal muscle which appears red in the fresh or living state, probably due to muscle hemoglobin and cytochrome, and which have a longer latency than white muscle; *syn.* slow twitch muscle; *opp.* white muscle

red-green blindness

see protanopia, protanomaly, deuteranopia, deuteranomaly, color deficiency, color blindness

redout

a condition in which vision appears blurred by a red mist due to centripetal (negative *g*) accelerations; also red-out

reduced comfort boundary

a set of limits representing the maximum exposure time for whole-body vibration in different frequency ranges

redundant

having more than one mechanism for accomplishing the same result; *n.* redundancy

reference lot

a lot consisting of selected components and used as a standard

reference standard

see secondary standard

referent power

the ability of management to gain support for a project because personnel are personally attracted to the manager or interested in the project

referred pain

a pain which is felt at one location, but which represents a disorder at another location

reflectance (ρ)

the ratio of the energy reflected from a surface to the energy incident on that surface; *syn.* luminous reflectance

reflected glare

that glare due to specular reflections from glossy or semi-glossy surfaces within the field of view

reflected light

that light which leaves an object or surface from the illuminated side

reflection

the return of that energy flux incident on an object or surface

reflection coefficient (α_r)

the ratio of the amplitude of a wave reflected from a surface or boundary to the amplitude of the wave incident on that surface or boundary; *syn.* coefficient of reflection

reflection law
see *law of reflection*

reflectometer
a photometer for measuring the reflectance of surfaces or materials

reflex
an innate biological stimulus-response mechanism

reflex angle
an angle between 180° and 360°

reflex arc
a neural chain consisting of a sensory receptor, an afferent neuron, an interneuron, a motor neuron, and an effector

reflexograph
a chart recorder for displaying graphically the magnitude of a musculoskeletal reflex

reflexometer
an instrument used to measure the force required to elicit a reflex

reflux esophagitis
a burning sensation in the thorax, generally due to stomach acid refluxing through the lower esophageal sphincter; *syn.* heartburn

refract
change the direction of an energy wave on passing from one medium to another in which the wave has a different velocity; *n.* refraction

refraction index
see *refractive index*

refractive index (n)
see *index of refraction*; also refraction index

refractory period
see *cellular refractory period, psychological refractory period*

regression
a statistical modelling technique for estimating the value of one variable from the value of one or more other variable(s) when a known correlation

exists between the sets of variables; see *linear regression, multiple regression, nonlinear regression*

regression coefficient
the weight or coefficient preceding each predictor variable in a regression equation

regression curve
a graphic curve or an equation for a non-linear function which may be used for predicting the value of an unknown variable from the known value of another

regression equation
an equation in which the value of a criterion variable may be predicted from one or more known predictor variable values

regression line
a graphic line or a linear equation representing a linear or approximately linear function which may be used for predicting the value of one variable from a set of known values

regression model
the use of regression techniques to explain projections

regular element
a job element which is performed at least once in each work cycle or operation; *syn.* repetitive element; *opp.* irregular element

regular reflection
see *specular reflection*

regular sampling
the continuous or intermittent (at a relatively fixed frequency) taking of a representative portion of the material being sampled

regular transmission
see *specular transmission*

regular transmittance
see *specular transmittance*

rehabilitate
use one or more forms of treatment in an attempt to restore some or all loss of capacity or develop residual capabilities to give an individual a maximal

state of independence; *n.* rehabilitation

rehearsal

the process of recycling information in working memory to maintain it within working memory or to store it in long term memory

Reid's base line

an imaginary line defined by the location of the auricular point and the lowest point of the orbit on the same side of the head, with an anterior-posterior extent

reinforcement

a meaningful reward or punishment after a response which results in a strengthening or weakening, respectively, of that response; *see positive reinforcement, negative reinforcement*

Reissner's membrane

see vestibular membrane

reject allowance

a special time allowance provided a worker for processing rejects from a process or for rework

relationship

an interdependence between two individuals, activities, or entities

relationship chart

a table which details what the response for carrying out some task should be for several possible situations

relationship diagramming

the process of examining various plant layouts with the intent to optimize according to the closeness of relationships within an organization

relative biological effectiveness (RBE)

a factor used for comparing the radiation damage for different types of ionizing radiation, typically

  1 X-rays, gamma rays, beta particles
  2 thermal neutrons
 10 fast neutrons, protons
 20 alpha particles

relative humidity (RH)

a dimensionless number representing that amount of water vapor present in the air relative to the water vapor ca-

pacity of air at a specified temperature

relative luminosity

see spectral luminous efficiency

relative luminosity factor

see spectral luminous efficiency

relative maximum

the highest value within a restricted region of a curve; *see also absolute maximum*

relative minimum

the highest value within a restricted region of a curve; *see also absolute minimum*

relative mode

an operational state in which cursor movement is a function of its original position

relaxation allowance (RA)

see rest allowance

relaxed reach

a reach which can be attained without having to strain or requiring maximal flexion/extension of limbs

relearning time

the time required for a previous user to re-achieve a previous level of competence following a period of non-use of a skill or training

release load (RL)

a work element in which an object is released

relevance tree

a means of organizing and presenting interrelated variables or problems in graphic form for better visualization through the use of nodes and branches

reliability [1]

the likelihood that a part, piece of equipment, or a system will perform an intended function under certain conditions for a specified period of time or at a specified time

reliability [2]

a measure of the likelihood of obtaining similar results on repeated testing or observations

reliability coefficient

the correlation coefficient between two successive performance samples; *syn.* coefficient of reliability

rem

see roentgen equivalent man, rapid eye movement; also REM

REM sleep

see rapid eye movement sleep

remedial maintenance

see corrective maintenance

remnant [1]

that portion of raw material, part, or component remaining after some process

remnant [2]

that portion of an actual system output which is unaccounted for or unexplained by the system model

　　Comment: usually referring to linear models

remote control

the control of a machine or process from a position not located physically at or near the actual system; *see also teleoperation*

remote handling

the transference of manipulative skills from the human operator to a more distant region via some electromechanical linkage

remote indicator

an instrument or display for showing conditions at a point some distance away from the sensing device

remote monitoring

see telemetry

remote operation

the monitoring and/or control of a system performing some function at a distance from an operator or control center

renal

pertaining to the kidney

renege

that customer or user behavior in which an individual leaves the queue or line after waiting for some period of time

rep

see roentgen equivalent physical

repair

restore that which is inoperable or operable only at reduced capability to full capability by replacement of components, assemblies, or sub-systems

repairability

having a system which is capable of being restored to operating status within a given repair time interval

repeat

perform again; *n.* repetition; *adj.* repetitive

repeat rate

the number of times a function is performed within a specified period of time

　　Comment: as with a keyboard key when continuously depressed and the number of characters which are input per second

repeated measures design

an analysis of variance technique in which each subject is exposed to more than one condition; *syn.* within-subjects design, treatments-by-subjects design

repetition

see repeat

repetitive element

see regular element

repetitive motion disorder

see repetitive motion injury

repetitive motion injury (RMI)

any of a class of pathologies created through excessively frequent use of a particular joint or tissue, especially in combination with awkward positioning, inadequate or no rest periods, or excessive loads; *syn.* cumulative trauma disorder, repetitive strain injury, repetitive trauma disorder, repetitive stress injury, overuse syndrome

repetitive strain injury (RSI)

see repetitive motion injury

repetitive stress injury

see repetitive motion injury

repetitive time method
see repetitive timing

repetitive timing
a work measurement time study technique in which the duration of each work element is measured and recorded in sequence as it is performed, then the timing device is immediately reset to zero to begin timing the duration of the next element; *syn.* repetitive time method, flyback timing, snapback method, snapback timing

repetitive trauma disorder (RTD)
see repetitive motion injury

repetitive work
a work activity in which the work or task elements are continuously repeated over a prolonged period of time

replace
substitute one unit for another; return to its original position; *n.* replacement

replacement therapy
the use of synthetic substances or substances recovered from natural sources to substitute for an organ, gland, or other body structure which is no longer present or has ceased functioning

replicate
conduct an experiment which uses the same methodology but is independent of a previously-conducted experiment; *n.* replication

report generator
see report writer

report writer
software which can transform database, spreadsheet, or other information into a hardcopy format which is understandable by someone not familiar with the software; *syn.* report generator

reportable accident
see OSHA reportable accident

representative worker
an average worker in terms of skill and performance for the group under consideration

reproducibility
see reliability

reproduction rate
the ratio of the total number of births from women of reproductive age to the number of women within that age group in the population

required evaporation rate
the amount of water from sweat or other sources which must evaporate from the body surface into the atmosphere per unit time to maintain the body's heat balance

required sweat rate
the volume of sweat per unit time which must be secreted to assure adequate body cooling

requirements contract
a legal agreement for a supplier to provide and for a buyer to purchase one or more types of products or materials for a specified period of time, usually at a specified price and delivery terms

research
(v) investigate, ideally using accepted scientific techniques, with the intent of discovering previously unknown facts, relationships, and laws; *n.* research

research and development (R & D)
the process of attempting to find new knowledge for commercial use in creating a new product or improving current products

reserve
that amount of energy, strength, heart rate, or other quantity which is available to an individual but which is not being used in the current activity

residual air
see residual volume

residual hearing
that capability for hearing retained by an individual with hearing loss

residual volume
the volume of air remaining in the lungs after a maximal expiration; *syn.* residual air

resilience
the ratio of energy given up on a recovery from deformation to the energy

required to produce the deformation in an elastic structure

resist-dyeing
a textile coloring technique in which a substance which prevents dyeing is applied to certain threads before dyeing to produce coloring patterns

resistance
see *behavioral resistance, electrical resistance, mechanical resistance*

resistance arm
that portion of a lever arm from the fulcrum to the point at which the resistance is applied

resistance heating
the use of electrical resistance to provide heat, normally within a relatively confined or localized volume

resistance strain gauge
a transducer constructed of a material which changes electrical resistance under stress or deformation; also resistance strain gage

resistance thermometer
a device containing a sensing element whose electrical resistance varies as a known function of temperature and thus can provide a temperature measurement

resolution
the number of horizontal and vertical pixels which are available for display on a screen at any given time

resolution acuity
the ability to distinguish small separation distances of separate stimuli as two or more stimuli, rather than one; see *two-point threshold, minimum resolution angle*

resolution angle
see *minimum resolution angle*

resolve
detect two distinct, separate sensory entities within the same modality as being separate; *n.* resolution

resonance
a phenomenon in which an external forced oscillation imposed on a physi-cal-mechanical system causes a maximal oscillatory response amplitude in that system; *adj.* resonant; see *resonant frequency*

resonant frequency
the frequency at which resonance occurs; also resonance frequency

resonate
cause to be in a state of resonance

resource
any entity which may be used or called upon in performing an activity, generally including some or all of money, land, structures, raw materials, machines/equipment, and personnel

respiration [1]
see *respire*

respiration [2]
the physical-chemical exchange of gases across tissues

respirator [1]
a device or system designed to protect the user from inhaling a hazardous atmosphere
Comment: usually a gas mask

respirator [2]
a device for helping an individual breathe who cannot adequately perform that function by himself due to disease or injury

respiratory capacity [1]
see *vital capacity*

respiratory capacity [2]
a measure of the ability of oxygen to combine with blood in the lungs and carbon dioxide with blood in the tissues for return to the lungs

respiratory coefficient
see *respiratory quotient*

respiratory disease
any pathological condition resulting from exposure to toxins, particulates, or biological agents in the respiratory tract

respiratory frequency
the number of respiratory cycles per unit of time; the inverse of the respiratory period

respiratory hazard

any airborne entity which may result in some form of respiratory disability either immediately or over time

respiratory irritant

any substance which irritates the respiratory system

respiratory period

the time interval between the beginning of two successive inspirations; the inverse of the respiratory frequency

respiratory quotient (RQ)

the value of the ratio of carbon dioxide volume excreted to the volume of oxygen consumed by an organism within a specified time interval; *syn.* respiratory coefficient

respiratory system

those structures and organs of the body involved in gaseous exhange with the environment, including the nose/mouth, pharynx, larynx, trachea, the lungs, and the diaphragm

respiratory tract

that part of the respiratory system through which air normally passes, the nasal cavities, the pharynx, the larynx, and the lungs

respire

breathe, inhale/inspire and exhale/expire atmospheric gases; *n.* respiration; *adj.* respiratory

respirometer

an instrument for measuring such respiration parameters as breathing rate and the force involved in breathing

response

an action or output of a system, usually occurring as a result of some stimulus or input

response flatness

having a constant output for a given input over a range of frequencies

response inventory

that entire set of kinesiological, physiological, and psychological responses available to an individual when presented with one or more stimuli; *syn.*

reactions inventory

response orientation

the ability to discriminate between two or more different stimuli and rapidly initiate the appropriate physical response

response time [1]

the temporal interval between a stimulus onset and the completion of that motor activity required for the response; *see also reaction time*

response time [2]

the temporal interval between the input of a signal to and the presentation of an output on a display from a computer system

response time [3]

the period of time required for an individual or organization to receive a service request, dispatch a worker, and arrive at the servicing location

responsibility

an obligation incurred by or assigned to an individual or organization to perform at a certain level and/or within a certain time

responsibility analysis

a determination of with whom the responsibility for making certain decisions regarding performance, use of resources, and other aspects should lie

rest allowance

that amount of time added to the basic time for completion of a task to permit the worker to recover from fatigue due to the task or working conditions and to attend to personal requirements; *syn.* relaxation allowance

rest for overcoming fatigue (R)

a work element (therblig) in which the worker is allowed to rest to overcome fatigue effects

rest pause

see rest period

rest period

a short interval of time, generally 5–15 minutes, during a work shift which is allocated to reducing or preventing fatigue and for which the worker is paid; *syn.* rest pause, coffee break, break time

resting potential

that voltage difference across a cell membrane under normal conditions without any exciting stimulus

Comment: generally a standard value for a given type of cell

restraint

any harness or other mechanical device intended to prevent or restrict unintended movement of some object, body part or the body as a whole, usually in response to vibration, or rapid acceleration or deceleration; *see active restraint, passive restraint*

restricted element

see externally-paced element

restricted work

see externally-paced work

retain

maintain information placed in memory; *n.* retention

retaliation

any act by an employer against an employee in response to some undesired action outside the workplace by the employee

retention

see retain

retina

the membranous lining on the interior of the posterior eyeball which contains the rods and cones for light transduction; *adj.* retinal

retinal disparity

see retinal image disparity

retinal illuminance

the luminous flux incident on the retina per unit area

retinal image

that portion of the field of view which is focused on the retina

retinal image disparity

any difference existing between the images formed in the two eyes when an object is viewed binocularly; *syn.* retinal disparity

retinal image size

that length and/or width of an external object as represented on the retina

retinal rivalry

see binocular rivalry

retrieval buffer

a temporary storage location from which the user may retrieve information after some action which would normally have deleted it

retro-torsion

twisting backwards

retroactive inhibition

that disruptive effect on the ability to recall information from a task by the imposition of an additional learning activity between the end of the primary learning task and the test for recall; *see also proactive inhibition*

retroflex

a consonant sound represented by the "r" as in read or beer

retrograde amnesia

the loss of memory for events preceding some event which caused the present condition

reverberant room

an enclosed volume which has highly sound reflective interior surfaces; *syn.* reverberation chamber

reverberation

the persistence of sound energy in a closed space due to multiple reflections after the source is no longer active

reverberation chamber

see reverberant room

reverberation time (T)

the temporal interval required for the average sound energy density of an abruptly terminated sound to decay 60 dB in a closed environment

reverse video

a highlighting technique in which the foreground and background colors are reversed for a segment of text or other portion of a display

rework

(v) reprocess to correct a defect or deficiency in a product, either before or after inspection

rheumatism

any disease with pain referred to the musculo-skeletal system, most commonly in the area of a joint, accompanied by stiffness

rho (ρ)

the correlation coefficient in the Spearman rank-order correlation test; *see* **Spearman rank-order correlation**

$$\rho = 1 - \frac{6 \sum D^2}{N(N^2 - 1)}$$

where:

D = the difference between the scores within a pair

N = number of pairs of scores

rhodopsin

a red photosensitive protein pigment in the rods of the eye and involved in light transduction under low light level conditions; *syn.* visual purple

rhyme test

any test, usually multiple choice, in which an individual's task is to select from a list of possibilities the word he believes was presented auditorily

rib

one of a set of bones which connect the vertebrae and the sternum, and which collectively enclose the thorax

ribonucleic acid (RNA)

a large cellular molecule, consisting of ribose, a phosphate group, and a cyclic organic base, which is involved in information transfer from DNA and protein synthesis within the cell

Ricco's law

a rule that, for small targets, the threshold intensity for detecting a target varies inversely with the size of the target

ride

see **ride quality**

ride quality

a measure of the amount of motion experienced in a vehicle by a crewmember or passenger; *syn.* ride

rider's bone

a calcium deposit in the adductor muscles of the leg in horse riders due to prolonged pressure of the thigh against the saddle

right- and left-hand process chart

see **two-handed process chart**

right-of-way [1]

the legal priority of one vehicle, pedestrian, or other object to proceed over another in the event of possible conflict

right-of-way [2] (ROW)

that land adjoining highway, roadway, or other public property or which provides a passage for vehicles, pipelines, and power or communication lines

right-to-work

pertaining to any designation, law, or legislation permitting employment by other than labor union members

rigid body

a modeling structure which is not capable of deformation on the application of forces

ring finger

digit IV, adjacent to the little finger, often referring specifically to the left hand

Rinne test

a method for detecting conductive hearing loss in which a vibrating tuning fork is brought into contact with the skin over the temporal bone behind the ear, then, when the sound can no longer be heard via bone conduction, the tuning fork is moved just anterior to the external auditory canal; *see also* **tuning fork test**

Comment: an individual with middle ear conductive hearing loss will not hear the tone after the change in position, while an individual with normal hearing will

rise angle

the angle which a stair makes with the ground or reference level; determined by the ratio of rise height to tread depth

$$\theta = \tan^{-1}\left(\frac{riser\ height}{tread\ depth}\right)$$

rise time
that temporal interval required for a signal or system output to reach a specified percentage of its peak amplitude; *see also time constant*

riser
the vertical portion of a stair step

risk
that uncertainty of attaining a desired goal or potential for experiencing some loss in undertaking some action

risk analysis
the process of determining what risk factors are present, their characteristics, and estimating the probability of success or failure

risk factor
any substance, disability, or environmental condition which increases the probability of an accident, disease, or failure

risk management
the process of regulating exposure to hazards, deciding which risks are acceptable, selecting alternative approaches under a given set of circumstances, and considering the consequences

Risley prism
a serial mounting of two thin, equivalent prisms with apposing faces which can be rotated individually about their common longitudinal axis
Comment: used for testing ocular convergence

ritual
a set of stereotyped actions which are believed to bring about some desired result

riveting hammer
a mechanical pounding device used for driving rivets

robot
an electromechanical device which may be equipped with sensing and reacting instrumentation and/or equipment, some calculating ability with a set of preprogrammed responses determined by the calculation results, some form of

mobility, and the ability to operate autonomously, at least for short periods of time

robotics
the study, design, manufacturing, development, and/or the use of robots

rocker switch
a manually-activated two-position toggle switch in which two faces are separated by an obtuse angle, and where one of the faces is normally depressed

rod [1]
a rod-shaped photosensitive cell which predominates in the peripheral retina and is used in scotopic vision

rod [2]
an ideal long, flat, thin structure which forms part of a lever system

roentgen (R)
a unit of exposure to X-rays or gamma rays equivalent to the absorption of 0.000258 coulombs per kilogram of air under standard conditions; also röntgen; *see Sievert, gray, radiation absorbed dose*

roentgen equivalent man (rem)
the equivalent dose from any type of ionizing radiation which produces a biological effect of 1 roentgen of X-rays or gamma rays in man

roentgen equivalent physical (rep)
a unit of ionizing radiation corresponding to that amount which results in soft tissue absorption of 93 ergs per gram

roentgenography
see radiography

roll
have or cause rotation about a longitudinal or fore-aft axis of a vehicle

rollbar
a heavy metal tube formed to approximate the cabin boundary of a vehicle to prevent crushing of the cabin for occupant protection in the event of the vehicle rolling over; also roll-bar

roller ball
see trackball

rolling ball
　see **trackball**

room criterion curve (RC)
　a sound pressure level guide for heating, ventilation, and air conditioning systems designed to produce a bland, non-disturbing sound

room surface dirt depreciation (RSDD)
　the reduction of reflected light from room surface walls, ceilings, and floors due to accumulated dirt; a recoverable light loss factor

root mean square [1]
　the square root of the square of the average of a time-varying signal, usually taken over one cycle if a periodic waveform; syn. effective value

root mean square [2] (rms)
　the positive square root of the average of the square of a data set

$$rms = \sqrt{\frac{\sum_{i=1}^{N} X_i^2}{N}}$$

　where:
　X_i = data point
　N = sample size

root mean square sound pressure
　see **effective sound pressure**

rotary pedal
　a foot-operated device which operates with a pair of pedals moving in a common direction about a shaft capable of continuous rotation

rotary selector switch
　see **rotary switch**

rotary switch
　a switch which operates by being turned about a central shaft; syn. rotary selector switch

rotate
　cause or experience an angular change in position of a non-axial point about one or more axes of an object; n. rotation; adj. rotary, rotating

rotating shift
　a work schedule that has one or more

individuals who work one shift for a period of time, then another shift, etc. in a cyclic manner; syn. shift rotation; see **clockwise rotating shift**, **counterclockwise rotating shift**

rotational acceleration
　see **angular acceleration**

rotational axis
　a defined line about which instantaneous angular motion takes place; also axis of rotation

rotational dynamics
　the study of the causes of rotation

rotational inertia (I)
　see **moment of inertia**

rotational kinematics
　the description of rotational motion, without regard to cause; see also **rotational dynamics**

rotational velocity
　see **angular velocity**

rotator
　any muscle which moves a bone around its longitudinal axis

rotator cuff
　a combination of the three muscles involved in rotating the arm and shoulder, the subscapularis, supraspinatus, and teres minor

round shoulders
　a posture in which the shoulders are drooped forward and the thoracic spine has increased convexity

round window
　a round-shaped osseous opening in the temporal bone which is covered by the secondary tympanic membrane, forming the terminus of the scala tympani of the cochlea at the junction with the middle ear

routine maintenance
　the performance of one or more maintenance functions after a certain period of time or after a certain amount of use; syn. scheduled maintenance, planned maintenance

row
　the horizontal set of numbers in a ma-

trix; the horizontal alphanumeric characters in a table

ruby

a gemstone consisting of a high quality red corundum

ructus

a condition in which gas is belched from the stomach

ruggedized

capable of withstanding a certain level of mechanical shock or vibration without damage to the unit or its components

rugitus

a condition in which the intestines make a rumbling sound

rule

an established guide or procedure for action

rule of 80-20

see *Pareto's law*

rule-based behavior

a cognitive operating mode in which the individual consciously attempts to perform some task in a situation for which clearly pre-established rules exist

rump

(sl) the region of the body near the bottom of the spine, including the buttocks

run

a type of gait in which both feet may be off the ground simultaneously within a stride cycle, generally consisting of the following phases: (a) support phase: foot strike, midsupport, takeoff; (b) recovery phase: follow-through, forward swing, foot descent; see also *gait*

run-in phase

that time in which a machine or system is operated either when new or after a period of maintenance and in which the probability of failure is highest; *syn*. infant mortality phase; see *life characteristic curve*

run-out time

the time period required by machine tools between the point at which cutting completion occurs and the point at which the tool and materials are free of interference so the next operational sequence can begin; also runout time

ruptured disk

a condition in which the nucleus pulposus of the intervertebral disk protrudes through the surrounding fibrous tissue; also ruptured disc; *syn*. herniated disk

rutherford (rd)

a unit for radioactive decay; equals 10^6 disintegrations per second

S

sabin
a unit for the acoustic absorption capability of a material

saccade
a rapid movement of the eyes from one fixation point to another; *syn.* saccadic eye movement

saccadic eye movement
see saccade

saccule
an expanded chamber within the vestibular apparatus

sacral spine
see sacrum

sacrospinalis
a collection of muscle groups located in the back and aligned parallel to the spine; *syn.* spinal erector
Comment: for maintaining erect posture, extending the spine, and bending the spine to the side

sacrum
the triangular bone near the base of the spine formed by the fusion of the five sacral vertebrae; *syn.* sacral spine; *adj.* sacral

safe
relatively free from the risk of danger, injury, or damage; *ant.* unsafe

safety
the practice of eliminating or minimizing and/or the freedom from conditions which may cause injury or death to personnel, damage to or loss of equipment or property, and/or loss of time

safety belt
any strap-like object worn about the waist region which attaches an individual to a secure structure and is intended to reduce or prevent injury in the event of an accident

safety coupling
a friction coupling set to slip at a controlled torque, thereby protecting the remainder of a system from overload

safety cut-out
a device to protect from overload in an electrical circuit

safety education
the transmission of information, skills, and attitudes dealing with environmental safety requirements to interested parties with the intent of producing favorable behavior changes

safety engineering
that engineering discipline concerned with system components and processes to achieve an optimal safety situation for both personnel and property

safety factor
see factor of safety

safety glass
any glass which is shatterproof or which breaks into granules rather than sharp strands

safety hat
see safety helmet

safety helmet
a hard, rigid headgear designed to protect a worker's head from impact, electric shock, and/or flying particles; *syn.* safety hat, hard hat

safety job analysis
see job safety analysis

safety lock
any lock which can only be opened by using its own key

safety margin
see margin of safety

safety professional
any individual who has achieved professional status in the safety field,

whether due to specialized knowledge, skills, and/or educational background; see **Certified Safety Professional**

safety rule

a written requirement stating personal protective gear to be worn, safe behavior, or other safeguards to be taken in certain activities

safety shoe

any of a variety of shoe types with safety features specific to certain environments, such as steel-toes, steel-soles, or spark-preventing design

safety stock

that inventory comprised of reserves in the case of unanticipated events which can prevent resupply

safety tongs

a gripping device with an extended arm/reach for placing objects into or removing them from a hazardous area

safety training

any training associated with the safety aspects of the home, job, workplace, or other aspects of living or working, possibly specifically including any potential hazards and their relationships to a particular individual or group

Saffir/Simpson Hurricane Scale

a five category scale for estimating the expected property damage and flooding associated with a hurricane; *syn.* Hurricane Scale

1 – winds of 74–95 mph, storm surge 4–5 feet above normal
2 – winds of 96–110 mph, storm surge 6–8 feet above normal
3 – winds of 111–130 mph, storm surge of 9–12 feet above normal
4 – winds of 131–155 mph, storm surge of 13–18 feet above normal
5 – winds of greater than 155 mph, storm surge greater than 18 feet above normal

sagittal

pertaining to any plane parallel and lateral to the midsagittal plane

sagittal arc

the surface distance over the top of the head from glabella to nuchale

Comment: measured over the hair with compression and with the scalp muscles relaxed

sailor's skin

that skin, especially on the back of the hands which has extensive pigmentation and may lead to squamous cell carcinoma; *syn.* farmer's skin

saliva

a secretion of various glands in the mouth which aid in the digestion of food

salt [1]

sodium chloride; *adj.* salty

salt [2]

any reaction product in the neutralization between an chemical acid and base

saltatory conduction

the jumping of the neural impulse along a myelinated axon from one node of Ranvier to the next

Comment: greatly increases the speed of neural impulse conduction

salty

having a taste resembling salt or sodium chloride

sample

(n) those units selected from the total population for study

sample

(v) select a sample, whether from a population, environment, or other entity; *ger.* sampling

sample size

the number of cases required or used within a sample

sampling error

the difference between the mean of a population and a sample mean

sanitize

reduce the number of microbial contaminants below some level

sarcolemma

the cell membrane of a muscle fiber

sarcomere

the basic longitudinal structural unit of a muscle cell

sarcoplasm
the cytoplasm of a skeletal muscle cell

saturation
see color saturation

scala tympani
the inferior portion of the osseus spiral of the cochlea

scala vestibuli
the superior portion of the osseous spiral of the cochlea

scalar
a quantity which has only a magnitude; *see also vector*

scale [1]
some proportion of size with respect to full- or normal-size

scale [2]
a device for measuring weight

scale [3]
a set of marks at measured distances

scalenus anterior syndrome
a sensation of weakness, numbness, or pain in the arm, due to compression of nerves and blood vessels supplying the arm by the scalenus anterior muscle; *syn.* scalenus anticus syndrome, thoracic outlet syndrome

scalenus anticus syndrome
see scalenus anterior syndrome

scan [1]
convert text and/or graphics from hardcopy form to electronic form; *n.* scanner

scan [2]
search a region of something for one or more particular details

Scanlon Plan
a system for encouraging productivity improvement as measured by the increase in the ratio of the total payroll cost divided by the total sales value of the items being produced and redistributing the gains in some proportion to the employees and the company

scaphoid bone
see navicular bone

scaphoid tubercle
a lateral protuberance on the scaphoid

bone which serves as an anatomical reference point in locating the wrist joint

scapula
a large, triangular-shaped flat bone, the superior-lateral portion of which forms part of the shoulder joint; *syn.* shoulder blade

scar
(v) set up or engineer/design a system/component for the installation of additional items in the future

scar
(n) a residual mark in a tissue after an injury or other invasive event

scatter diagram
a plot showing the locations of the individual data points within a coordinate system; *syn.* scatter plot

scatter plot
see scatter diagram

scheduled maintenance
see routine maintenance

sciatic nerve
a large spinal nerve innervating the thigh, lower leg, and foot

sciatica
pain along the sciatic nerve
Comment: often caused by herniated disk in lumbar or sacral spine

scintillation [1]
see wander

scintillation [2]
rapid variations in the brightness of a distant self-luminous object viewed through the atmosphere

scintillation [3]
a flash of light produced in a phosphor by a penetrating ionized particle or photon

scintiscan
a film or hardcopy output resulting from the scanning of intact internal tissues or organs for radioactive tracer

score
a quantitative indication of performance on some test or other measurement technique

scotoma
a totally or partially blind region in the visual field surrounded by a region of normal vision

scotopic
pertaining to relatively low illumination levels at which vision is mediated primarily by the rods; *opp.* photopic

scotopic adaptation
see dark adaptation

scotopic luminosity function
see scotopic spectral luminous efficiency function

scotopic spectral luminous efficiency function $(V'(\lambda), V_{\lambda'})$
a mathematical representation or curve describing the relative sensitivity of the eye to the wavelengths of light at low light levels; *opp.* photopic spectral luminous efficiency function

scotopic vision
that vision which occurs in low light levels or after dark adaptation using the retinal rods; *syn.* night vision, low light level vision; *opp.* photopic vision

screen [1]
(n) the face of a visual display monitor

screen [2]
(v) give one or more tests to an individual and determine whether he meets certain evaluation criteria

screen dump
direct the contents of the screen to a printer or file

scroll
advance vertically line-by-line or horizontally column-by-column under operator control to view text, alphanumeric information, or graphics on a screen; *see vertical scroll, horizontal scroll*

scrotale
the junction point of the posterior scrotum and the perineum

scrotale – cervicale, rear, sitting
the surface distance from scrotale up the back to cervicale
Comment: measured with the individual sitting erect

scrotale – cervicale, rear, standing
the surface distance from scrotale up the back to cervicale
Comment: measured with the individual standing erect

scrotale – midshoulder, frontal, sitting
the surface distance from scrotale up the front of the torso to midshoulder
Comment: measured with the individual sitting erect

scrotale – midshoulder, frontal, standing
the surface distance from scrotale up the front of the torso to midshoulder
Comment: measured with the individual standing erect

scrotale – midshoulder over buttock, sitting
the surface distance from scrotale, over the buttock and up the back to midshoulder
Comment: measured with the individual sitting erect

scrotale – midshoulder over buttock, standing
the surface distance from scrotale, over the buttock and up the back to midshoulder
Comment: measured with the individual standing erect

scrotale – midshoulder, rear, sitting
the surface distance from scrotale up the back to midshoulder
Comment: measured with the individual sitting erect

scrotale – midshoulder, rear, standing
the surface distance from scrotale up the back to midshoulder
Comment: measured with the individual standing erect

scrotale – scye level, frontal, sitting
the surface distance from scrotale up the front of the torso to the scye height
Comment: measured with the individual sitting erect

scrotale – scye level, frontal, standing
the surface distance from scrotale up the front of the torso to the scye height

Comment: measured with the individual standing erect

scrotale – scye level, rear, sitting
the surface distance from scrotale up the back to the scye height
Comment: measured with the individual sitting erect

scrotale – scye level, rear, standing
the surface distance from scrotale up the back to scye the height
Comment: measured with the individual standing erect

scrotale – suprasternale, sitting
the surface distance from scrotale up the front of the torso to suprasternale
Comment: measured with the individual sitting erect

scrotale – suprasternale, standing
the surface distance from scrotale up the front of the torso to suprasternale
Comment: measured with the individual standing erect

scrotale – waist level, frontal, sitting
the surface distance from scrotale up the front of the torso to the seated waist height
Comment: measured with the individual sitting erect

scrotale – waist level, frontal, standing
the surface distance from scrotale up the front of the torso to the standing waist height
Comment: measured with the individual standing erect

scrotale – waist level over buttock, sitting
the surface distance from scrotale over the buttock and up the back to the sitting waist height
Comment: measured with the individual sitting erect

scrotale – waist level over buttock, standing
the surface distance from scrotale over the buttock and up the back to the standing waist height
Comment: measured with the individual standing erect

scrotale – waist level, rear, sitting
the surface distance from scrotale up the back in the midsagittal plane to the seated waist height
Comment: measured with the individual sitting erect

scrotale – waist level, rear, standing
the surface distance from scrotale up the back to the standing waist height
Comment: measured with the individual standing erect

scrotum
the pouch in the male containing the testes and other tissues

scye [1]
the lower level of the axilla, represented by the highest point of the axillary fold

scye [2]
the armhole of a garment

scye circumference
the surface distance around the shoulder over acromion and through scye
Comment: measured with the individual standing or sitting erect and the arms hanging naturally at the sides

search (Sh)
a therblig for locating an article

seasonal affective disorder (SAD)
a disorder in which an individual suffers symptoms such as depression, lethargy, sleep disturbances, and weight gain during the winter months when daytime hours are reduced

seat
any structure for assuming a sitting or reclining posture which consists of a seatpan and a seatback which have a seatback angle of about 90° or more and is elevated off the floor or local other reference surface

seat angle
the angle of the seatpan plane above a horizontal reference with the origin at the seat reference point

seat belt
a strap or similar restraint about waist level for restraining a person at or below the waist in a seat; *syn.* lap belt

seat depth

the linear distance in the plane of a seatpan from its intersection with the seatback plane to the front-most edge of the seatpan of a chair; *syn.* chair depth

seat reference point (SRP)

the midpoint of the line formed by the intersection of the seatpan and the seatback

seatback

the back of a seat structure, for accommodating the human back; also seat back

seatback angle

the angle between the seatpan and the seatback; also seat back angle; *syn.* backrest-to-seat angle

seatback plane

that geometrical plane established by the seatback

seated

see sitting

seatpan

that portion of a seat on which the buttocks and thighs rest when sitting; also seat pan

second [1] (s, sec)

the time required for 9,192,631,770 radiation periods corresponding to the cesium-133 atom transition between the two ground state hyperfine levels

second [2]

an item not up to all specifications, but which may satisfy most uses

second quartile

see median

second shift

a late afternoon-evening work shift of about 8 hours duration, extending approximately between 3 P.M. and 1 A.M.; *syn.* evening shift, B shift

second-class lever

a lever system in which the fulcrum is at or near one end, the effort near the other end, and the resistance is located between them

secondary motor vehicle controls

those controls not critical to moving or stopping a motor vehicle

Comment: e.g., radio, turn signals, lighting

secondary positioning movement

that part of a positioning movement which brings the body member into an exact relationship with the point of aim; *syn.* corrective positioning movement, fine adjustment

secondary radiation

that ionizing radiation produced from the interaction of primary radiation with other materials; *see primary radiation*

secondary standard

any standard prepared by direct comparison to a primary standard; *syn.* reference standard

secondary task

a task which must be performed in addition to an individual's primary task; *see loading, non-loading secondary task*

secondary tympanic membrane

that membrane which covers the round window of the cochlea; *syn.* round window membrane

secondary viewing area

that portion of a computer display, screen, or window which contains information such as system function status or messages

secondary work

any activity which is not directly related to on-the-job productivity, but which must be performed by direct labor personnel to support the primary job

secular trend

a change in some parameter of a population that occurs gradually over long periods of time

sedentary occupation

an occupation involving a great deal of sitting and/or having a gross metabolic cost of not more than 65 calories per square meter of body surface per hour

sedentary work

that work which is normally accomplished in a sitting posture

seed

an inclusion impurity which resembles a tiny embedded pebble in glass

segmentation method

a technique for finding the center of gravity of the body by subdividing the body into segments/links and using their weights, lengths, and relative locations in space

select (St)

a therblig in which a choice is made from a set of possible options

selected element time

see selected time

selected time

that time value chosen as representative or expected for a set of work elements; *syn.* selected element time

selective attention

the ability to consciously or willfully focus on a restricted set of desired inputs, to the exclusion of the remaining concurrently impinging sets; *syn.* focused attention

selective transmittance

the transmission of certain wavelengths of electromagnetic energy through a transparent or translucent medium

self-contained breathing apparatus (SCBA)

a respiration system which includes a face mask, an air supply, and a regulator/delivery mechanism

self-paced job

a job which is under complete control of the worker

self-paced work

that work which is performed manually by a worker using simple tools or machines which are controlled by him such that the output rate/performance level is solely determined by the worker; *syn.* internally-paced work, man-paced work, effort-controlled cycle, unrestricted work; *opp.* external pacing

self-pacing

pertaining to self-paced work

self-protectivity

the capacity of a robotic or telerobotic system to protect itself from damage caused by its own activities

sellion

the point of greatest surface indentation at the base of the nose between the eyes

sellion height

the vertical distance from the floor to sellion; *syn.* nasal root height

Comment: measured with the individual standing erect, looking straight ahead, and the scalp and facial muscles relaxed

sellion to back of head

the horizontal linear distance from inion to sellion; *equiv.* sellion to back of head

sellion to top of head

the vertical distance from sellion to the vertex level

Comment: measured with the individual standing erect

sellion to wall

the horizontal linear distance from a wall to sellion; *equiv.* sellion to back of head

Comment: measured with the individual standing or sitting erect with his head against the wall and his scalp muscles relaxed

Selspot

an active motion tracking system which uses flashing infrared LEDs and optoelectronic sensors

semi-interquartile range (Q)

that range represented by one half the range from the first quartile to the third quartile

$$Q = \frac{Q_3 - Q_1}{2}$$

semicircular canal

any of three bony tubular structures making up part of the inner ear which are filled with perilymph and oriented at approximately right angles to each other

semicircular duct

any of the three tubular structures lying within the semicircular canals which are filled with endolymph and form part of the membranous labyrinth of the inner ear

semilunar notch

see trochlear notch

semimental basic element

any of a set of work elements which involve a mental activity component as well as a physical component

sensation

the detection of some stimulus by one or more sensory receptors

sense

any system through which information is acquired about the environment, normally referring to human or robotic abilities; *adj.* sensory

sensible

capable of being perceived by the senses

sensing time

the temporal interval required for a human operator or other controller to become aware of a signal

sensitivity

the ability to detect a specific condition/stimulus, or distinguish differences between conditions/stimuli

sensitivity analysis

a systematic study of those changes in results or output when one or more inputs are changed

sensitivity training

that training to improve an individual's appreciation for and consideration of his environment, especially with regard to the people in it and their feelings

sensitize

promote a greater response to succeed-

ing stimuli than was exhibited by the original stimulus; *n.* sensitization

sensor

that part of a transducer, including a human sensory receptor, which responds to incoming excitation

sensorimotor

pertaining to both sensory and motor capabilities, activities, or other interactions; *syn.* sensory-motor

sensorimotor integration

that central nervous system processing of sensory information and the resulting direction to motor regions which culminates in a motor response for the given input

sensorineural hearing loss

a form of hearing impairment caused by excessively loud sounds

sensory

see sense

sensory adaptation

an adjustment in the sensitivity of a sensory structure to compensate for the intensity or quality of a stimulus and attempt to maintain sensory effectiveness

sensory deprivation

the reduction in intensity or elimination of stimuli which an organism would normally receive, usually from the external environment

sensory disability

any disability involving the visual, auditory, olfactory, tactile, gustatory, proprioceptive, kinesthetic, or other bodily receptor apparatus

sensory feedback

any information received by a sensory system which may be used to indicate the quality of performance of a voluntary act and enable adjustments if desired or necessary

sensory homunculus

a representation of the human body on the surface of the cerebral somatosensory cortex whose distribution is proportional to the density of innervation in various parts of the body; *opp.* motor homunculus

sensory load

a measure of the number, rate, and variety of stimuli which must be perceived

sensory memory

a brief form of memory resulting directly from sensory input, and typically having a duration of a few seconds but which depends on the modality; *syn.* sensory storage; *see iconic memory, echoic memory*

sensory nerve

see afferent nerve

sensory overload

a condition in which the sensory load is too great for the human to process it effectively; *see load stress, speed stress*

sensory receptor

any neuron or specialized portion or a neuron which is capable of detecting some particular aspect of the environment

sensory-motor

see sensorimotor

septum

a dividing wall between two bodily cavities

sequence of use principle

a concept in equipment design that controls and displays should be placed such that their physical locations correspond with the order in which they are used; also sequence-of-use-principle

sequencing

the process of specifying or performing a series of tasks in a certain order

serial behavior

an integrated sequence of acts, usually leading to some goal or conclusion

serious injury

any disabling injury or a non-disabling injury of one of the following types: (a) an eye injury requiring treatment by a physician, (b) fractures, (c) any injury requiring admission to a hospital for

observation, (d) loss of consciousness, (e) any injury requiring treatment by a physician, (f) any injury which requires a work restriction or assignment to another job

serious injury frequency rate

the number of serious injuries per one million employee-hours of exposure in the working environment

$$SIFR = \frac{no.\ serious\ injuries \times 1,000,000}{no.\ employee\ exposure\ hours}$$

Comment: an old NSC measure

serum

that fluid remaining when the cells and fibrinogen are removed from blood plasma

servicability

a measure of the ease with which a machine or system can be serviced

service [1]

(v) replenish consumables

service [2]

(n) a labor commodity or product in which an individual or group performs some function for another

service test model

see mockup

service time

that period of time actually required for customer/user service

servomechanism

any device which senses the difference between a true state and some desired state and operates to bring a system toward the desired state

sesamoid bone

any of a class of bones which are small, have a generally rounded appearance, and are enclosed by tendons or other tissues at joints; *see patella*

Comment: may not be consistent across individuals

set [1]

the predisposition or readiness to perceive and/or respond only to certain situations or in certain ways to a given stimulus; *syn.* mental set

set [2]

any defined collection of objects or data

set up

prepare a workstation, facility, or work item for carrying out a specific job or task; *syn.* make ready; *n.* setup; *opp.* teardown

setup allowance

a special time allowance to cover a worker for the time required to prepare equipment or machinery for some process; *syn.* make-ready allowance; *opp.* teardown allowance

severity rate [1]

a measure for comparing severity of industrial injuries, based on the number of total days charged, based on a standard schedule, for work injuries per one million employee-hours of exposure

$$SR = \frac{no.\ total\ days\ charged \times 1,000,000}{no.\ hours\ of\ employee\ exposure}$$

Comment: an old NSC measure

severity rate [2]

see accident severity rate

Severity Index (SI)

a guideline for estimating the likelihood of injury from sudden accelerations; equals the integral over time of the acceleration curve duration (impulse)

sex

a classification category for the male and female of a species; *adj.* sexual; *see also gender*

sex ratio

the number of males per 100 females in the population

shade [1]

(v) to place in a shadow or shield from light

shade [2]

(v) apply a color to in order to darken

shade [3]

(n) an area or region shielded from direct light, especially sunlight

shade [4]

(n) any color which appears darker than a medium gray

shadow [1]

a literal copy of something, as a database

shadow [2]

that dark image of an object created by placing the object between a source of direct or specularly reflected light and a background

shadow [3]

a font in which the appearance of shading is present

shadow controller

a human operator who views the same display and uses the same type of hand controls as the real controller, but who has no connection to the active controls

shape coding

the use of different shapes on control surfaces for distinguishing both visually and tactually between control devices

shared time

see job sharing

Shaver's disease

see bauxite fume pneumonoconiosis

Sheldon somatotype

a body type classification system, in which men are divided into three basic groups: endomorph, mesomorph, and ectomorph

shielding

any physical structure, device, or other mechanism which can be used to deflect, absorb, or otherwise attenuate an undesirable entity

shift [1]

(n) that period of time within a given day when an individual is scheduled to be at work, usually about eight hours in length

shift [2]

(v) change from one location, position, level, or posture to another, often with respect to some fixed or reference point

shift maladaptation syndrome

any one or a combination of physical complaints of illness, injury, or fatigue resulting from an inability to adjust to a particular shift

shift rotation

see rotating shift

shift work

that work performed primarily by other than those working first shift, generally between the hours of 5 P.M. and 8 A.M., or that work performed by those working a rotating or other shift schedule; also shiftwork

shift worker

an employee who performs shift work

shin

the anterior portion of the lower leg

shin splint

a repetitive motion injury, typically from running or jogging, which results in pain in the anterior lower leg

shinbone

see tibia

shipping

those activities involved in distributing finished products to the markets

shock

see cardiovascular shock, electrical shock, psychological shock, mechanical shock

shoe

a form of footwear which has a hard sole, covers the majority of the foot surface, and whose uppermost portion is generally lower than the lateral malleolus height

shop rule

any of a set of regulations established either by an employer, by collective bargaining, or by a union constitution; *syn.* working rule

short-term

a brief period of time, which may range from seconds to minutes to days, depending on the magnitude of the referenced time scale

short-term exposure limit (STEL)

the OSHA-established maximum time weighted average concentration of certain specific hazardous substances in the working atmospheric environment to which a worker should be exposed for a specified period of time during a work day, typically 15 minutes

short-term memory

see working memory; also short term memory

shoulder

the joint which connects the arm to the trunk/body and all local tissues associated with that region

shoulder belt

see shoulder harness

shoulder blade

see scapula

shoulder breadth

see bideltoid breadth, biacromial breadth

shoulder circumference

the surface distance around the torso and shoulders at the level of the greatest lateral protrusions of the deltoid muscles

Comment: measured with the individual standing or sitting erect and the upper arms hanging naturally at his sides

shoulder – elbow length

the vertical distance from acromiale to the bottom of the elbow

Comment: measured with the individual sitting or standing erect, the upper arm vertical, and the elbow bent 90° so the forearm is horizontal

shoulder – fingertip length

see acromion – dactylion length

shoulder – grip length

the horizontal linear distance from a wall to a pointer held in a clenched fist

Comment: measured with the individual standing or sitting against the wall, with the elbow fully extended anteriorly and the arm horizontal

shoulder harness
a strap, belt, or other means of restraint which crosses the shoulder region, usually for the purpose of holding the torso, shoulders, and/or back against a rigid object to prevent or minimize injury in the event of impact; also shoulder strap, shoulder belt

shoulder height, sitting
see *acromial height, sitting*

shoulder height, standing
see *acromial height, standing*

shoulder length
the surface distance laterally from acromiale in a medial direction to the junction of the shoulder and the neck at which the tissue angle between the two is 45°

shoulder – neck junction
see *neck – shoulder junction*

shoulder strap
see *shoulder harness*

shutdown allowance
a special-case time allowance in those situations where a worker has to turn off or otherwise shut down equipment or machines at the end of a work period; *opp.* startup allowance

SI Units
see *Système International d'Unites*

sick leave
an employee benefit providing a certain amount of time in which a worker is permitted to be absent due to illness or accident without loss of job, seniority, and usually pay

siderosis
a pneumoconiosis caused by chronic exposure to airborne dust containing iron compounds; *syn.* arc-welder's disease

sidetone
the feedback signal from a speaker's microphone which is returned to the speaker through his earphones

sievert (Sv)
the SI unit for radiation dosage; equivalent to the dose from a 1 mg point source of radium in a 0.5 mm thick platinum container at a sampling distance of 1.0 cm in one hour's time; *syn.* millicurie-of-intensity-hour

sight
see *vision*

sight time
the temporal interval from the point at which a visual stimulus is recognized to that point at which a response must be made

sign test
a non-parametric statistical test involving the use of only plus and minus signs for analysis; see also *Wilcoxon matched-pairs signed-ranks test*

signal
that information conveyed via audible, electromagnetic, mechanical, or other means

signal detection
the observation and/or reporting of a signal; see *signal detection theory*

signal detection theory
a psychophysical model in which decisions as to whether a signal is judged to be present against a background are based on an individual's evaluation of the risks involved and the signal-to-noise ratio

signal-to-noise ratio (S/N, SNR)
the magnitude of a signal relative to that of any accompanying and/or background noise; also signal-noise ratio, S/N ratio

signature
any characteristic pattern or information generated by an individual or system which is repeatable under similar circumstances and can be used to recognize that individual or system

signed-ranks test
see *Wilcoxon matched-pairs signed ranks test*

significance
see *statistical significance*

signpost
any highway signage conveying information to drivers

silicosis
a pulmonary disease precipitated by the chronic inhalation of dust containing silicon dioxide, leading to fibrosis in the lungs and shortness of breath

similarity estimate
a judgment as to which two of three given items are more similar along one or more dimensions

similarity index
the reciprocal of the mean square average difference of standard scores between individuals for all the measured variables

$$SI = \frac{1}{\sqrt{\dfrac{\sum d_z^2}{N}}}$$

where:
d_z = difference in standard scores
N – number of variables measured

simo chart
see *simultaneous motion chart*

simple harmonic motion (SHM)
that motion with reference to an axis in which the perpendicular displacement of an object from a given point on the line is proportional to its acceleration toward that point and a sinusoidal function of time; *syn.* harmonic motion, harmonic vibration

simple linear regression
see *linear regression*

simple random sample
see *random sample*

simple reaction time
the temporal interval required to initiate a single predetermined response to a single predetermined stimulus

simple reflex
a motor reaction to an external stimulus where the neural signal is sent directly to the effector by a spinal motor neuron without input from the brain

simple sound source
a point source of sound energy which radiates uniformly in all directions

simple tone
see *pure tone*

simulate
carry out certain test conditions in an attempt to duplicate as much as possible actual field operations or conditions as a model for training or experimental purposes; *n.* simulation

simulator
any device or apparatus used to simulate one or more conditions

simulator fidelity
a measure of the degree of representation accuracy which a simulator has for the real-world system

simultaneous color contrast
see *chromatic contrast*

simultaneous contrast
an apparent change in the intensity or other quality of one stimulus due to the presence of an adjacent stimulus which differs from the reference along some dimension such that the differences appear to be magnified; see *simultaneous color contrast, simultaneous lightness contrast*

simultaneous lightness contrast
an alteration in the apparent lightness or brightness of one stimulus due to the presence of a different lightness stimulus

simultaneous masking
that masking of an auditory stimulus provided by a stronger signal at a different frequency; *syn.* frequency masking

simultaneous motion
the performance of two or more elemental motions concurrently by different body members

simultaneous motion chart
a chart for recording work involving simultaneous motion, and containing a minimum of columns for a therblig or motion symbol abbreviation, time value, and the body member involved; *syn.* simo chart, simultaneous motion cycle chart, micromotion data

simultaneous motion cycle chart
    see *simultaneous motion chart*

sine
    a trigonometric function, represented
    by the value of the ratio between the
    side opposite an acute angle of a right
    triangle to the hypotenuse of that tri-
    angle

single shift
    an operating mode in which all work-
    ers are working the same shift, usually
    the first shift

single underline
    a highlighting technique in which one
    line is drawn below a set of text

sinistral
    having a left-handed preference in
    motor activity

sinoatrial node
    a small strip of specialized muscle tis-
    sue in the heart which generates the
    normal heart rhythm; also sino-atrial
    node; *syn.* S-A node, sinus, sinuatrial
    node, cardiac pacemaker

sinuatrial node
    see *sinoatrial node*

sinus [1]
    see *sinoatrial node*

sinus [2]
    any cavity in the frontal portion of the
    skull

sinus arrhythmia
    an irregular heart stimulation rhythm
    by the sinoatrial node

sinus block
    the failure of one or more sinus cavities
    to equalize pressure due to blockage of
    the duct leading to the sinus cavity

sinusoidal
    varying over time with approximation
    to a sine wave

sitting
    pertaining to a posture in which the
    torso is approximately vertical, the hips
    are flexed about 90°, and the knees are
    flexed between about 45° and maxi-
    mum flexion; *syn.* seated

sitting bone
    (sl); see *ischial tuberosity*

sitting height
    the vertical linear distance from the
    sitting surface to the vertex plane
        Comment: measured with hair com-
        pression, the individual sitting erect,
        looking straight ahead, the knees
        flexed about 90°, and with a non-
        compressible seat at approximately
        the politeal height

sitting height, relaxed
    the vertical distance from the sitting
    surface to the vertex plane
        Comment: measured with the indi-
        vidual sitting relaxed, looking
        straight ahead, with a non-com-
        pressible seat at approximately the
        politeal height, and the knees flexed
        about 90°

sitting thigh clearance height
    see *thigh clearance height (sitting)*

situation-caused error
    see *situational error*

situational error
    a human error attributable primarily
    to a faulty design of the working envi-
    ronment; *syn.* situation-caused error

situational factor
    any job characteristic which is not di-
    rectly a part of a job performance itself,
    but are associated with the job, through
    management pressures, personalities,
    or other variables

size coding
    the use of different sized controls to
    indicate different functions

skeletal configuration
    see *posture*

skeletal link
    the straight-line distance between any
    two joint centers of rotation

skeletal muscle
    that muscle type having a heavily stri-
    ated appearance and which is gener-
    ally voluntary and attached to or in-
    volved in moving bone or cartilage

skeleton

the bones of the body configured in their normal relationships; *adj.* skeletal

skelic index

the value of the ratio of leg length to torso length

skewness

a measure of the departure from symmetry of a normal distribution, based on the third moment about the mean; *see negative skew, positive skew*

skill

an organized or coordinated activity pattern comprising a task at which some training or practice is normally required to become proficient; *see also ability*

skill-based behavior

an operating mode in which the individual performs some highly-practiced task without the need for conscious intervention following initiation of the activity

skin

the organ covering the surface of the body

skin conductance response (SCR)

a rise in skin surface conductance following some triggering event; *see also skin resistance response*

skin resistance response (SRR)

a decrease in the electrical resistance measured between two points on the skin surface within seconds after some triggering event; *syn.* galvanic skin response, galvanic skin reflex; *see also skin conductance response*

skin rule

a task design guideline that pressure from objects should not be concentrated on small areas of the skin

skin wettedness

a measure of heat stress, based on the amount of sweating; *syn.* index of skin wettedness

skinfold

a pinch of surface body tissue consisting of a double fold of skin and its associated subcutaneous fatty tissue for determining the amount and distribution of body fat; *syn.* fatfold

 Comment: usually selected from standardized sites

skinfold caliper

a spring-loaded caliper which exerts a standardized pressure per unit area and is designed to provide an estimate of skinfold thickness for body fat determinations

skinfold measurement

the process of obtaining or that data obtained from the use of a skinfold caliper in measurement of the thickness of a pinch of skin in specified body areas for estimating the percentage of body fat

skull

the collection of bones making up the head

sleep

a periodically-occurring behavior in which the eyelids are closed, the body is generally relaxed and inactive, consciousness appears reduced, and the EEG produces a distinctive waveform cycle

sleep debt

that amount of additional sleep which is required to feel properly rested or not fatigued due to sleep deprivation

sleep deprivation

a condition in which an inadequate amount of sleep is obtained

sleep disturbance

any interference with normal sleeping habits, regardless of cause

sleep period shifting

the process of advancing or delaying the sleep period for some length of time to accommodate a work schedule or other activities

sleep-wake cycle

a circadian rhythm in which normally about two-thirds of the cycle is spent awake and about one-third of the cycle asleep

sleepiness

a state of fatigue in which an individual tends to fall asleep easily

sleeve

that portion of an article of clothing normally intended to cover all or part of the arm

sleeve inseam

the linear distance from scye past the elbow to the level of the base of the thumb; *syn.* inseam

Comment: measured with the elbow fully extended

sleeve length – posterior

the surface distance along the lateral edge of the arm over olecranon from a point on the shoulder in the scye circumference plane to the distal end of the ulna

Comment: measured with the individual standing erect and the proximal phalanges of the fist placed on the hip

sleeve length segment – spine to scye

the surface distance from the midline of the spine to scye

Comment: measured with the individual standing erect, the arms held in a horizontal plane at shoulder level, the elbows flexed as required, and the proximal phalanges of the fists from each hand touching

sleeve length segment – spine to elbow

the surface distance from the midline of the spine to the tip of the elbow

Comment: measured with the individual standing erect, the arms held in a horizontal plane at shoulder level, the elbows flexed as required, and the proximal phalanges of the fists from each hand touching

sleeve length segment – spine to wrist

the surface distance from the midline of the spine to the ulnar styloid process

Comment: measured with the individual standing erect, the arms held in a horizontal plane at shoulder level, the elbows flexed as required, and the proximal phalanges of the fists from each hand touching

slew

see primary positioning movement; ger. slewing

slider bar

a computer display graphic for use in direct manipulation to select a value from a continuous range of values for some parameter

sliding caliper

a caliper which has a vernier mechanism in which one portion slides with respect to another, fixed portion to provide the measurement desired

sling psychrometer

a psychrometer which is capable of being spun about its handle to cause air flow over the wet bulb

slipped disk

(sl); *see* **herniated disk**

Sloan letter chart

a chart containing capital letters graded by size in equal logarithmic steps chosen to be of equal difficulty to each other and to the Landolt ring for use at specified distances from the observer for testing visual acuity

slope (m)

the tangent to a line or curve at a given point

$$m = \frac{\Delta y}{\Delta x} = \tan \theta$$

where:

$\Delta y =$ the vertical distance from the curve to a specified point

$\Delta x =$ the horizontal distance from the curve to a specified point

Comment: slope is undefined if the tangent to the line is vertical

slope intercept form

an equation for a straight line in which the slope and ordinate intercept are variables

$$y = mx + b$$

where:

m = slope

b = intercept

slot velocity
the linear flow rate through the openings of a slot-type hood

slouched
pertaining to a variant standing posture in which the shoulders and upper back are rotated anteriorly, with the neck slightly flexed

slow twitch muscle
see *red muscle*; *opp.* fast twitch muscle

slow wave sleep
that sleep phase during which delta frequency EEG waves predominate, with little muscle activity and eye movements and regular, slightly lower than the normal awake heart and respiration rate

slug
a unit of mass in the English system, equal to that mass whose acceleration by a force of one pound is one foot/sec

slumped
pertaining to a near sitting posture in which the shoulders and back are rotated anteriorly, with the neck slightly flexed; *syn.* slumping

small saphenous vein
a superficial leg vein extending from the ankle to the popliteal region in the posterior leg; *syn.* short saphenous vein

Smedley hand dynamometer
a hand-operated mechanical dynamometer with a rotary read-out dial for measuring grip strength

smoke
an airborne dispersion of small particles and/or droplets

smooth muscle
a normally involuntary, non-striated muscle tissue which is involved in visceral and other internal bodily movements

snap reading method
see *activity sampling*

snap reading technique
see *activity sampling*

snapback method
see *repetitive timing*

snapback timing
see *repetitive timing*

snapping finger syndrome
a stenosing tenosynovitis of a finger

Snellen acuity
a measure of visual acuity, referenced to a Snellen chart at 20 feet, expressed either as a decimal number or as the distance at which a given row of letters capable of being read by the individual being tested subtends 1 minute of arc

Snellen chart
a standardized chart containing rows of letters or numbers of the same size within a row, but of different sizes across rows for measuring visual acuity; *syn.* Snellen letter chart

Snellen test
a test using a Snellen chart to determine visual acuity

snow blindness
a normally temporary visual impairment, possibly accompanied by actinic keratoconjunctivitis, due to reflection of sunlight from surrounding snow; *syn.* solar photopthalmia

snowball survey
the gathering of initial responses on some issue from a selected sample of people, followed by a presentation of those responses to a second, different sample for revision, then a review by the first group, and finally a revision by the entire group

snubber
a mechanical device for increasing the stiffness of an elastic system when the displacement becomes larger than some specified value

soap
a fatty acid salt, usually with a hydrocarbon chain of about 12 units

sociology
the study of interpersonal and group behavior, and the influences of society and culture on such behavior; *adj.* sociological

soft light
any type of luminaire or the diffuse

illumination from a luminaire which gives shadows without sharp definition

soft palate

the posterior extension from the hard palate, consisting of muscles and a covering mucous membrane; *see also pala-tine velum*

soft water

water having low levels of disolved calcium and magnesium salts; *opp.* hard water

software

any or all of the complete set of programs, flowcharts, disks, tapes, and documentation associated with processing by computer systems

software engineer

a computer programmer or program designer

software engineering

the practice of computer programming

solar particle event (SPE)

the eruption of a large flare on the sun which results in radiation consisting of large numbers of high energy protons and alpha particles

solar photopthalmia

see snow blindness

solar retinopathy

damage to eye retinal tissue from looking directly at the sun without eye protection

solid model

a three-dimensional model in which the edges, surfaces, and volume of an object are represented

somatic

pertaining to the body

somatogram

a graphed profile consisting of girths at several body locations for classifying fat patterning, muscular distribution, or other body proportions

somatotype

(v) classify individuals into particular categories by body characteristics

somatotype

(n) any classification within a system describing individuals according to body characteristics; *see Heath-Carter somatotype, Sheldon somatotype, Kretschmer somatotype*

Comment: used historically as an attempt to correlate body type or structure with human character or personality traits

sonar

a sound system, most commonly used under water, for detecting, locating, or communicating with other objects

Comment: originally an acronym for SOund Navigation And Ranging

sone

a unit of loudness; the reference of 1 sone is the intensity of a pure 1000 Hz tone 40 dB above a listener's absolute auditory threshold

sonic

pertaining to sound at frequencies within the normal human audible range

sonics

the use of sound in measurement, control, or processing

sonogram

an image resulting from sonography

sonography

the use of sound energy and its reflections to image structures; *syn.* echography; *see also ultrasonography*

sonometer

an instrument for measuring hearing acuity; *see audiometer*

Comment: an outdated term

sonorant

a class of consonant produced by narrowing the vocal tract slightly, but not enough to cause turbulence

sorbent

any material which removes toxic gases and vapors from air which is to be inhaled

sound [1]

any mechanical vibration which normally results in an auditory sensation

sound [2]

a mechanical oscillation in pressure, stress, particle displacement, or other similar characteristics within a medium allowing internal forces such as elasticity or viscosity

sound absorption

the transduction of sound energy into another form, usually heat

sound analyzer

any device for measuring sound pressure level as a function of frequency

sound energy flux density

see **sound intensity**

sound intensity

the rate at which sound energy passes a unit normal area from a specified point; syn. sound energy flux density, sound power density, sound intensity level

$$I = \frac{p_{rms}^2}{\rho c}$$

where:

p^2 = mean square sound pressure
ρ = atmospheric density
c = sound velocity in that atmosphere
Comment: units in W/m^2

sound intensity level (L_I)

see **sound intensity**

$$L_I = 10 \log(\frac{I}{I_{ref}})$$

where:

I = measured sound intensity
I_{ref} = reference sound intensity
Comment: units in W/m^2; proportional to sound power level

sound level

see **sound intensity level**, **sound pressure level**, **sound power level**

sound level meter

an instrument for measuring sound pressure level; syn. sound meter, noise meter

sound localization

the ability to identify the direction from which a sound source is emitting relative to the observer

sound meter

see **sound level meter**

sound power density

see **sound intensity**

sound power level (PWL, W)

a measure of the acoustic power radiated by a sound source, specified to some reference level

$$PWL = 10 \log(\frac{W}{W_{ref}})$$

where:

W = power level
W_{ref} = reference power level
Comment: usual reference power is 1 pW

sound pressure

see **effective sound pressure**

sound pressure level (SPL, L_p)

a logarithmic pressure measurement of acoustical vibrations with respect to some specified reference pressure level; see **sound power level**

$$SPL = 10 \log \frac{p_{rms}^2}{p_{ref}^2} = 20 \log \frac{p}{p_{ref}}$$

where:

P_{ref} = reference pressure level
P_{rms} = measured root mean square pressure level
Comment: usually A-weighting against a reference pressure of 0.0002 microbar (20 µPa)

sound protective helmet (SPH)

any piece of headgear which both cushions against impact and attenuates noise

sound radiation

the conduction of acoustic energy through a solid, liquid, or gaseous medium; syn. radiation

sound transmission class (STC)

a single number rating representing a measure of the ability of a material to absorb sound or block sound transmission from one region to another

soundproof room

a room or chamber which insulates to some stated level against external noise penetration

South Atlantic Anomaly

a region over the south Atlantic ocean where high radiation levels are encountered at a lower altitude than elsewhere for earth orbiting vehicles

space[1]

the region surrounding a specified point or individual on earth

space [2]

the universe outside the earth's atmosphere

space adaptation syndrome (SAS)

a temporary disability encountered early in a spaceflight by some astronauts/cosmonauts which resembles motion sickness on earth, having symptoms including nausea, dizziness, and headache; *syn.* space sickness, space motion sickness

space motion sickness

see space adaptation syndrome

space planning

the development and/or use of information regarding operations for a specific location in determining a better or optimum use of the area or volume available for those functions

space sickness

see space adaptation syndrome

spacecraft maximum allowable concentration (SMAC)

a maximum concentration level of certain substances permitted by NASA in manned spacecraft

spaced practice

see distributed practice

span

the horizontal linear distance between the fleshy tips of right and left middle (or longest) fingers; *syn.* arm span, total span

Comment: measured with the individual's arms maximally extended laterally while standing erect and the body weight equally distributed on both feet

span akimbo

the horizontal linear distance between the the most lateral points of the two elbows; *syn.* akimbo span

Comment: measured with the individual standing erect and the elbows flexed sufficiently to permit the following position: the forearms held horizontal at chest level with the palms down, the fingers extended and adducted, the tips of the middle fingers of each hand touching, and thumbs touching the chest

span of reach

see reach envelope

spasm

a sudden, often unexpected, strong involuntary contraction of one or more muscles

spasmodic dysphonia

a spasm disorder fo the laryngeal muscles which results in an inability to produce speech sounds

spasticity

a movement disorder of central nervous system origin which is characterized by muscle hypertonia and a sudden increased muscle resistance to passive stretching, usually followed by relaxation

spatial correspondence

a condition in which the actuator(s) on a robotic device mimic the motion of the operator's controls

spatial facilitation

see spatial summation

spatial frequency

the reciprocal of the spacing between equally-spaced elements of a repeating visual pattern

spatial orientation

an awareness and ability to express

one's position and location in relation to other objects

spatial perception

the ability to acquire information about the direction, distance, form, and size of physical objects in the visual environment

spatial summation

the additive effect of signals from different neurons acting to cause a greater effect than would be achievable by a single neuron; *syn.* spatial facilitation

spatio-temporal anthropometry

the changing of mass distribution, surface area, and other measures over time, within an individual

speaker identification

the ability of a system or individual to distinguish a particular person's voice, whether spoken directly or synthesized

speaker verification

the ability of a system to correctly decide through word or speech recognition techniques whether or not a speaker is who he claims to be; the process of making such a decision

Spearman rank-order correlation

a significance test for the correlation of two variables based on differences in rank-ordered data; *syn.* rank-difference correlation; *see Spearman rank-order correlation coefficient*

Spearman rank-order correlation coefficient (ρ)

see rho

specialization

the concentration of effort, skills, and/or resources within a narrow scope of activity

specific heat (c)

the ratio of the quantity of heat required to raise the temperature of a unit mass of a substance by 1°C compared to that required to raise an equal mass of water 1°

Comment: units are cal/gm-°C

specification

a description of specific details as to how requirements for manufacturing a product will be implemented, including materials, dimensions, and other relevant considerations; sl. spec

specification by reference

see incorporation by reference

specificity

the percentage of variance accounted for by other than the independent variable being considered; *opp.* generality

spectral analysis

the breakdown or decomposition of a time series into frequency bands; *syn.* spectrum analysis

spectral density

see power spectral density

spectral fluorescent radiance factor (β_F)

the ratio of the radiance from a fluorescent sample to that produced by an identically-irradiated perfect reflecting diffuser for a specified wavelength

spectral locus

see spectrum locus

spectral luminous efficiency

the value of the ratio of the relative effectiveness of the conversion of light in the retina of a second wavelength of light to that of a reference wavelength which yields the greatest effectiveness; *syn.* relative luminosity, relative luminosity factor

spectral luminous efficiency for photopic vision

see photopic spectral luminous efficiency function

spectral luminous efficiency for scotopic vision

see scotopic spectral luminous efficiency function

spectral luminous efficiency function

a mathematical function or graphical curve describing the relative sensitivity of the eye to various wavelengths of light; *syn.* eye sensitivity curve, spectral sensitivity, CIE standard observer response curve, luminous efficiency function, luminosity function; *see photopic spectral luminous efficiency func-*

tion, scotopic spectral luminous effi-ciency function

spectral reflected radiance factor (β_R)
the ratio of the radiance produced by the reflection from a sample to that produced by an identically-irradiated perfect reflecting diffuser for a speci-fied wavelength

spectral sensitivity
see spectral luminous efficiency func-tion

spectral tristimulus value
see tristimulus value

spectrogram
a graphic record of the power spectral density

spectrometer
an optical instrument for measurement of the wavelengths of a light or for measurement of the radiant intensities of light at selected wavelengths

spectrophotometer
an optical instrument for the measure-ment of emitted, transmitted, or re-flected light as a function of wave-length for determining color or comparing luminous intensities

spectrophotometric curve
a graphical plot or mathematical func-tion of the amount of light emitted, reflected, or transmitted by an object at all wavelengths within the visible spec-trum

spectrophotometry
the use of a spectrophotometer to mea-sure light intensities at selected wave-lengths; *adj.* spectrophotometric

spectroscope
a device for enabling the eye to exam-ine the output of a spectrophotometer

spectrum
a graph of a continuous range of time-series data for amplitude, phase, power, intensity, or other information against its component frequencies or wave-lengths; *adj.* spectral

spectrum analysis
see spectral analysis

spectrum colors
those colors normally produced from the decomposition of white light by a prism, generally red, orange, yellow, green, blue, and violet

spectrum density
see power spectral density

spectrum level
see power spectral density

spectrum locus
that slanted, inverted-U-shaped boundary line on a chromaticity dia-gram on which all the colors within the visible region of the electromagnetic spectrum are located; *syn.* spectral lo-cus

specular
having a highly reflective finish for light or sound

specular angle
that planar angle between a specularly reflected ray of light and the perpen-dicular to a surface from which the reflection originates

specular glare
see reflected glare

specular gloss
shininess

specular reflectance
the value of the ratio of the departing flux from a surface due to specular reflection to the incident flux; *syn.* regu-lar reflectance; *opp.* diffuse reflectance

specular reflection
a reflection in which electromagnetic or other radiation travels in a specific direction and the specular angle is nu-merically equal to the angle of inci-dence; *syn.* regular reflection; *opp.* dif-fuse reflection

specular surface
a surface providing primarily specular reflections

specular transmission
the passing of an incident energy flux through a medium without diffusion; *syn.* regular transmission; *opp.* diffuse transmission

specular transmittance

the value of the ratio of that specularly transmitted flux having passed through a medium to the incident flux; *syn.* regular transmittance; *opp.* diffuse transmittance

specularly reflected light excluded (SPEX)

a light/color measurement condition involving the use of gloss traps to eliminate specularly reflected light

specularly reflected light included (SPINC)

a light/color measurement condition in which specularly reflected light is not eliminated

speech

a series of meaningful sounds produced by air flow through various structures (a) in the head, neck, and upper torso for human-generated speech, or (b) through various physical cavities for artificial speech

speech articulation index (AI)

a quantitative measure of speech intelligibility in the presence of background noise, ranging from 0 to 1 for impossible voice communication to excellent voice communication, respectively; *syn.* articulation index

speech clarity

the ability to convey spoken verbal information clearly such that an average listener will understand what is being said

speech compression

any form of reduction in the time or bandwidth required to convey essentially the same amount of information in speech

Comment: may include a modulation technique for transmitting over a narrower frequency band, reduction in redundancy, or reduction in the interval between words

speech enhancement unit (SEU)

a real-time unit for eliminating commonly found communication channel interference with only minor degrad-

ing of the speech signal itself

speech frequency band

those frequencies generally between 200 Hz and 4000 Hz, which are predominant in human speech

speech hearing

the ability to perceive and comprehend human speech

speech intelligibility

those characteristics or combinations of emitted speech sounds which enable the normal listener to comprehend what is being spoken; *see also speech intelligibility score*

speech intelligibility score

the percentage of spoken material which is understood by the listener

speech interference level (SIL)

the average sound pressure levels of noise (in dB) in three octave bands from 0.6–1.2 KHz, 1.2–2.4 KHz, and 2.4–4.8 KHz; *see also preferred speech interference level*

Comment: an older measure

speech perception test

an examination which measures hearing acuity by the administration of a standard list of words with performance evaluated against the average performance of persons with normal hearing

speech quality

a subjective judgement regarding the overall naturalness or acceptability of spoken words and the ability to recognize the speaker from voice cues alone

speech quality scale

a 5-point Likert scale used for rating the quality of speech, ranging from 1 (unsatisfactory) to 5 (excellent)

speech reception threshold (SRT)

the masking noise level which results in a 50% correct response of presented words or sentences

speech recognition

the use of a computer or other device to compare the syntax, semantics, and spectra of more than just a few spoken

315

words, from more than one person, with those algorithms and spectra of speech sounds in memory to perform a match; *syn.* continuous speech recognition

speech synthesis
the generation of meaningful speech sounds through the use of a computer or other equipment using previously coded data

speech transmission index (STI)
a number ranging from 0 to 1 which represents the quality of a speech communication channel, based on dynamically determined measures of the signal-to-noise ratio for speech in a number of frequency bands

speed [1]
the rate at which something is accomplished or an object or control moved; *syn.* rate

speed [2]
the magnitude aspect of the velocity vector; *syn.* rate

speed of closure
see rate of closure

speed rating
see performance rating

speed stress
a form of sensory overload in which the rate of information presentation is too high for adequate human processing

sphenoid bone
a highly irregular-shaped bone comprising a portion of the central base of the skull

spherical aberration
that image degradation occurring in an optical system when electromagnetic rays at different distances from the optical axis are refracted through a lens or reflected from a spherical mirror and cross the axis at different points

spherical wave
a wave which radiates outward equally in all directions from a central point source

sphincter
a muscle which surrounds a soft tissue opening and whose function on contraction is to decrease the size of that opening

sphyrion
the most distal point on the medial tibia at the ankle

sphyrion height
the vertical distance from the floor or other reference surface to sphyrion
    Comment: measured with the individual standing erect and his weight balanced equally between both feet

spice
an aromatic vegetable product for flavoring food

spike
a type of electrophysiological event generally characterized by a rapid, large, short duration change in amplitude, followed by a return toward the baseline

spin table
a platform on which organisms, objects, or equipment can be oriented in various positions for rotation testing

spinal column
see spine

spinal cord
that elongated portion of the central nervous system enclosed by the vertebral canal of the spine, and from which spinal nerves leave and enter to innervate the body

spinal erector
see sacrospinalis

spinal nerve
any of the 31 pairs of nerves containing sensory and/or motor components which exit from/enter the spinal cord between vertebrae

spine
the sequence of vertebral bones in the neck and back; *syn.* backbone

spinous process
the primary posterior prominence of a vertebra

spiral organ of Corti
see organ of Corti

spirant
see fricative

spirometer
an instrument for measuring the air capacity of the lungs

spirometry
the measurement of air volumes inspired or expired from the lungs; *adj.* spirometric

split shift
a work shift in which one or more individuals work less than about 7 hours, are released for more than an hour for other than a meal break, then return for a regularly-scheduled additional work period; *syn.* broken shift

split-halves reliability method
a testing reliability measurement technique in which a correlation coefficient is computed between performance on a test which has been divided into two equivalent groups of questions; *syn.* split-half reliability method

spondylosis
a stiffness in a portion of the spine

sponginess
the inexact response of a teleoperator to controls due to the use of a compressible fluid as a transmission medium

spontaneous ignition
a fire resulting from a condition in which there is sufficient oxygen to permit a slow chemical oxidation to generate heat, insufficient air circulation to dissipate the heat buildup, and the combustion temperature of the materials involved is reached

spot cooling
body cooling maintained within a restricted area

spot heating
body heating within a resticted area

sprain
an injury, typically occurring at a joint, in which the ligaments are stretched

and/or torn

spreading caliper
a caliper which has two tips on curved rods which are separable and a scale which is used to measure the separation distance between those tips

spreadsheet
a matrix in which the values of certain cells may be determined/computed from the values of other cells

spring element
a modeling structure characterized by deformation and resilience

spring-loaded control
a switch or lever which remains in the active position only as long as force is maintained due to resistance by some type of spring mechanism

spur
a growth projecting from a bone

squatting height
the vertical distance from the floor or other reference surface to the highest point on the head
Comment: measured with the individual in the squatting position, with the trunk, neck, and head erect and balanced on the toes of both feet

squeeze
the failure or inability to equalize pressure in gas-filled spaces within the body during changes in environmental pressure

squelch
see noise suppressor

stabilimeter
a device for measuring body sway; also stabilometer

stability ¹
that condition in which a position is maintainable, tends to be maintained, or is returned to after some movement; *adj.* stable

stability ²
the tendency of an individual or workforce to remain in the same geographical region or in the same employment

stabilize [1]

insure that an individual will not be adversely affected by external forces while performing some function; *n*. stabilization

stabilize [2]

treat a substance such that it is not itself or in combination with other substances capable of adversely affecting the general environment; *n*. stabilization

stabilize [3]

reduce undesirable vehicle motion either through active or passive mechanisms or devices; *n*. stabilization

stabilized retinal image

that type of vision in which, through artificial means, the target image does not move on the retina when the eye moves, thus exposing the retina to fixed stimulation

stabilizer

see **fixator**

stabilometer

see **stabilimeter**

stadiometer

a device for measuring height, consisting of a vertical rod with an attached rule and an adjustable horizontal headboard that can be moved vertically

staggered shift

a work schedule established by management which involves work hours displaced in time by some portion of the workday; *syn*. staggered work hours

staggered work hours

see **staggered shift**

staggers

a CNS-involved form of decompression sickness in which motor function is adversely affected, giving the appearance of staggering movements when walking

staircase procedure

a method of limits technique in which the value of a stimulus is increased or decreased on a given trial based on the result of the observer's response on the previous trial or group of trials; *syn*. up-and-down methodology

stairway

a stair and the region immediately surrounding it

stairwell

a stairway which is enclosed on at least two sides

stamina

the ability of the various bodily systems being utilized in a given task to sustain efficient performance for long periods of time

STAMINA

a computer model for predicting the level of highway noise to determine whether or not acoustic barriers are needed

stammer

(n) a speech impairment in which an individual involuntarily speaks hesitatingly, usually making multiple attempts to say a speech element or transposing speech elements; *see also* **stutter**

stamping gait

a type of gait in which the individual stamps one or both of his feet

stance [1]

that phase of a gait cycle where at least one foot is in constant contact with the ground

stance [2]

any static body position or posture where at least one foot is in constant contact with the ground or other base of support

standard

any established value for comparison purposes or accepted procedure for measurement or testing; *see* **consensus standard, mandatory standard, voluntary standard, primary standard, secondary standard, reference standard, working standard**

standard allowance

any allowance established in advance by calculation, arbitrary setting, or negotiation to provide for specified working conditions

standard atmosphere

an atmospheric pressure of 1.01325 N/m^2 with a density of 13.5951 gm/cc

Standard Colors of Textiles (SCOT)

a variant of the Munsell color ordering system for the textile industry

standard conditions

see *standard temperature and pressure*

standard cost

the estimation of the cost of a system, product, or activity based on standard times for similar work

standard coverage

the number of jobs, personnel, or total hours which are covered by standards during the reporting period; *syn.* actual coverage

standard data

a structured collection of normal time values for work elements arranged in some readily accessible form; *syn.* standard time data

standard daylight

any of a variety of standard illuminants which define the spectra for different types of daylight

Standard Deviate Observer (SDO)

a derived value which is typical of those color differences which occur on metameric matches made by real observers with normal color vision

standard deviation

a measure of the dispersion of scores in a sample (sd) or population (σ); equivalent in the normal distribution to the positive square root of the variance where:

$$sd = \sqrt{\frac{\sum X_i^2 - \frac{(\sum X_i)^2}{N}}{N-1}} \quad (computational)$$

$$sd = \sqrt{\frac{\sum (X_i - \overline{X})^2}{N}} \quad (theoretical)$$

X_i = data point or score
$\overline{X}$ = mean of the data set
N = number of data points in the set

standard element time

the standard time for a given work element; *syn.* standard elemental time

Standard Ergonomic Reference Data System (SERDS)

a proposed standardized reference system for ergonomic/human factors data

standard error

see *standard error of the mean*

standard error of estimate

a measure of the deviation of the measured values from the predicted value over repeated samples

$$\sigma_{yx} = \sqrt{\frac{\sum (Y_m - Y_p)^2}{N}}$$

where:
Y_m = measured value
Y_p = predicted value
N = number of samples

standard error of the mean (SEM)

a measure of the variability of the distribution of sample arithmetic means with respect to the theoretical population standard deviation

$$\sigma_{\overline{X}} = \frac{\sigma}{\sqrt{N}}$$

where:

$\sigma_{\overline{X}}$ = standard error of the mean

σ = population standard deviation
N = number of cases in the sample

standard hour

the production quantity required from an operator to meet a one hour quota

standard hour plan

an incentive plan in which a worker is paid for standard hours instead of the actual work hours

Standard Industrial Classification (SIC)

a well-recognized numerical coding system used to identify industry types by a number

standard illuminant
any of a set of specified, but not necessarily physically realizable, radiant light sources having a defined spectrum

standard luminous efficiency
see *spectral luminous efficiency function*

standard man
see *standard worker*

standard method
see *standard practice*

standard nine scale (stanine)
a measure of dispersion scale having a range of nine, a mean of 5.0, and a standard deviation of 1.96
Comment: an older scale

standard normal deviate
see *standard score*

standard observer
see *CIE standard observer*

standard of living
the ability to provide oneself and his immediate family with those necessary and desirable things in life

standard operator
see *standard worker*

standard output
that quantity of work completed by a standard worker or group of standard workers using a specified method or standard practice over a given work period

standard performance
that performance of a qualified individual worker or group of workers which meets a standard output

standard practice
a standardized work method, whether recorded or not, for the various steps involved in some operation; *syn.* standard method; *see also* **written standard practice**

standard practice sheet
a form which is used to provide the written standard practice information for an operator; *syn.* operation card; *see also* **operator instruction sheet**

standard rating
that rating which corresponds to a motivated, qualified worker, adhering to a specified method and working at an average pace

standard score
the expression of a point within a sample distribution as some multiple of the unit standard deviation to indicate its direction and distance from the mean; *syn.* z score, standard normal deviate
Comment: the mean is always 0; the standard deviation is always 1

standard source
a manufacturable or otherwise physically realizable generator of radiant visible light having a defined spectrum

standard stimulus
a stimulus whose value is fixed in a given dimension, and which is used in experiments to determine difference thresholds

standard system
a coded time-and-motion data set which is regarded as authoritative for a given plant or location

standard temperature and pressure (STP)
the condition of a volume of gases at $0°$ C and a pressure of 1 atmosphere; *syn.* standard conditions

standard temperature and pressure, dry (STPD)
the conditon of having a water vapor free volume of atmospheric gases at $0°$ C and 1 atmosphere pressure

standard time [1]
the total expected time, as determined by work measurement techniques, to complete a given task when working at a standard rating

Standard Time [2]
the mean solar clock time during the winter months; *opp.* Daylight Savings Time

standard time data
see *standard data*

Standard Time System

an incentive plan in which worker awards are determined by the number of units produced per unit time

standard workweek

the 5-day, approximately 40-hour work schedule within one calendar week

Comment: typically on the first shift

standardize

establish a common set of terms, procedures, criteria, or other appropriate aspects for a particular item or activity; *n.* standardization

standardized death rate

that number of deaths per 1000 population which would normally occur in a given group having a known age-specific death rate

Standardized Tests for Research with Environmental Stressors (STRES)

a test battery selected by AGARD as having evidence for a good psychometric basis for standardization in the evaluation of human performance

standards audit

one or more work measurement studies designed to examine whether existing standard times and methods are still proper

standby

an operational mode in which a worker, crew, or set of equipment is not actively engaged in generating work output, but is prepared for whatever action may be required

standby time

a time interval or the total amount of time spent in a standby mode

standing

pertaining to a posture in which the torso is approximately vertical, the hips and knees are extended, and the body weight vectors are directed through the ankles to the feet

standing height

see stature

standing wave

a periodic wave with nodes and antin-odes in a fixed position in space, and which occurs as the result of interference between progressive waves of the same frequency/wavelength and type; *syn.* stationary wave

stanine

see standard nine scale

stapedius

that muscle in the middle ear which inserts into the stapes

startle response

a strong psychophysiological response, triggered by a sudden, intense stimulus, which prepares the body for possible physical action; *syn.* defensive response; *see also* **orienting response**

startup allowance

a special-case time allowance for any waiting or time involved where a worker has to turn on or otherwise check out equipment or machinery at the beginning of a work period; *opp.* shutdown allowance

startup curve

the learning curve applied to a new job, process, individual, or group to allow for the longer than standard initial work times

state

the condition of matter — solid, liquid, gas, or plasma

static

motionless or unchanging; *syn.* stationary; *opp.* dynamic, kinetic

static anthropometry

the study of the bodily dimensions of an individual in a given fixed posture; *syn.* structural anthropometry; *opp.* dynamic anthropometry

static contraction

see isometric action

static display

a display containing one or more screen structures which remain the same for long periods of time; *opp.* dynamic display

static equilibrium

the ability to maintain body posture or

balance through a sense of position or motion of the head with respect to gravity from the integrated involvement of the utricular macula, vision, and the cerebellum and muscle tension; *see* *dynamic equilibrium*

static friction
that friction acting between surfaces with no relative motion between them; *opp.* kinetic friction

static muscle contraction
see *isometric action*
Comment: an older term

static muscle work
see *static work*

static strength
the force generated by a maximal voluntary isometric muscular exertion in a brief period of time; *syn.* static ultimate strength

static ultimate strength
see *static strength*

static work
that manual work performed when muscles are isometrically contracted, but no readily observable motion occurs; *syn.* static muscle work, isometric work, isometric muscle work

statics
the study or use of forces resulting in equilibrium, causing body parts or the body as a whole to be at rest; *opp.* dynamics

stationarity
a condition in which time-series data are stationary

stationary ¹
see *static*

stationary ²
pertaining to a condition or function where the mean, spectral density, and probability distribution are independent of time; *opp.* non-stationary

stationary time series
a stochastic time series whose characteristics are unchanged by an integral increase in the time axis; *opp.* non-stationary time series

stationary wave
see *standing wave*

statistic
a data point, estimate, or other descriptive property of a sample; *see also* *parameter*

statistical analysis
the use of statistical techniques to summarize, describe, or draw inferences from data

statistical effect
see *statistical significance*

statistical energy analysis (SEA)
a computer modeling and analysis package for estimating vibroacoustic impacts in the design of complex systems

statistical inference
the use of randomly sampled data to draw conclusions about a population

statistical process control
see *statistical quality control*

statistical quality control
the use of statistical principles in the manufacturing environment for quality control purposes; *syn.* statistical process control

statistical significance
having a sufficiently low probability that an given result is probably not due to chance; *syn.* statistical effect, significance
Comment: the probability level used should be stated

statistical standard time
a standard time derived from the statistical analysis of past time data; *syn.* historical time

statistics
the field of applied mathematics which is concerned with the analysis, presentation, and deriving conclusions from data; *adj.* statistical

statograph
a lever system for determining the body's center of mass/gravity

statokinetic
pertaining to the body's posture and

balance during standing and locomotion

stature

the vertical distance from the floor or other reference surface to the top of the head; *syn.* standing height; *see also stature (maximum)*

Comment: measured with the individual standing erect, the lower limbs vertical, and looking straight ahead

stature as reported

the individual's stated value for his stature

stature, maximum

the greatest vertical distance attainable from the floor or other reference surface to the top of the head

Comment: measured with the individual having taken a deep breath maximally extending himself vertically while keeping both feet flat on the floor, and having his weight equally balanced on both feet

steadiness

a measure of the ability to maintain a fixed posture with a minimum of tremor

steady-state [1]

pertaining to a condition that overall is unchanging with time, as being in equilibrium

steady-state [2]

that physiological condition in which oxygen uptake by the lungs and delivery to bodily tissues by the circulatory system equals the oxygen requirement of the tissues for a particular activity

steering wheel

a circular control device which is connected to a guiding mechanism and may be turned by hand to control the course/direction of a vehicle

stenosing tenosynovitis

a partial reduction in the flexion or extension of a joint due to an inflammation and thickening/swelling of the tendons or its sheaths of the muscles providing action about that joint; *see*

also **snapping finger syndrome**, **trigger finger syndrome**

step test

see **Harvard step test, Master's two-step test**

steppage gait

a type of locomotion with an exaggerated flexion of the hip and knee resulting in a high-stepping gait with a flopping of the foot which tends to drag

steradian (sr)

a unit of measure for solid angles, with one steradian equal to that solid angle of a sphere which has its vertex at the center and intersects an area on the surface of that sphere with four equal-length sides, each equal to the radius in arc length

stereogram [1]

a two dimensional graphic which gives the impression of depth using contour lines or shading

stereogram [2]

a pair of separate two-dimensional views which, when positioned properly and presented separately to the right and left eyes using a stereoscope, cause the visual system to integrate them into a single view appearing to be three-dimensional or to have depth

stereographic anthropometry

see **stereometric anthropometry**

stereometric anthropometry

a form of non-contact anthropometric measurement using stereophotogrammetric techniques to determine surface distances, angles, areas, and other appropriate measures; *syn.* stereographic anthropometry

stereophotogrammetry

the determination of positions in space from video or film involving either two cameras aligned along different axes in the same plane or a single camera and mirrors

stereophotography

a photographic technique which simulates stereoscopic vision by using two separate cameras separated by some distance in space or a stereoscopic camera; *syn.* stereoscopic photography

stereopsis
see stereoscopic vision

stereoscope
an optical device capable of giving an impression of depth when presenting an appropriate visual display to each eye

stereoscopic acuity
the ability to perceive the three-dimensional aspect of physical space, or depth, by the use of two eyes or vision sensors separated in space; *syn.* stereoacuity

stereoscopic photography
see stereophotography

stereoscopic vision
the capability of perceiving depth and distance in the region near the fixation point by the use of two eyes or video sensors slightly separated in space; *syn.* stereopsis

stereoscopy
the study or use of three-dimensional or depth aspects, effects, or techniques; *adj.* stereoscopic

sterilize
reduce living microbial life forms to below some specified quantity or render them incapable of reproduction

sternocleidomastoid
a bilateral, voluntary muscle located approximately vertically from the sternum and clavicle to the lower part of the posterior temporal bone

sternum
the bone articulating with the ribs and clavicle bones in the midline of the anterior torso, and consisting of three segments — the xiphoid process, body, and manubrium; *adj.* sternal; *syn.* breastbone

sternum height
see substernale height

Stevens' power law
a psychophysical relationship between sensation and stimulus physical intensity, consisting of a power function of the form

$$S = k(I_s)^n$$

where:
S   =   sensation strength
I_s =   stimulus intensity
k   =   a constant depending on the units of measurement
n   =   an exponential value which varies depending on the subject and type of stimulus/sensory modality

stick
(sl); *see control stick, joystick*

stickman
a simple figure for modeling human posture or motion in which straight lines are used to approximate the various body links or segments

sticky keys
a feature which enables sequential use of keyboard keys rather than simultaneous use in certain multi-key operations for allowing disabled individuals to operate the computer

stiction
that friction which tends to prevent relative motion between two movable objects at a neutral position

stiffness [1]
the rigidity of a teleoperated system

stiffness [2]
the ratio of the force/torque applied to the corresponding change in translational/rotational displacement of an elastic element

stigmatism
a condition in which the refractive system of the eye causes light rays to be accurately focused on the retina; *opp.* astigmatism

stilb (sb)
a CGS unit of luminance; equals 1 cd/cm^2
Comment: an older term

Stiles-Crawford effect
the reduced effectiveness in stimula-

tion of the retina by a light ray entering the eye near the periphery of the pupil compared to that light entering at or near the center of the pupil

Stilling test
a color vision test involving the use of a set of plates, each containing a colored digit embedded within a set of colors easily confused by those with color vision deficiencies; *syn.* hidden digit test

stimulus
any type of cue presented to an organism, whether internal or external
Comment: typically referring to a cue strong enough to be consciously perceived

stimulus generalization
the production of a response due to the presentation of a similar, but not identical, stimulus to that which originally produced the response

stimulus-onset interval
see interstimulus-onset interval

stochastic
random; describable only by using statistical processes

stochastic process
see random process

stochastic variable
see random variable

stoker's cramp
see heat cramp

stomion
the point of contact between the upper and lower lip of the mouth in the mid-sagittal plane

stomion to top of head
the vertical distance from stomion to the vertex plane
Comment: measured with the individual sitting or standing erect and facing straight ahead

stop [1]
see limit stop

stop [2]
a point which should not be passed prior to performing some type of op-

eration

stop [3]
a consonant sound whose production requires a brief, complete cessation of air flow through the closing off of one or more of the cavities in the vocal tract, followed by opening and the release of pressure

stop [4]
the useful opening of a lens

stopclock
an electrical or electromechanical timing device for measuring time intervals

stopping power
a measure of the ability of a substance or material to reduce the velocity or energy of an entity

stopwatch
a portable electrical, mechanical, or electromechanical timing device

stopwatch time study
the measurement of short time intervals in repetitive operations using a stopwatch or stopclock

strabismus
a disorder in which the eyeballs are uncoordinated due to the lack of control over the extrinsic eye muscles, resulting in the visual axes not intersecting at the desired point

straight back rule
a task design guideline that the back and neck should remain straight at all times during performance of a task

straight wrist rule
a task design guideline that a flexed or extended wrist be avoided when grasping, squeezing, or otherwise executing hand movements requiring any significant strength application

strain [1]
an injury or disability involving the overuse, overextension, compression, or twisting of a muscle, ligament, or joint

strain [2]
the biomechanical, physiological, and /

or psychological effects from one or more stressors on an individual; *see also stress, stressor*

strain [3]

a change in one or more dimensions of some object due to elongation, contraction, or shear stressors

strain gauge

an electrical device which uses a change in resistance on deformation to measure the amount of force applied

strain propagation

see stress transmission

strain synthesis

a simulation of work strain by a combination of work stressors

strained reach

the reach capability under conditions of maximal joint extension, applying pressure to any restraints

strap length

the distance from one bra tip over the back of the neck to the other bra tip
Comment: measured with the individual standing or sitting erect, without following the body contour

stratified sampling

a subject selection procedure in which the population is divided into different strata, each having one or more common characteristics, then randomly drawing samples from each strata in proportion to that group's representation in the population

stray light

any undesired scattered or reflected light within a specified volume; *syn.* stray luminance

stray luminance

see stray light

strength [1]

the maximum capability of an individual to exert a brief force using only his muscles and body segments under specified conditions; *syn.* muscular strength; *see also endurance, breaking strength*

strength [2]

see breaking strength, yield strength

strength assessment

any determination of an individual's strength under a given set of conditions

strength-duration curve

a curve indicating the relationship between the time duration of stimulation and current flow in artificially exciting a muscle or nerve

stress [1]

the collective mental and physical conditions resulting when an individual experiences one or more biomechanical, physiological, or psychological stressors above comfortable levels; *see also stressor, strain*

stress [2]

the resistance by an object to application of an external mechanical force which tends to produce a deformation
Comment: usually quantified as the force per unit area

stress equivalent

a quantitative biomechanical relationship between physiological outputs and physical workload

stress incontinence

an involuntary urination due to shock or a startle response

stress reduction

the use of any of several techniques, such as deep muscle relaxation, meditation, cognitive restructuring, and biofeedback, in an attempt to reduce stress levels

stress transmission

the transfer of physical or psychological stress (a) from the environment to the individual, (b) from the individual to the environment, or (c) from one part of the body to another

stressful work conditions

having an excessive amount of one or more stressors in the workplace

stressor

a stimulus or set of stimuli which is

perceived to create discomfort; *see also* *stress, strain*

stretch

draw out or elongate an elastic entity

stretch out

a reduction in the delivery rate or increase in the length of time to deliver a product, without any decrease in the total number of products delivered

stretch reflex

the contraction of a muscle following a sudden longitudinal stretching of that muscle

stria

a visible line inherent in certain materials or tissues such as skeletal muscle or imperfect glass; *pl.* striae

string diagram

a model of a plant or facility on which a thread or string has been used to track the flow path of employees, materials, or equipment during some operational sequence

strobe light

a flash tube which may be capable of adjustable-frequency, rapid flashing rates over an extended period of time; *syn.* stroboscopic lamp

stroboscope

a device which flashes a bright light or opens a shutter intermittently to make moving objects visible

stroboscopic effect

the apparent immobilization of an object when the object is illuminated by periodic pulses of bright light

stroke ¹

the motion which depresses one key when using a keyboard or keypad

stroke ²

a single motion of a pen or cathode ray tube gun

stroke ³

a condition in which blood flow to some portion or all of the brain is severely reduced or eliminated

stroke volume

the amount of blood ejected from the left ventricle of the heart into the arterial system from one cardiac contraction cycle

stroke width

the width of a drawn or displayed line

stroke writing

the construction of an entity or object on a cathode ray tube directly by an electron gun; *syn.* vector graphics

Strong Interest Inventory

a commonly used self assessment test for aid in job seeking

Stroop color-word test

the presentation of conflicting color and word stimuli

strophosphere

a reach envelope for the hand/arm combination or the leg/foot combination in which any translational or rotational motion of the limb or its terminal segment is permitted; *see also* **kinetosphere**

structural anthropometry

see static anthropometry

structure-borne noise

that noise transmitted vibroacoustically through some structure

Student's t-test

see t-test

study session

an interaction between interested participants to obtain or learn new information and evaluate current information

stutter

(n) a speech impairment in which an individual speaks hesitatingly due to difficulty in saying certain syllables; *see also* **stammer**

stylion

the most distal point on the lateral margin of the radius styloid process at the wrist

stylion height

see wrist height

styloid process

a long, spine-like projection from a bone

stylus

a pen- or pencil-shaped computer input device, usually used in conjunction with a digitizing tablet for drawing or marking input locations

sub-

(prefix) below, beneath, or under

subarachnoid space

a region between the arachnoid layer and the pia mater surrounding the brain, which is filled with cerebrospinal fluid

subassembly

the combination of two or more parts forming a unit which is a component of a larger product or system

subcontract

an agreement with a party other than the prime contractor or the original contracting customer to perform services and/or provide one or more products

subcontractor

a party having a subcontract

subcutaneous emphysema

an accumulation of gas beneath the skin surface

subdural hematoma

an accumulation of blood between the dura mater and arachnoid layer covering the brain

subgravity

see *hypogravity, microgravity*

subischial height

the vertical linear distance from the floor or other reference surface to the height of the lowest point of the ischial tuberosity; *syn.* lower extremity height
Comment: generally assumed to represent the length of the lower extremities; estimated by subtracting sitting height from stature

subject

a member of a specified population or sample who is selected according to some specified methodology and from whom a researcher intends to obtain data; *syn.* experimental subject

Comment: normally an informed, voluntary consent must be obtained

subjective

pertaining to some internal measure, state, or aspect which is not directly observable or verifiable by more than one person except as the product of an individual's verbal or other reporting means; *opp.* objective

subjective brightness

see *brightness*

subjective report

any appropriate form of expression by an individual regarding some effect, experience, or other phenomenon which cannot be independently verified or quantified

subjective vertical

that direction which an individual perceives as vertical

Subjective Workload Assessment Technique (SWAT)

a method for determining mental workload, in which ratings on three scales (time, mental effort, and psychological stress) are combined or examined separately to provide the workload measure

subliminal

pertaining to a stimulus having an intensity below the perceptual or responsive threshold

sublingual

pertaining to or a structure which lies beneath the tongue

submandibular

below the mandible

submarine

(v) slide under a lap safety belt or other object intended for restraint

subnasale

the junction of the base of the nasal septum and the philtrum

subnasale – sellion length

see *nose height*

subnasale to back of head

the horizontal linear distance from the base of the nasal septum to inion; *equiv.* subnasale to wall

Comment: measured with the individual standing erect and facing straight ahead

subnasale to top of head
the vertical distance from the base of the nasal septum to the horizontal vertex plane
Comment: measured with the individual standing or sitting erect

subnasale to wall
the horizontal linear distance from a wall to the base of the nasal septum; *equiv.* subnasale to back of head
Comment: measured with the individual standing erect with his back against the wall

subrogation
the legal process in which one party attempts to recover the amount paid under a policy to an insured from a third party when the latter may have been responsible for the situation causing the loss

subscapular skinfold
the thickness of a skinfold taken at an angle about 45° to horizontal just below the inferior angle of the scapula
Comment: measured with the individual standing comfortably erect and the arms hanging naturally at the sides

subscapular skinfold, recumbent
the subscapular skinfold measure, but using a skinfold from the inferior angle of the scapula pointing toward the elbow
Comment: measured with the individual lying on one side, the acromial processes of the shoulders aligned vertically, and the arm positioned along the side of the body with the palm against the thigh

subscript
an alphanumeric character or symbol placed just to the right and below another character or symbol, often of smaller point size; *opp.* superscript

subsonic
traveling at a velocity less than that of

sound in a given medium; outdated *syn.* infrasonic; *opp.* supersonic

substernale
the most inferior point on the xiphoid process in the midsagittal plane

substernale height
the vertical distance from the floor or other reference surface to substernale in the midsagittal plane; *syn.* sternum height
Comment: measured with the individual standing erect and his weight balanced equally on both feet

substitution analysis
an examination of the rate at which new equipment or technology is projected to replace that existing in the present economy

subsystem
any portion of a system which itself performs a specific, essential subset of the activity of that system; *see* **system**

subtask
a set of task elements which comprise a logical, describable unit within a task

subtracted time
the time representing the period required for completion of one or a group of work elements obtained from the difference between successive stopwatch or stopclock readings when using a continuous timing technique

subtractive color mixing
the addition and integration of one or more colored substances or materials to an existing set; *opp.* additive color mixing

sudoriferous gland
see **sweat gland**

suffusion
the spreading or flow of any bodily fluid through interstitial spaces into surrounding tissue

sugar
any of a class of carbohydrates, having a chemical formula of the type $C_nH_{2n}O_n$ or $C_nH_{2n+2}O_{n-1}$, and generally having a sweet taste

sulfur dioxide (SO_2)
a combustion product of sulfur burning in oxygen or air

summation
the additive effects in neural, muscular, or mental activities

sunburn
a discoloration or inflammation of the skin due to excessive exposure to the ultraviolet light from the sun; *adj.* sunburned, sunburnt

sunlight
that radiation from the sun within or near the visible spectrum

Supercockpit
a USAF program for developing a virtual workspace having many advanced display/control technologies for pilot/crew interaction

superhigh frequency (SHF)
that portion of the electromagnetic spectrum consisting of radiation frequencies between 3 GHz and 30 GHz

superhighway
a multi-lane, limited access highway for high-speed surface vehicular traffic, such as a turnpike, freeway, or expressway

superior
located above relative to another structure

superior levator
a flat extraocular muscle which raises the upper eyelid

superior oblique muscle
a voluntary extraocular muscle principally for rotating the upper part of the eyeball medially about the optical axis

superior rectus muscle
a voluntary extraocular muscle parallel to the optical axis along the upper eyeball for looking/pitching the eyeball upward

superior thigh clearance, maximum
see **thigh clearance height, sitting**

superscript
an alphanumeric character or symbol normally placed just above and to the right of another character or symbol, and which may be smaller in point size; *opp.* subscript

supersonic [1]
faster than the velocity of sound; *opp.* subsonic

supersonic [2]
see **ultrasonic**
Comment: an older term

superstition
an unsubstantiated belief that a cause-effect relationship exists between two or more events

supervisor
one who oversees and directs the work activities and is involved in any personnel actions of a subordinate

supervisory control
having computer hardware and software at either or both ends of an operator – teleoperator loop to aid decision making

supinate
rotate the forearm clockwise about its proximal-to-distal longitudinal axis, as viewed from the shoulder; *n.* supination; *opp.* pronate

supinator
any muscle which is involved in a supinating motion; *opp.* pronator

supine
pertaining to a posture in which the anterior portion of the body faces upward, the torso is aligned parallel to a reference surface, and the hips and knees are extended; *see also* **reclining**

Supplemental Data System
a Bureau of Labor Standards program involving the national collection and distribution of worker's compensation data for the purpose of aiding in the determination of accident causes

supplementary lighting
any lighting supplied in addition to that general lighting which is normally available at a given location to provide a certain quality or quantity

supplied-air suit
a closed suit which is impermeable to most particulate and gaseous contaminants and which provides the wearer with an adequate supply of breatheable air

suppuration
the formation of pus

supra-
(prefix) on or above

suprachiasmatic nucleus
a group of cells above the optic chiasm in the hypothalamus which receives input directly from the retina and is believed to be involved as a pacemaker in biological rhythms

supracondylar
a structure located superior to a condyle

suprailiac skinfold
the thickness of a skinfold directed antero-medially and downward at an angle of 45° on the midaxillary line just superior to the level of the iliac crest
    Comment: measured with the individual standing comfortably erect, the body weight distributed equally on both feet, and the abdominal muscles relaxed

supraliminal
pertaining to a stimulus intensity above the perceptual threshold; *opp.* subliminal

suprapatellar skinfold
the thickness of a vertical skinfold on the lower thigh, 2 cm above the patella
    Comment: measured with the individual standing comfortably erect, the weight equally distributed on both feet, and the leg muscles relaxed

suprasternal notch
the depression on the superior surface of the manubrium; *syn.* jugular notch

suprasternale
the lowest point in the suprasternal notch on the superior edge of the manubrium

suprasternale height
the vertical distance from the floor or other reference surface to suprasternale
    Comment: measured with the individual standing erect and his weight equally balanced on both feet

suprasternale height, sitting
the vertical linear distance from the upper seat surface to suprasternale; *syn.* trunk height
    Comment: measured with the individual sitting erect, on a non-compressible seat; used to provide a measure of trunk length

surface
the exterior of an object; the most superficial layer or part of a structure

surface acoustic wave touchscreen
a display having a surface layer which emits ultrasonic energy and which indicates a touch location through absorption of the energy by the water content of the finger

surface active agent (surfactant)
any of a class of chemicals used in cleaning which act to emulsify oils, grease, and attached dirt by reducing the interface tension of the substances involved

surface area
the area contained by a surface

surface distance
a measurement representing the distance when following the general surface contour of some structure

surface model
an image in which only the edges and surfaces of objects are displayed

surface tension
that force acting on the surface of a liquid substance which tends to minimize the surface area

surfactant
see surface active agent

surge
a sudden, short-lived increase in energy output

surround brightness
the brightness of the immediate background near the work area

surround inhibition
see *lateral inhibition*

survey
the process of subject selection, the development of a set of verbal questions or a questionnaire, presentation of the questions/questionnaire to the subject(s), and an analysis of the results obtained; *syn.* survey study

survival ratio
the ratio of the number of individuals surviving a situation to the number present prior to that situation

susceptor
one or more aluminum strips within a microwave cooking dish which help brown the enclosed food

suspension [1]
a temporary dismissal from work for improper activities in the workplace, which may be with or without pay

suspension [2]
that mechanism used to improve ride quality by isolating a land vehicle from the shock and vibration experienced by movement on a surface

sustained hold
the maintenance of a position for an indefinite or long period of time; *opp.* momentary hold

suture
(v) close a wound using sewing techniques

suture
(n) a type of skeletal joint in which adjacent bone surfaces are essentially fused together; *syn.* cranial suture

swatch
a piece of cloth for comparing color or pattern in the textile industry

swayback
(sl) a greater than normal lordosis of the lumbar spine
Comment: may be accompanied by increased compensatory kyphosis of the thoracic spine

sweat
(v) secrete sweat; *syn.* perspire

sweat
(n) the transparent, colorless, water-based fluid consisting of fats, salts, carbohydrates, and other materials secreted from the sweat glands; *syn.* perspiration
Comment: cools the body on evaporation

sweat gland
a coiled tube-shaped gland in the skin which secretes sweat; *syn.* sudoriferous gland; see *apocrine gland, eccrine gland*

sweat test
see *Minor's sweat test*

swimmer's ear
see *otitis externa*

swing
the phase of a gait cycle during which the foot is not in contact with the ground

swing shift [1]
a work shift which overlaps two other shifts, usually on an around-the-clock operation

swing shift [2]
that work shift which a crew works following and in additon to its regular shift in order to rotate shifts by eight hours in an around-the-clock operation
Comment: this crew works two shifts (about 16 hours) on the day of the shift

swing shift [3]
that work shift which is manned by a crew which rotates to work all three shifts within a week on a 7-day, around-the-clock operation in order to provide days off to workers on each of the other shifts

swinging arm rule
a task design guideline that any movements of the arms should follow a natural arc, and that a barrier or stop should bring the motion to a halt rather than muscular activity

switch
(n) any mechanical or electrical device which may close or open a path, or change the direction of an entity traveling along a path

swivel
(n) the ability of a chair or other object to rotate either direction about a central vertical axis

Sydenham's chorea
a usually temporary form of chorea associated with rheumatic heart disease in children; *see also* **Huntington's chorea**

symbiosis
a relationship in which different species of organisms coexist in close proximity and contribute to each other's benefit; *adj.* symbiotic

symbol
any graphical character or other representation which is intended to: (a) stand for something else, (b) communicate a use for an object/structure, or (c) communicate what should or should not be done at a given time or location

symbolic control
the use of symbols/graphics as input to exercise a controlling function

symbology
the study or use of a set of symbols for communication

symmetry
having corresponding similar components or appearance on either side of an imaginary point, axis, or plane; *adj.* symmetric

sympathetic
pertaining to the division of autonomic nervous system originating from the thoracic and lumbar sections of the spinal cord, has its ganglia located near the spinal column, and which generally opposes the parasympathetic division by actively responding in stressful conditions; *syn.* thoracolumbar; *opp.* parasympathetic

sympathy
a shared feeling or identification with another individual; *adj.* sympathetic

symphyseal height
the vertical distance from the floor or other reference surface to symphysion
> *Comment:* measured with the individual standing erect and weight equally distributed on both feet

symphysion
the lowest point of the pubic symphysis

symphysis
a normally non-movable skeletal joint in which bones are tightly joined by a cartilaginous plate; *see* **pubic symphysis**

synapse
the electrical and/or chemical junction between two or more neurons at which information can be passed from one cell to another; *adj.* synaptic

synchronization allowance
see **interference allowance**

synchronizer
see **entraining agent**

syncope
a temporary loss of consciousness due to hypoxia of the brain; *syn.* fainting

syndrome
a set of signs and/or symptoms which together indicates a particular pathological condition

synergism
see **synergy**

synergist
an entity which acts to assist another entity when their efforts are combined, referring especially to a muscle aiding another muscle in performing its action; *adj.* synergistic

synergy
a condition in which the interaction of two agents or systems produces an effect which may be greater than the effect produced by equal amounts or efforts of the individual agents or systems acting separately; *syn.* synergism; *adj.* synergistic

synesthesia
experiencing a sensation in another sensory modality which accompanies a primary sensation evoked by a stimulus in the primary modality

synkinesia
the involuntary motion of limbs which coincides with voluntary movement of another part of the body; *syn.* accessory movement

synovial fluid
a clear, viscous fluid contained within certain joints, bursa, and tendons which provides lubrication for movement of those structures

synovial joint
a joint between two or more bones which has a synovial cavity, articular cartilage, and an articular capsule

synovial membrane
that tissue which encloses a synovial cavity at a synovial joint

synovitis
an inflammation of a synovial membrane

synthesis
the collection and summation of the individual work element times obtained from synthetic data or from other, previously-studied jobs containing those elements to define a performance level for a new job

synthetic basic motion times
a set of standard times assigned to individual motions and groups of such motions via synthesis

synthetic data
any value(s) obtained from established tables or formulas, not empirically from the actual situation(s) to which they are relevant; *see synthesis*

synthetic standard data
standard times obtained from synthetic data

synthetic time standard
see synthetic standard data

synthetic vision
the use of millimeter radar waves to see through clouds, haze, or certain other non-visually transparent media to create images of what lies within or beyond that media

system
an integrated collection of facilities, parts, equipment, tools, materials, software, personnel, and/or techniques which make an organized whole capable of performing or supporting some function

system anthropometry
a representation of the human body in three-dimensional coordinate space, describing all body links and joint angles; also systems anthropometry

system dynamics
the interactions within a functioning system

system engineering
the application of engineering principles to concept formation, requirements and specifications development, hardware/software design and development, testing, and verification of a system, including all supporting documentation for development and use; also systems engineering

system response time
the elapsed time from the signal to begin a command to the notification that the command has been executed

system safety
the use of system engineering principles to provide a specified level of safety given the tradeoffs involving cost, time, and the operations involved

system safety engineering
the use of system safety techniques to identify any hazards within a system

system status information
that information about a system's operating condition which is presented to the user

System for Aiding Man-Machine Interaction Evaluation (SAMMIE)
an interactive, three-dimensional computer modeling software package for

designing the physical aspects of man and workplaces

systematic observation

a non-random, representative, organized program for observing and recording the activities of individuals, systems, or events

systematic sampling

a subject selection procedure in which an ordered population exists and the units sampled are located at fixed or otherwise pre-defined intervals after a random starting point

Système International d'Unites (SI)

an international dimensional system, based generally on the metric system, which has been adopted for worldwide measurements; *syn.* International System of Units, Le Système International d'Unites

systems anthropometry

a representation of the human body in three-dimensional coordinate space, describing all body links and joint angles

systems engineering

see system engineering

systolic blood pressure

the maximum arterial blood pressure during the cardiac cycle, obtained during the heart contraction portion of the cycle; *opp.* diastolic blood pressure

T

t distribution

a distribution of sampling means obtained from a normally distributed population and having the probability distribution function

$$f(t) = G(v)\left[1 + \frac{t^2}{v}\right]^{-\frac{(v+1)}{2}}$$

where:

v = number of degrees of freedom

G_v = constant for a given n

Comment: approaches the normal distribution as sample size increases

t ratio

the ratio of the quantity of the obtained mean in a sample minus the expected mean, divided by the estimated standard error of the mean

$$t = \frac{\overline{X} - \mu}{\dfrac{s}{\sqrt{N-1}}}$$

t-test

a statistical test using the *t* distribution and *t* ratio with small sample sizes for determining the significance of differences between means: (a) of a sample and the population, (b) of two independent samples, or (c) of two related samples; also *t* test; *syn.* Student's *t*-test
Comment: usually used for small sample sizes and when the variances are unknown

table [1]

a structure generally containing rows and columns of information or data which are related and uniquely identified; *adj.* tabular

table [2]

a flat-topped furniture piece having small leg-like structures for support and on which one or more functions such as dining, games, or work are carried out

tablespoon (tbsp)

a measure of volume used primarily in cooking, corresponding to the level volume held in a large spoon or about 14.8 ml

tabular display

a display consisting of alphanumerics, words, and/or other symbols in a table format

tachistoscope

an instrument for presenting time-controlled exposures of visual stimuli

tachometer

a device for measuring and displaying rotational frequency or angular velocity per unit of time

tachycardia

a higher than normal heart rate; *opp.* bradycardia

tachypnea

an abnormally high respiration rate

Tactical Air Navigation (TACAN)

a relatively short range radio frequency aerial navigation system

tactile

pertaining to the sense of touch, mediated by sensors located in the skin

tactile coding

the use of vibratory mechanical stimuli for communication purposes

tactile control

having the possibility of control through some type of tactile feedback, usually via a distinctive surface texture, vibration, or a stop

tactile stimulus
any mechanical stimulus which activates the touch/pressure receptors

tactual
causing a touch sensation

tailbone
(sl) the sacral vertebrae of the spine

tailor's ankle
a growth over the lateral malleolus commonly found in tailors
Comment. generally caused by pressure from sitting on hard surfaces with crossed legs

takedown
see teardown; also take down

takedown time
see teardown time

talbot (T)
an SI unit for that amount of light delivered by a luminous flux in one second; *syn.* lumen-second

talc
a soft, white or gray hydrated magnesium silicate with a somewhat slippery feel; *syn.* talcum

talcosis
a pneumoconiosis from the chronic inhalation of talc dust

talent
a greater than normal ability, especially pertaining to the arts or sports

talus
the most superior bone of the foot, which articulates with the bones of the lower leg to form the primary aspect of the ankle joint; *pl.* tali; *adj.* talar; *syn.* ankle bone

tandem gait
a type of gait in which the individual places the heel of the advancing foot in line with and directly in front of the toes of the stationary foot

tangent
a trigonometric function; equal to the value of the ratio of the opposite side of a right triangle to the adjacent side

tapotement
the use of percussion movements in massage

target [1]
any object which is capable of reflecting a sufficient amount of a transmitted sonar or radar signal to produce a blip on detection/display equipment

target [2]
any object, point, or other entity toward which an object, activity, or energy is directed

target acquisition
the first appearance of an echo on a radar or sonar tracking system

target discrimination
the ability of a system to distinguish one or more targets within a noisy background

target gland
any gland affected directly by a hormone released from another gland or tissue

target glint
see wander

target organ
an organ affected directly by a hormone released from another gland, tissue, or organ

target scintillation
see wander

tariff
a schedule showing the number of each size of an item required to outfit the user population

tarsus [1]
the group of bones in the posterior foot, consisting of the talus, calcaneus, cuboid, navicular, and the three cuneiform bones; *pl.* tarsi

tarsus [2]
a plate of dense connective tissue which gives form and some rigidity to each of the eyelids

task
a logical, describable group of related subtasks which comprise a discrete component of a job and which are performed within a job classification

task allocation
the distribution of tasks or task elements between workers and machines

task analysis
a systematic breakdown of a task into its elements, specifically including a detailed task description of both manual and mental activities, task and element durations, task frequency, task allocation, task complexity, environmental conditions, necessary clothing and equipment, and any other unique factors involved in or required for one or more humans to perform a given task; *see also* **time study, job analysis**
   Comment: the end product is a detailed document

task assumption
any of a set of background or unstated ideas or concepts which underlie the performance of a task

task clustering
see job

task complexity
a measure of or statement about the number, variety, and difficulty of the separate task elements making up a task

task definition
see task description

task description
a written statement providing an overall label and a label for each of the task elements which must be performed to carry out a given task; *syn.* task definition, task statement

task duration
see task time

task element
the smallest work unit into which a task may be logically divided, typically comprised of several therbligs

task element time
the time interval required to complete a task element

task frequency
the number of times a task is/must be performed within a given time period

task hierarchy
a description of the manner in which lower level tasks are organized to form more complex tasks

task inventory
a listing of the tasks performed or required for a given occupation

task lighting
that illumination which is directed onto a localized workplace for a specific visual task

task load
see workload

Task Load Index (TLX)
a subjective rating technique developed by NASA for determining overall workload, and consisting of six dimensions (mental, physical, and temporal demands, effort, perceived performance, and frustration level), which are rated on the basis of low to high

task needs analysis
a determination of those tasks forming part of some function which an individual or group wishes to perform

task rating
a subjective rating of tasks using a relevant set of work habitability criteria

task statement
see task description

task time
the time interval allowed or required to complete a given task; *syn.* task duration

taste
a chemical sense involving the stimulation of sensory receptors located in the tongue and oropharynx for the four basic sensations of sweetness, saltiness, sourness, and bitterness

taste blindness
see ageusia

taste bud
an ovoid-shaped structure embedded just below the surface of the tongue which contains the sensory receptors for taste

tasteless

a characteristic of a substance in which it does not stimulate taste buds above threshold levels

taxon

a classification within a taxonomy; *pl.* taxa

taxonomy

an organizational structure for classification and description purposes

teach

provide knowledge, information, and/or concepts; *n.* teaching

tear strength

that force required to initiate a tear in a fabric or sheet of material

teardown

the process of partial or total dismantling of a facility or workplace to prepare for another job which uses the same facility or workplace; *syn.* takedown; *opp.* setup

teardown allowance

the teardown time applied as a special time allowance for a worker who performs a teardown operation; *syn.* dismantling allowance; *opp.* setup allowance

teardown time

that time period required or projected for a teardown; *syn.* takedown time

teaspoon

a volume of approximately 5 cc or 1/3 tablespoon

technical error of measurement

see measurement error standard deviation

technical mel scale

a scale of subjective auditory perceptual sensitivity, which is an approximately linear function to 1 KHz, and logarithmic above 1 KHz

Technique for Human Error Rate Prediction (THERP)

a procedure for applying tables consisting of human reliability estimates for certain steps in the analysis of tasks to determine the probable overall likelihood of a successful outcome

technological forecasting

the process of gathering and analyzing data in an attempt to predict the types and characteristics of future equipment and technologies

tectorial membrane

that gelatinous membrane overlying the hair cells of the organ of Corti in the cochlea

TEKTITE

a program conducted in the early 1970s involving two studies (I and II) in which people lived for up to 30 days in an underwater chamber

tele-

(prefix) at a distance

tele-existence

see telepresence

teleceptor

any sensory receptor sensitive to stimuli of distant or remote origin

telefactor

see teleoperator

    Comment: an older term

telemetry

the process of acquisition of a signal at a remote location and the transmission of the information contained in that signal to a more convenient location for presentation, storage, or analysis; also telemetering; *syn.* remote monitoring

teleoperation

the use and control of a teleoperator or telerobot, usually involving either a hardwired connection or communication using the electromagnetic spectrum, with a video display for the operator; also tele-operation

teleoperator

a general-purpose, dexterous device capable of sensing its environment and of direct, essentially real-time control by a human operator to perform tasks like grasping, moving, and other operations using its appendages; *syn.* telefactor

teleopsia
a visual perception disorder in which depth is not judged correctly

telepresence
a condition in which the operator of a remote system is provided with sensory information and/or feedback regarding the operation of that remote system; *syn.* tele-existence

telepuppet
see telerobot
Comment: an older term

telerobot
a remote, dexterous manipulator device capable of self-locomotion and pre-programming for periods of autonomous operation; *syn.* telepuppet

telescience
the direction of an individual or using teleoperation to perform scientific research remotely via video or other tele-communication means

telestereoscope
an instrument which effectively increases the interpupillary distance to produce the appearance of exaggerated depth

temper
a disposition toward a sudden display of anger

temperature (T)
a measure of the heat intensity possessed by an entity, commonly reported by reference to a standard scale; *see Celsius, Fahrenheit, Kelvin*

temperature color scale
see color temperature scale

temperature gradient
the change in temperature with distance in a medium (dT/dx); *syn.* thermal gradient

temperature sense
the ability to detect relative heat and cold

temple
that portion of the head anterior to the pinna and superior to the zygomatic arch; *adj.* temporal

temporal [1]
pertaining to the lateral direction of the head or skull, as opposed to nasal or medial; toward the temples or the temporal bone

temporal [2]
pertaining to time

temporal bone
a flat bone located on the inferior-lateral portion of the skull

temporal crest
a narrow bony ridge comprising the posterior extension of the zygomatic arch as part of the temporal bone on the side of the head above the ear

temporal lobe
that portion of the cerebrum which is located on the lower lateral region of the cerebral hemisphere

temporal muscle
that muscle located over the temporal bone region of the skull

temporal pole
the inferior and anterior projecting portion of the temporal lobe

temporal summation
the additive effect over time of multiple subthreshold neural inputs, with each subsequent input occurring before the previous potential has returned to baseline

temporary editing buffer
see clipboard

temporary rate
a production output rate or incentive wage rate based on a temporary standard

temporary standard
a standard time or output measure to be used for a limited period, generally to allow for an unusual job situation, revising a task, or developing a new task

temporary threshold shift (TTS)
a temporary reduction in sensitivity to stimuli in one or more sensory modes due to any cause, where essentially normal sensitivity is recovered with

the passage of time or on removal from those conditions; *opp.* permanent threshold shift; *see also **noise-induced temporary threshold shift***

temporary total disability
an injury classification in which, for some limited period of time subsequent to an injury, an individual cannot perform the normal duties of one's job

temporary work
that work performed for a short term, without any agreement by the parties involved of long-term or permanent employment

temporo-mandibular joint disorder (TMJ)
some condition involving the joint between temporal and mandibular bones which may cause symptoms of tooth, jaw, ear pain or headache; also temporomandibular joint syndrome

temporo-parietal
pertaining to or toward the side of the skull or head, or the junction between the temporal and parietal bones

tender
pertaining to a soft, easy to chew food

tendon
a tough connective tissue, often in the shape of a cord, which provides a junction between muscle and bone; *adj.* tendinous

tendon reflex
the contraction of a muscle due to tapping of its tendon

tendonitis
an inflammation of a tendon; *see also tenosynovitis*

tendovaginitis
see tenosynovitis

tennis elbow
an epicondylitis condition in the lateral elbow region, especially at the origin of the wrist extensor muscles on the humerus

tenosynovitis
a lesion or inflammation of a tendon and its sheath, typically due to chronic or acute overuse, direct impact, or infection; *syn.* tenovaginitis, tendo-

vaginitis; *see also **tendonitis***

tenovaginitis
see tenosynovitis

tensile strength
that maximum stress to which a material can be exposed prior to breaking or tearing

tensiometer
a mechanical spring-type device for measuring static strength

tension headache
a headache caused primarily by prolonged muscle contraction of the scalp muscles

tension movement
see controlled movement

tensor
any muscle which makes a portion of the body more rigid

tentorium
that portion of the dura mater which separates the cerebral hemispheres from the cerebellum

teratogen
any agent which can produce a perma nent alteration in the structure/function of cells, tissues, and/or organs of the embryo or fetus during gestation; *adj.* teratogenic

terminal threshold
see upper threshold

terminal velocity
the maximum velocity at which the body will free fall in the atmosphere at a given altitude

territoriality
the concept that an individual has a physical location or personal space that belongs to him, and resists or resents the entry of another person into that space

tesla (T)
a unit of measure for magnetic field density; equals 1 weber per meter^2

test
carry out a technique or procedure for determining a quantity or performance measure on one or more dimensions for an individual or product

test battery

a group of tests administered to one or more individuals to obtain a total or composite score for evaluation with regard to some fitness or aptitude

test conditions

the total circumstances or environment under which an individual, part, or system is tested

test-retest reliability

a measure of the consistency obtained in repeated administrations of the same test to the same group, separated by some time interval

Comment: represented by the correlation coefficient between performances on the test

testicle

see testis

testis

the ovoid-shaped male gland in the scrotum which produces sperm; syn. testicle; pl. testes; adj. testicular

tetanus

an acute, often fatal, disease characterized by painful tonic muscle contractions due a toxin from the presence of the bacterium Clostridium tetani in a wound not exposed to oxygen; adj. tetanic; syn. lockjaw

tex

a unit of measure for the fineness of a fiber, represented by the weight of 100 meters of thread in grams

text

any combination of alphanumeric character strings, usually having a structure comprised of words, sentences, and paragraphs

text editing

a form of text processing, which may consist of any or all of the following steps: (a) scanning for text errors, (b) locating a segment of text on a screen, (c) making the desired text changes/ corrections, and (d) formatting

text entry

the process of inputting text to a computer, through whatever method

text processing

one or more of the acts of: text entry, text retrieval or storage, text editing, printing, or similar actions performed in the production of text materials

text-to-speech system

a speech synthesis device in which text and small speech units are processed to yield sounds approximating human speech

textile

any material made from either manmade or synthetic fibers and intended for use in clothing, furniture coverings, or carpets

textile softener

see fabric softener

thenar

pertaining to the thenar eminence

thenar eminence

the protrusion on the radial/lateral side of the palm near the base of the thumb due to the underlying mass of muscle tissue

theoretical biomechanics

the use of knowledge from classical mechanics and the life sciences to generate models which predict the relationship of man, his body segments, or tissues to a particular biomechanical environment

theory

a concept based generally on accepted principles and which uses standard rules to predict or otherwise explain certain phenomena; adj. theoretical

Theory X

a belief that workers are naturally passive, self centered, require constant motivation, lack ambition, resist change, and that management must operate by active intervention

Theory Y

a belief that motivation, responsibility, and a willingness to work as a team member are human characteristics which management should provide conditions to bring out

therapeutic index

the value of the ratio of the median lethal dose of a drug to its therapeutic dose; *syn.* therapeutic ratio

therapeutic ratio

see ***therapeutic index***

therapist

one who practices a therapeutic technique

therapy

a technique intended to produce a cure or assist in managing a disability or disease; *adj.* therapeutic

therblig

a specific fundamental division of mental, sensory-motor activity, or the lack of activity, within a larger task; also Therblig; *syn.* basic division of work, basic division of accomplishment, Gilbreth basic element, elemental motion, fundamental motion, basic motion, basic element, work element

Comment: named by Frank B. and Lillian Gilbreth; therbligs are:

assemble
avoidable delay
disassemble
grasp
hold
inspect
plan
position
pre-position
release load
rest
search
select
transport empty
transport loaded
unavoidable delay
use

therblig chart

an operation chart subdivided into therbligs, with each therblig designated by its appropriate symbol or abbreviation; *see also* ***simultaneous motion chart***

thermal

pertaining to heat or temperature

thermal adaptation

the physiological adjustment of the body over time to cold or heat

thermal capacity

see ***heat capacity***

thermal comfort

satisfaction with the thermal environment; *opp.* thermal discomfort

thermal comfort zone

that range of dry bulb air temperature, mean radiant temperature, air velocity, and humidity within which some specified percentage of people are expected to be satisfied indefinitely when in a sedentary posture with suitable clothing

thermal conduction

see ***heat conduction***

thermal conductivity

a proportionality constant for the rate of heat transfer across a unit area for a given perpendicular temperature gradient

$$k = - \frac{\frac{dQ}{dt}}{A \frac{dT}{dx}}$$

where:

k	=	thermal conductivity
dT/dx	=	temperature gradient
dQ/dt	=	time rate of heat flow
A	=	area across which heat flow occurs

thermal convection

see ***heat convection***

thermal discomfort

that sensation of being outside the thermal comfort zone, but not to the extent at which the body would encounter significant thermal stress; *opp.* thermal comfort

thermal environment

the total combination of dry bulb temperature, radiant temperature, air velocity, humidity, physical activity, and clothing effects for an individual or group

thermal equilibrium [1]
a condition in which the rate of excess body heat production equals the rate of heat loss to the environment

thermal equilibrium [2]
a condition in which all objects in contact with each other or within the same closed thermal environment have the same temperature

thermal indifference zone
that range of thermal environments within which some specified percentage of active people are not expected to complain

thermal insulation
any material resistant to the flow of heat

thermal insulation value of air (I_a)
the resistance to heat transmission through the atmosphere immediately surrounding an individual

thermal insulation value of clothing (I_{cl})
that amount of thermal insulation provided by a clothing system as worn; *syn.* insulation value of clothing

thermal photograph
see thermogram

thermal radiation
see heat radiation

thermal resistance
a material's ability to prevent heat flow

thermal resistance value (R)
a measure of the resistance to heat flow provided by clothing or other thermal barrier; *syn.* R value

thermal resistivity
the inverse of thermal conductivity

thermal stress
any type of stress caused solely by deviations from comfortable temperatures or thermal environments for a given clothing system; *see heat stress, cold stress*

thermochromic effect
exhibiting a color change on exposure to heat

thermodynamics
the study of heat and its relationships

to other forms of energy

thermogenesis
heat production by the body or other means; *adj.* thermogenic

thermogram
a graphic representation of the temperature distribution on the surface of the body or other object; occasional *syn.* thermal photograph

thermography
the study or use of measuring surface temperatures of objects via emitted or reflected electromagnetic radiation

thermoluminescence
any emission of electromagnetic radiation within or near the visible range due to the heating of an object

thermoreceptor
a sensor responding to or capable of sensing temperature or temperature differences

thermoregulation
any or all of the processes used by living organisms to balance heat production, gain, and loss in an attempt to maintain a suitable body temperature; *syn.* thermal regulation

theta rhythm
an EEG frequency band consisting of frequencies from about 4 to 8 Hz

thigh
the upper portion of the leg, generally from the hip to the knee, consisting of the femur and the surrounding tissues

thigh circumference, distal
the surface distance in the horizontal plane around the upper leg just proximal to the femoral epicondyles
Comment: measured without tissue compression with the individual standing erect, his weight equally distributed on both feet, and no unnecessary muscular contractions

thigh circumference, midthigh
the surface distance around the upper leg in a horizontal plane at the vertical midpoint between the lowest point in the pubic crotch and tibiale; *syn.*

midthigh circumference

> Comment: measured without signifi-
> cant tissue compression with the
> individual standing erect, his body
> weight distributed equally on both
> feet, and no unnecessary muscular
> contractions

thigh circumference, proximal

the surface distance around the upper
leg in the horizontal plane at a level as
high in the pubic crotch as possible and
at the level of the gluteal furrow; *syn.*
upper thigh circumference, standing

> Comment: measured without tissue
> significant compression with the
> individual standing erect, his weight
> balanced on both feet, and no un-
> necessary muscular contractions

thigh circumference, sitting

the surface distance around the upper
leg in a plane perpendicular to the
thigh longitudinal axis and as high in
the pubic crotch as possible; *syn.* upper
thigh circumference, sitting

> Comment: measured without signifi-
> cant tissue compression and with
> the individual sitting erect, having
> no unnecessary muscular contrac-
> tions

thigh clearance

see thigh clearance height (sitting)

thigh clearance height, sitting

the vertical distance from the upper
seat surface to the highest point of the
thigh above the sitting surface; *syn.*
thigh clearance, maximum superior
thigh clearance; thigh height above seat,
sitting thigh clearance height

> Comment: measured with the indi-
> vidual sitting erect in a seat whose
> height is at approximately the pop-
> liteal height, the thigh longitudinal
> axis horizontal, the lower leg verti-
> cal, and without unnecessary
> muscle contractions

thigh height above seat

see thigh clearance height (sitting)

thigh length

the linear distance from the horizontal

midpoint of the inguinal crease to the
proximal border of the patella

> Comment: measured with both the
> hip and knee flexed 90°

thigh skinfold

the thickness of a vertical skinfold taken
along the midline of the thigh at the
mid-thigh point

> Comment: measured with the indi-
> vidual standing comfortably erect,
> the weight distributed equally on
> both feet, and the leg musculature
> relaxed

thigh – thigh breadth, sitting

the maximum horizontal linear distance from the most left lateral point of
the thigh spread across the thighs to
the most right lateral point of thigh
spread

> Comment: measured with the indi-
> vidual seated erect and thigh longi-
> tudinal axes parallel or as nearly
> parallel as possible

third octave band

see one-third octave band; syn. third
octave, 1/3 octave

third shift

a night work shift of about 8 hours
duration, between approximately 11
P.M. and 7 A.M.; *syn.* C shift, graveyard shift, night shift

third-class lever

a lever system in which the force or
effort is located between the fulcrum
and resistance

thirst

a desire for water due to a sensation of
dryness in the mouth and/or throat or
a general bodily need for water

thoracic

see thorax

thoracic cavity

that bodily cavity which contains the
heart and lungs, with their supporting
tissues

thoracic outlet syndrome

see scalenus anterior syndrome

thoracic spine

that portion of the spine associated

with the level of the thorax, vertebrae T1–T12

thorax
that portion of the trunk from the neck to the diapragm, containing the chest and its organs; *adj.* thoracic; *pl.* thoraces

THORAX
a finite element structural dynamic human thorax model

three-component theory
see Young-Helmholtz theory

three-dimensional
having extent in length, width, and depth; giving the impression of depth

three-dimensional anthropometry
the measurement of points on the human body with reference to a well-defined three-dimensional space

thresher's lung
an acute pneumonia which may be found in agricultural workers exposed to moldy grain or hay; *syn.* farmer's lung

threshold
that level on some dimension at which a stimulus or change in stimulus is just adequate to be detectable, effective, or elicit a certain response on a specified proportion of events; *syn.* detection threshold; *see also **absolute threshold, threshold dose, difference threshold, contrast threshold, threshold shift, discomfort threshold, upper threshold***

threshold contrast
see contrast threshold

threshold dose
that minimum amount of a drug, radiation exposure, or other agent which will cause a detectable biological effect

threshold illuminance
see illuminance threshold

threshold limit value (TLV)
a set of safety guidelines established by the American Conference of Governmental Industrial Hygienists (ACGIH) for exposure to toxic substances in the normal working environment in order

to protect most workers from known adverse effects; *see **threshold limit value – ceiling, threshold limit value – short term exposure limit, threshold limit value – time-weighted average***

threshold limit value – ceiling (TLV-C)
that airborne concentration of a toxic substance above which individuals should never be exposed

threshold limit value – short term exposure limit (TLV-STEL)
that maximal airborne concentration of a toxic substance to which most individuals may be exposed for 15 minutes without knowingly causing harmful effects such as irritation, tissue damage, or narcosis, under the condition that the daily TLV-TWA is not exceeded

threshold limit value – time-weighted average (TLV-TWA)
the time-weighted average airborne concentration of a toxic substance above which individuals should not be exposed repeatedly within a single day or week

threshold of audibility
the minimum effective sound intensity or pressure level at a specified frequency that will evoke an auditory sensation in a specified proportion of trials or in a specified proportion of people; *syn.* auditory absolute threshold, threshold of detectability, threshold of hearing, zero level

threshold of detectability
see threshold of audibility

threshold of discomfort
see discomfort threshold

threshold of feeling
that acoustic stimulation level at which sound begins to be perceived as pain

threshold of hearing
see threshold of audibility

threshold of pain
see pain threshold

threshold potential
that membrane potential in an excitable cell at which the potential becomes

unstable, leading to an action potential

threshold shift

a change in either an absolute or difference threshold; *see also* **permanent threshold shift, temporary threshold shift**

throat

the fauces, oropharynx, and laryngopharynx with their enclosing tissues

thrombocytopenia

having a below normal number of blood platelets

throughput

the output quantity of completed items from a machine or system within a specified time interval; also thruput, through-put

thrust

a rapid forward motion, especially of an upper limb

thumb

the first digit on the hand
> *Comment:* has only two phalanges instead of three

thumb crotch

that region of tissue between the thumb and the first finger (digit II); *see also* **interdigital crotch**

thumb crotch length

the linear distance from the level of the base of the crotch between the thumb and index finger to the level of the crotch between the first and second fingers (digits II and III)
> *Comment:* measured parallel to the long axis of the hand

thumb-tip reach

the distance from a wall to the tip of the thumb
> *Comment:* measured with the individual standing erect against the wall, his elbow/arm extended forward, and his thumb aligned with the index finger

thumb-tip reach, extended

the thumb-tip reach except that the measured shoulder is extended as far as possible while keeping the non-mea-sured scapula firmly against the wall

thumbwheel

a manual rotary control device, only a portion of which extends above a panel surface, usually having the numbers 0 through 9 displayed spaced at equal distances around the circumference; also thumb wheel
> *Comment:* may have individual, discrete detents for each number or a continuous rotary motion

thyroarytenoid

a skeletal muscle in the larynx which is involved in shortening and relaxing the vocal cords; *opp.* cricothyroid

thyroid

a bi-lobed endocrine gland located in the neck attached to the thyroid cartilage which is involved in regulating body metabolism

thyroid cartilage

the large, most superior cartilage of the larynx
> *Comment:* the anterior prominence of this cartilage is larger in the male than the female and commonly referred to as the Adam's apple

tibia

the larger, medial bone of the lower leg; *syn.* shinbone
> *Comment:* bears most of the body weight when standing

tibiale

the most superior point on the medial border of the tibia's medial condyle

tibiale height

the vertical distance from the floor to the proximal medial margin of the tibia
> *Comment:* measured with the individual standing erect and his weight balanced between both feet

tic

any movement disorder involving involuntary, brief, rapid contractions of related groups of muscles

tidal air

see **tidal volume**

tidal volume

the volume of air inhaled and exhaled

during each respiratory cycle; *syn.* tidal air

tie [1]

a condition in which two individuals, objects, or other entity have the same value or score

tie [2]

an article of clothing, usually long and narrow, for tying around the neck

tight rate

see **tight standard**

tight standard

a time standard which allows less than the normal time for a given task to be done by a normal operator working at the normal pace; *syn.* tight rate

tile

place two or more windows on a display such that they abut one another, but do not overlap; *ger.* tiling

timbre

that auditory attribute by which a listener can discriminate between two sounds having the same loudness and pitch, but different tonal quality

time allowance

see **allowance**

time and materials contract

a legal agreement providing for materials at the contractor's cost and labor hours at a specified rate, including both direct and indirect costs, overhead, and profit

time and motion study

the observation, measurement, and analysis of the operations involved in the performance of a job or task to determine standard times; *syn.* time-motion study, motion and time study

time balancing

the redistribution of those work elements or assignments performed by workers having the greatest workloads among workers who have less time-consuming workloads in an attempt to reduce the total job time

time constant

the time required for a physical signal

to rise to or fall from to a specified percentage of its normal operating level

Comment: typically an exponential relationship: $= 1 - 1/e\ (63.2\%)$

time delay

that time required for a signal to be transmitted from one point and received at another point, separated by some physical distance

time domain

the expression of a function in terms of time; *opp.* frequency domain

time magnification

the ratio of film/video camera sampling frequency over film/video projection frequency, both normally in units of frames/sec

time measurement unit (TMU)

an MTM time interval equal to 10^{-5} hour

time perception

the ability to accurately judge the duration of time intervals; *syn.* time sense

time sense

see **time perception**

time series

a set of time domain data which has been sampled over some period of time, usually at regular intervals

time standard

see **standard time**

time standards maintenance

the periodic verification and updating of standard times to keep pace with technology, methodology, and other changes

time study

the use of appropriate sampling, measurement, and analysis techniques to determine a standard time, including time allowances, for an operator to perform a given task or job using specified methods and under the prescribed working conditions; *see also* **task analysis**

time-lapse photography

the use of one or more still, motion picture film, or video cameras operating at less than normal frame rates to record long-duration operations or processes for viewing later at normal frame rates to determine times

time-weighted average (TWA)

a calculated average of an individual's exposure to some environmental condition over a period of time when the levels or intensity of that condition vary throughout the period of interest; also time weighted average

$$TWA = \frac{\sum_{1}^{n} L_i T_i}{\sum_{1}^{n} T_i}$$

where:

L_i = level of exposure (intensity, concentration, etc.)

T_i = length of time at exposure level i

n = number of exposure intervals used

timeline

a schedule of expected or projected events, with the dates and/or times at which each is to occur and its duration; also time line

timeline chart

a graphical or symbolic representation of a timeline

timeshare

integrate performance on two or more tasks using divided attention, parts of which are performed either in parallel or sequentially, all within a given period of time; also time share

tinnitis

a ringing, buzzing, or hissing sound heard in either one ear or both ears

tint

any color lighter than median gray

tintometer

an instrument using a combination of colored glass filters to estimate the intensity of colors

tissue

an aggregation of cells having a similar function with their supporting connect-

ing and other intercellular materials; *see also **cell**, **organ***

tissue rheology

the study of the deformation and flow of tissue under external stress

title

an identifier which names a display and is distinguishable from other display structures

TKA line

*see **trochanter – knee – ankle line***

toad skin

(sl) a dry, rough skin texture correlated with a deficiency of vitamin A

toe

any of the digits of the foot

toeboard

a type of guard which may be installed around flywheels, open pits, and overhead catwalks

toenail

the elastic tissue covering the dorsal portion of any of the terminal phalanges of the foot; *see also **fingernail***

toggle switch

a two- or three-position switch, either electro-mechanical or a graphic display image, which can be flipped either from side to side or from a central position to either side for control operation

tolerance [1]

a measure of an entity's ability to endure stressful conditions or workloads without injury or damage; *adj.* tolerant

tolerance [2]

*see **engineering tolerance***

tolerance [3]

*see **drug tolerance***

tolerance limits

*see **engineering tolerance***

tolerance specification

*see **engineering tolerance***

tomography

the study or practice of sectional radiography; *adj.* tomographic; *see also **computerized axial tomography***

ton
> a unit of weight, equal to 2000 pounds; *syn.* short ton; *see also* **long ton**

tonal discrimination
> *see pitch discrimination*

tonal gap
> a sound frequency band in which an individual has low or non-existent auditory sensitivity and has normal or better sensitivity on both sides of that band

tonal island
> a sound frequency band in which an individual has significant hearing, but has little no auditory sensitivity on either side of that band

tondal
> the unit of force which will accelerate a long ton 1 foot per sec^2

tone [1]
> *see tonus*

tone [2]
> a sound wave capable of giving the auditory sensation of pitch

tone [3]
> a shade or variation in shade of color

tone deafness
> the relative inability to make a fine discrimination between tones close together in pitch; *adj.* tone deaf

tongue
> the muscular structure attached to the posterior floor of the mouth and which is involved in mastication, speech, swallowing, and taste

tonicity
> the level of tension or contraction in a static muscle or group of muscles

tonus
> that degree of tension in a static muscle; *syn.* tone, muscle tone

tool
> any device, piece of instrumentation, or machine intended to perform an operation or aid in the performance of an operation

tool allowance
> a time allowance for a worker to adjust and/or maintain his tools

tool design
> the part of engineering involved with the design of tools, especially of hand tools

top light
> illuminate one or more individuals or a scene from directly above

topography
> the study or description of the body surface or its parts; *adj.* topographical

torque (τ)
> the effective perpendicular component of a force (or the effective sum of forces) applied to an object at some distance from a point representing an axis about which rotation can occur, inducing or tending to induce an angular acceleration; *syn.* moment of force; *see also* **moment arm, torsion**

$$\vec{\tau} = \vec{r} \times \vec{F}$$

torr
> a unit of pressure; equals 1/760 atmosphere or 1 mm of mercury under standard conditions; also tor

torsion
> the twisting of a rigid structure about an internal axis due to an applied torque

torsional vibration
> that motion which tends to rotate a rigid body about its axis, alternating in direction of rotation

torso
> that part of the human body generally between the neck and lower limbs, including the thorax, abdomen, pelvis, and the thoracic, lumbar, and sacral regions of the spine, and specifically excluding the head, neck, and limbs; *syn.* trunk

total body electrical conductivity
(TOBEC)
> a technique for estimating/measuring total body fat and lean body mass by placing the body in an electromagnetic field and observing the change in the coil's impedance

total body fat (TBF)
the sum of all the fat deposited in the body; *see body fat*

total body water (TBW)
an estimated value for all the water resident in the body; the sum of both the intracellular and extracellular water

total bottom time (TBT)
the length of time an individual was or has been working at depth underwater; that total time an individual has been rehabilitated at the maximum working pressure in a hyperbaric chamber

total color blindness
see color blindness

total disability
any disability short of death which prevents an individual from following any gainful employment, or which includes the loss of use of: (a) both eyes, (b) one eye plus a hand, arm, leg, or foot, (c) any two members (hand, arm, foot, or leg) not on the same limb

total float
that additional time available for performing an activity beyond its actual duration

total light loss factor
the mathematical product of all recoverable and non-recoverable light loss factors

total lung capacity
the volume contained within the lungs after a maximal inspiration

total metabolic cost
see gross metabolic cost

total organic carbon (TOC)
a measure of the amount of carbon existing in organic molecular form within a water sample

Total Quality (TQ)
a technique intended generate a culture which is quality oriented and provide for continual improvement through developing an environment in which employees gain a better understanding of work processes, employee-management communication is encouraged, and employees are empowered to aid in enhancements; *see also Total Quality Management*

Total Quality Management (TQM)
a management style which is intended to be conducive to total quality

total reaction distance
that distance traveled by a vehicle during the operator's total reaction time

total reaction time
that time required for the operator of a vehicle to make contact with the brake pedal or other stopping mechanism or to begin an evasive maneuver once it is recognized that such an action is necessary

total solids
the sum of all dissolved and suspended solids in water

total span
see span

total stopping distance
that distance required for a vehicle to come to rest from the moment at which the necessity for braking is required; equal to the sum of the total reaction distance, brake reaction distance, and braking distance

total thermal insulation value of clothing (I_T)
the overall thermal insulation value of a clothing system, represented by the sum of the thermal insulation value of clothing and a factor for the insulation value of air divided by a factor for the surface area of the clothing; *syn.* effective thermal insulation value of clothing

$$I_T = I_{cl} + \frac{I_a}{f_{cl}}$$

where:
I_{cl} – thermal insulation value of clothing
I_a = thermal insulation value of air
f_{cl} = clothing area factor

touch

the sensation arising from light to moderate stimulation of the pressure receptors in the skin

touch feedback

see *force feedback*

touch tablet

a pressure-sensitive computer input device consisting of a flat surface which is capable of converting local position to position on a display for cursor or other control

touch temperature

the temperature of an object while possible for an individual to come into contact with it

touch zone

that area within a display on a touch screen which can respond to activation by pressure

touch-sensitive display

see *touchscreen*

touch-tone

pertaining to a numeric keypad on which each key, when depressed, outputs a pair of tones unique to that key

Comment: for use with telecommunications equipment or as a computer input device

touchscreen

a display, at least some portion of which is sensitive to pressure or the position of an object on the face of the screen for input or direct manipulation; also touch screen; *syn.* touch-sensitive display

toxemia

a condition in which the blood contains poisonous products; *adj.* toxemic

toxic

poisonous; see *toxin*

toxic substance

see *toxin*

toxic tort

a lawsuit resulting from exposure to toxic substances

toxic unit

any established unit of toxic activity, often expressed in terms of the minimal dose which will be lethal

toxicant

see *toxin*

toxicity

a measure of the harmful effect by a substance of either biological or chemical origin on a biological mechanism under specified conditions

toxicologist

an individual who is qualified and practices in toxicology

toxicology

the branch of medicine dealing with the study of the nature and effects of poisons on living organisms, including detection, analysis, mechanisms, diagnosis, and therapy; *adj.* toxicological

toxin

any chemical substance capable of producing disease, injury, or death in a living organism; *adj.* toxic; *syn.* toxicant, toxic substance

trace element

any chemical element present in small quantities in living organisms, but essential to normal health, growth, and development

trace minerals

those essential inorganic chemicals found or required in small quantities in living organisms, generally providing the trace elements

traceability

having the capability to verify the original requirement, measurement, standard, calibration, or other statement/process by suitable reference or demonstration

trachea

that tubular portion of the respiratory system between the larynx and the division into the two bronchi which is ribbed by pieces of circular cartilage; *adj.* tracheal; *syn.* windpipe

track [1]

(v) attempt to or follow a moving target, while minimizing the error, on a performance test

track ²

(v) assign to a group according to perceived abilities for educational performance

track ³

(n) the trace left by a moving target on a display or hardcopy

track ⁴

(n) the tendency of individuals to retain their rank order within a population on a given variable over time

trackball

a computer input device consisting of an upward-facing sphere enclosed in a housing with transducers for converting rotational motion caused by the hand or fingers into translational motion of a cursor or other object on a display; *syn.* rolling ball, roller ball

trade

(sl); *see tradeoff*

trade name

a name by which a product is known in normal industry or commerce

trade study

see tradeoff study

trade test

see worksample test

tradeoff

the giving up of some desired characteristics or parts of those characteristics for other desired characteristics when not all desires can be available due to cost or other considerations; also trade-off; *syn.* trade

tradeoff study

any study intended to determine which characteristics of a product should be sacrificed in order to include or retain other characteristics; *syn.* trade study

traditional anthropometry

see classical anthropometry

traffic

the flow of people and/or vehicles along a defined route

traffic density

the average number of vehicles occupying a specified length of road

Comment: usually a mile or kilometer

traffic diagram

a chart or figure to illustrate the traffic flow within a certain region

traffic flow

a measure of the quantity of people or vehicles passing a specified point in a given period of time

tragion

the deepest point at the notch just above tragus

tragion height, sitting

the vertical distance from the seat surface to the level of tragion

Comment: measured with the individual sitting erect, looking straight ahead

tragion height, standing

the vertical distance from the floor to tragion

Comment: measured with the individual standing erect and the body weight distributed equally on both feet

tragion to back of head

the horizontal linear distance from tragion to inion; *equiv.* tragion to wall

Comment: measured with the individual standing erect, looking straight ahead

tragion to top of head

the vertical distance from tragion to the horizontal vertex plane

Comment: measured with the individual sitting or standing erect, looking straight ahead

tragion to wall

the horizontal linear distance from a wall to tragion; *equiv.* tragion to back of head

Comment: measured with the individual standing erect, looking straight ahead, and with his back and head against the wall

tragus

the piece of cartilaginous tissue just anterior to the entrance to the external auditory meatus; *pl.* tragi; *adj.* tragal

trailer

see *trailing hand*

trailing hand

that hand which tends to lag the leading hand when using both hands in a synchronized operation because it is not the center of focus; *syn.* trailer

train

impart one or more particular skills to an individual via some combination of information, instructions, demonstration, and directed activity under controlled conditions leading with the intent of leading directly to the practical use of those skills; *ger.* training

trainability

the capacity for being trained on a given task within a reasonable period of time; *adj.* trainable

training aid

any device or item developed, acquired, or used primarily for assistance in training

training allowance

a compensation in time allowance, performance expectations, or rate of pay due to an experienced employee taking time to train an inexperienced or new worker

training time

that total amount of time involved in training a new worker or a worker being taught a new task

trait

any psychological or physical characteristic of an individual or group

trajectory

the ballistic path taken by the body as a whole, by a point on the body, or of an object released from the body

transarthral

across a joint

transcutaneous electrical nerve stimulation (TENS)

the electrical stimulation of nerves and muscles from electrodes placed on the skin

transducer

a device for converting one form of signal or energy to another form of signal or energy

transfer function

a mathematical relationship between the input and output of a system

Comment: usually a function of frequency

transfer of learning

see *transfer of training*

transfer of training

the phenomenon in which the training, knowledge, and/or information acquired previously on one task affects an individual's ability to be trained later on another task; see *positive transfer, negative transfer*

transform

systematically modify a set of values or an equation to change the form of the relationship; *n.* transformation

Comment: often used to convert power or logarithmic functions to linear functions

transformation

a change in view, perspective, or distance of an object in computer modeling

transient

brief

transient response

a brief phenomenon caused by a sudden change in system conditions

transillumination

the indirect illumination of an object or structure via light transmitted through a translucent object; *syn.* translumination

translate [1]

move from point to point along some path; *n.* translation

translate [2]

convert some text/language, entity, or format from one type into another; *n.* translation

translational motion

see *rectilinear motion*

translatory pedal

a foot-operated device which operates

by simple linear motion in an in-and-out pattern

translucent

allowing a portion of the light incident on a material to pass through as diffused light

translumination

see ***transillumination***

transmissibility

the ratio of the response amplitude to the excitation amplitude in a steady-state, forced vibration system

transmission [1]

see ***neuromuscular transmission***

transmission [2]

the process of passing through a medium without a change in frequency/wavelength

transmission coefficient

the ratio of the transmitted wave intensity to the incident wave intensity at a boundary or discontinuity

transmission lag

a temporal delay due to processing in which the output signal emerges identical to the input except for the time shift

transmission loss

the decrease in amplitude, intensity, or other measure of energy as that form of energy passes through some structure or medium

transmission time

the length of that temporal interval from the transmission of a signal to reception of that signal

transmissivity

see ***transmittance***

transmit

send a message or other information via some communication system; *n.* transmission

transmittance (T, τ)

the value of ratio of the energy flux transmitted through a medium to the incident flux on that medium; *syn.* luminous transmittance, transmissivity

$$T = \frac{I_t}{I_o}$$

where:

I_o = incident intensity
I_t = transmitted intensity

transmitted light

that light emerging from the surface of an object on a non-illuminated side

transmitter

see ***neurotransmitter***

transonic

pertaining to those velocities/speeds near that of sound

transparent [1]

permitting the passage of most incident light through an object without significant diffusion

transparent [2]

an operation or processing sequence which the user doesn't observe when using a system

transport empty (TE)

a work element in which a hand or container is moved without contents

transport loaded (TL)

a work element in which a hand or container is moved with some contents

transportation

the movement of goods or people

transverse [1]

perpendicular to the longitudinal axis of the body or of a body segment
Comment: syn. horizontal if the body is standing erect or the body segment is vertical

transverse [2]

perpendicular to the ground or reference surface; *syn.* horizontal

transverse g

an acceleration vector directed perpendicular to the frontal plane of the body in a reclining or lying posture

transverse pelvic breadth

see ***biiliocristale breadth***

transverse plane [1]

a plane perpendicular to the longitudi-

nal axis of the body or body segment
> Comment: *syn.* horizontal plane if the body is erect or the body segment vertical

transverse plane [2]
a plane parallel to the ground or reference surface; *syn.* horizontal plane

trapezium
one of the distal bones in the wrist; *syn.* greater multiangular bone

trapezius
a broad, flat muscle in the upper back and posterior neck

trapezoid bone
one of the distal bones in the wrist; *syn.* lesser multiangular bone

trapezoidal approximation
the use of the trapezoid area formula for computing the area under a small portion of a curve in integration

trauma
any physical or mental injury induced by an external force or agent; *adj.* traumatic

travel chart
a set of quantitative data, arranged in tabular form, on the movements of workers, materials, and/or equipment between workplaces; *syn.* trip frequency chart

travel time
that time required for personnel, materials, equipment, or hardcopy to move or be moved from one location to another

tread
the horizontal step on a stair

tread depth
the depth of an individual step on a stair, including the overhang

treadmill
a device consisting of a motor- or human-driven belt on which an individual walks, jogs, or runs at certain velocities and/or inclination angles

treble
pertaining to or enhancing the higher auditory frequencies in a sound system; *opp.* bass

tremor
an involuntary movement in which continuous rhythmic oscillations, usually of a smaller amplitude and higher frequency than volitional movements, occur between opposing, normally voluntary muscles at one or more points of the body

trend
a directional tendency for a data set relative to time or other variable

trend analysis
the use of statistical techniques in the evaluation of historical or other data for the determination and/or quantification of any periodicities, correlations, or predictors

trial [1]
a single event during an experimental session in which one or more subjects are presented with a stimulus set and the response set is recorded

trial [2]
a temporary period of testing or evaluation of an individual or product prior to a formal commitment

trial and error
pertaining to a blind, initially random, uninformed search for the correct solution or a path to that solution; also trial-and-error

trial and error learning
the process of narrowing a wide range of possible responses through feedback to the one response or the set of responses which is most appropriate for a given situation; also trial-and-error learning

triangular bone
one of the proximal row of bones in the wrist; *syn.* triquetral bone

triceps
a large, three-headed, voluntary muscle on the lateral posterior upper arm

triceps skinfold
the thickness of a vertical skinfold on the midline of the posterior arm halfway between acromion and the lower tip of olecranon, the point determined

as for the arm circumference measure
Comment: measured with the individual standing erect and the arms hanging naturally at the sides

triceps skinfold, recumbent
the reclining triceps skinfold measure
Comment: measured with the individual lying on his side such that the right and left acromial processes are perpendicular to the bed/table surface and the arm is lying against the side of the body

trichromat
an individual with normal color vision; *see also* **trichromatopsia**

trichromatic theory
a color vision theory based on the concept that any hue may be derived from an appropriate mixture of three primary colors

trichromatopsia
the condition or state of having normal color vision; *see also* **trichromat**

trigeminal nerve
the fifth cranial nerve, containing motor fibers, which are involved in hearing and mastication, and sensory fibers, which convey touch, pain, and temperature from the facial region

trigger finger syndrome
a stenosing tenosynovitis of the index finger

triglyceride
a chemical combination of three fatty acids and a glycerol molecule

trigonometry
the study and measurement of triangles and the relationships of their components; *adj.* trigonometric

trill
a consonant produced by the rapid vibration of one or more articulators

trim
remove some proportion of the values at the extremes in a dataset before processing the data

trip ^{1}
the process of falling forward after the

toe or some other portion of the foot strikes an object above the normal surface elevation which impedes the foot's forward motion

trip ^{2}
that travel from one point to another

trip frequency
the number of instances in which a trip is made from one point to another within a unit time

trip frequency chart
see **travel chart**

trip hazard
any object projecting above the typical level surface over which an individual may trip when walking; *syn.* tripping hazard

tripodal grasp
a type of grasp where an object is held by the combination of the index finger, middle finger, and the thumb, with the object possibly extending toward or touching the palm

tripping hazard
see **trip hazard**

triquetral bone
see **triangular bone**

tristimulus colorimeter
a colorimeter which provides tristimulus values on measuring a color

tristimulus value
the amount of each of the three primary colors, represented by magnitudes of X, Y, and Z, which are used by a CIE Standard Observer to match a given color; *syn.* spectral tristimulus value, color matching function

tritanomaly
a form of color vision deficiency involving a reduced ability to discriminate blue colors within stimuli

tritanope
one who has tritanopia

tritanopia
a form of color blindness involving an inability to discriminate blue colors due to the absence or nonfunction of the blue cone in the retina; *syn.* blue blindness; *adj.* tritanopic

trochanter, greater
   see **greater trochanter**

trochanter – knee – ankle line (TKA line)
   an imaginary straight line from the
   trochanter, through the knee, to the
   ankle, which approximates the body
   weight support axis when standing
   erect

trochanter, lesser
   see **lesser trochanter**

trochanteric height
   the vertical linear distance from the
   floor or other reference surface to
   trochanterion
      Comment: measured with the indi-
      vidual standing erect and his body
      weight distributed equally on both
      feet

trochanterion
   the most superior lateral point on the
   greater trochanter of the femur

trochlea
   a bone structure, rounded and depressed in the middle and high on both
   sides, at the anterior distal end of the
   humerus for articulating with the ulna
   at the elbow

trochlear nerve
   the fourth cranial nerve, containing
   both motor fibers, which innervate the
   superior oblique muscle of the eye, and
   sensory fibers, which convey proprioception from that eye muscle

trochlear notch
   a large, crescent-shaped depression in
   the proximal ulna for articulation with
   the humerus; syn. semilunar notch

troffer
   a long, recessed lighting unit, typically
   placed so the opening is flush with the
   ceiling surface

troland (td)
   a unit of retinal illuminance; equal to
   the value of the ratio of the luminance
   of a surface or light source in cd/m^2 to
   the area of the pupil in mm^2; syn. luxon
      Comment: originally called a photon
      — now obsolete usage, since pho-
      ton has another meaning

tropical climate
   a climate having generally high temperatures and humidity through much
   of the year, with high rainfall amounts
   during at least part of the year

true value
   that theoretically correct measurement
   of some characteristic of a system

truncal depth
   the maximum horizontal linear depth
   of the trunk, not including the bust in
   women, at whatever level it occurs;
   also trunk depth; see **abdominal depth,
   chest depth, waist depth**
      Comment: measured with the indi-
      vidual standing erect; note the level
      at which the measurement is taken

truncal height, sitting
   see **suprasternale height, sitting**; also
   trunk height, sitting

trunk
   see **torso**; adj. truncal

tuberosity
   a relatively large protuberance on a
   bone

tubular bone
   see **long bone**

tuning fork test
   any of several simple hearing tests performed with a tuning fork; see **Weber
   test, Rinne test**

tunnel
   an enclosed passageway, usually artificial and underground

tunnel vision
   a reduced ability to see toward the
   periphery of the normal visual field

turbidity
   cloudiness, due to suspended particles
   or dirt, as in water or air; adj. turbid

turnaround time
   the period of time required to check
   out an item, service it if necessary, and
   return it to operating status; also turnaround time

turnkey
   pertaining to a system which is delivered and installed to be fully opera-

tional, such that the user has only to turn on the system to operate it

turnover

the process of having one or more employees leave of their own will and having to replace them to maintain the desired operational performance level

tweeter

a loudspeaker designed for reproducing the higher audible sound frequencies, generally above 3 KHz

twilight

that intermediate lighting level between daylight and dark during the period of each day before sunrise and after sunset

twilight vision

see *mesopic vision*

twill weave

a weave pattern in which lines run diagonally to the length of the fabric

twist

rotate a body segment about its longitudinal axis or some portion of the entire body about its (normally vertical) longitudinal axis

twisting moment

that combination of torques produced by contraction of various muscle groups and ground reaction forces to yield a twisting motion of some portion or all of the body

two-alternative, forced choice paradigm

an experimental design in which a subject is presented with one of two stimulus alternatives on each trial of an experiment, and he must indicate which stimulus he believes was present

two-and-a-half-D model

the stacking of several two-dimensional components, each of which is composed of a cross-section of a three-dimensional object, to give a semblance of depth in the image

two-hand controls

a safety interlock control technique which requires the operator to have both hands on the control system, thus

insuring that the hands will not be in the operational area while the machine is functioning

two-handed normal working area

see *normal working area (two handed)*

two-handed process chart

a special case of a multiple activity process chart detailing by symbols, text, and/or graphics the individual and relative motions of the two hands or limbs in a workplace operation, generally for a repetitive task; *syn.* right- and left-hand process chart, workplace chart, operation chart

two-interval, forced choice paradigm

an experimental design in which a stimulus is presented to the subject in only one of two sequential time periods, and the subject must indicate which period he believes contained the stimulus

two-man rule

a policy or procedure used in certain highly critical or hazardous situations in which two authorized individuals must be present at all times to ensure use of the correct procedures, to help each other, or to enable one to remove the other from the situation in the event of some accident; *syn.* two-man concept, two-man policy

two-point threshold

that minimal tactile separation distance on a given region of the skin which can be distinguished as two stimuli instead of one when two distinct stimuli are applied; *syn.* resolution acuity

two-sided test

see *two-tailed test*

two-step test

see *Master's two-step test*

two-tailed test

a non-directional statistical test of significance in which the null hypothesis should be rejected if the sample value is either greater or less than pre-established critical values; *syn.* two-sided test; *see also* **one-tailed test**

tympanic membrane

a thin, semitransparent tissue layer separating the outer ear from the middle ear and which is involved in the transduction of airborne sound in hearing; syn. eardrum, tympanum

type I error

an error in statistical judgement in which a true hypothesis is falsely rejected; also Type 1 error

type II error

an error in statistical judgement in which a false hypothesis is not rejected; also Type 2 error

typeface

a particular style or size of letter in printing or displays

typhoon

a hurricane in or adjacent to the Pacific Ocean

typing stick

see **mouth stick**

U

U function

see *convex function*; *opp.* inverted U function

ulna

the medial bone in the forearm; *adj.* ulnar

ulnar deviation

a movement or position of the longitudinal axis of the hand toward the ulnar/little finger side of the forearm

ulnar nerve

a spinal nerve innervating generally the medial part of the forearm and hand

ultimate factor of safety

that number by which the load limit is multiplied to yield the ultimate load

ultimate load

the product of the load limit and the ultimate factor of safety

ultradian

pertaining to periodicities having a period of less than 24 hours

ultradian rhythm

a biological rhythm having more than one cycle per day, or a period less than a day in length

ultrahigh frequency (UHF)

that portion of the electromagnetic spectrum consisting of radiation frequencies between 300 MHz and 3 GHz

ultrasonic

see *ultrasound*

ultrasonics

the use or study of ultrasonic sound energy

ultrasonography

the use of pulsed ultrasound and echo recording for diagnostic purposes in the human body or elsewhere

ultrasound

that acoustic energy with frequencies higher than those the human ear can normally hear, above about 20 KHz; *adj.* ultrasonic

ultraviolet (UV)

pertaining to that region of the electromagnetic spectrum having wavelengths between about 10 nm and 400 nm; *see also* **ultraviolet A, ultraviolet B, ultraviolet C, far ultraviolet, middle ultraviolet, near ultraviolet**

ultraviolet A (UV-A)

that portion of the ultraviolet spectrum between about 315 nm and 400 nm; partial *syn.* near ultraviolet

ultraviolet B (UV-B)

that portion of the ultraviolet spectrum between about 280 and 315 nm; *syn.* actinic ultraviolet

ultraviolet C (UV-C)

that portion of the ultraviolet spectrum between about 100 and 280 nm; partial *syn.* far ultraviolet

ultrawide band radar

a radar system emitting a broad range of frequencies, from megahertz to gigahertz

umbilicus

the residual scar on the inferior, anterior abdomen from the removal of the umbilical cord; *syn.* navel

umbilicus height

see *omphalion height*

umbra

the small central core of the greatest ionization within an ionization track of tissue or other material

unavoidable delay (UD)

a therblig representing any delay in an ongoing process or operation which is beyond the control or responsibility of a worker

unavoidable delay allowance
see *delay allowance*

unburden
ease the human physical or mental workload through computerization, mechanization, or some other means

undependability
that measurement error due to physiological variation over time; see also *unreliability, imprecision*

undergarment
a single piece of underwear

underline
a highlighting technique in which one or more straight lines are drawn below a line of text; see also *single underline, double underline*

underwear
that minimal clothing worn adjacent to the body and which one does not normally wear exposed when in public
Comment: usually referring to pants and bra for women, shorts and t-shirt for men

undo
re-establish the condition on a display prior to execution of the last command

unilateral
pertaining to only one side of a body or structure

unilateral teleoperator
a type of teleoperator system capable only of one-way force and motion transmission — from the operator to the teleoperator

uniocular
see *monocular*

unit strain
the change in length due to stress divided by working length

unnecessary cost
any cost in money, time, materials, energy, or other asset which does not contribute to the quality, usefulness, life, appearance, or features of a product

unoccupied time
that time which may result from a worker doing neither internal work nor taking a rest allowance due to machine-controlled time or a team effort in which his immediate participation is not required

unoccupied time allowance
that allowance made for unoccupied time

unrecoverable light loss factor
see *non-recoverable light loss factor*

unrestricted element
see *manual element*

unrestricted job
see *self-paced job*

unrestricted work
see *self-paced work*

unsafe act
any departure from mandated, accepted, or standard practice, whether of commission or omission, which either has caused or has the potential for causing personal injury or property damage

unsafe condition
see *hazard*

unvoiced sound
a sound produced by air turbulence under pressure at a constriction point in the vocal tract; opp. voiced sound

up-and-down methodology
(sl); see *staircase procedure*

upper arm circumference
the surface distance around the upper arm at the level of the midpoint between the lateral projection of acromiale and the inferior aspect of the olecranon process of the ulna
Comment: measured with the individual standing erect and the upper arms hanging naturally at the sides with the elbow flexed 90°

upper arm circumference, recumbent
the upper arm circumference in a reclining position
Comment: measured with the arm

lying beside the body and the elbow raised slightly by a small pillow or other pad

upper explosive limit (uel)

see upper flammable limit; opp. lower explosive limit

upper extremity

the hand, wrist, and arm, with its junction to the shoulder

upper flammable limit (ufl)

that concentration at a given temperature of a flammable gas or flammable liquid vapor in air above which flame propagation will not occur; *syn.* upper explosive limit; *opp.* lower flammable limit

upper thigh circumference, sitting

see thigh circumference (sitting)

upper thigh circumference, standing

see thigh circumference (proximal)

upper threshold

the maximum stimulus intensity which will produce a specific type of sensory experience or elicit a specific response; *syn.* terminal threshold; *opp.* absolute threshold

uptime

the period of time during which a system or system element is either performing or capable of performing

urethra

the tubular structure which extends from the bladder to the body surface for excretion of urine

urinal

a receptacle into which urine may be directed and which is normally connected to appropriate plumbing

urinate

eliminate urine from the body via the urethra; *n.* urination

urine

the liquid excretion from the kidneys (normally containing various salts, urea, bodily metabolites, and some solids) which is passed to the exterior through the bladder and urethra

usability

a measure of the ease with which one may use a product or learn how to use a new product

use (U)

a therblig representing the implementation of some device's intended function under the control of one or both hands

useful life

that period of time in the existence of a machine or system following any run-in phase and prior to the wear-out phase in which it is generally functionally stable in its operation; *see life characteristic curve*

user guidance

any prompt or feedback which assists the user in performing a computerized task

user interface development system (UIDS)

any set of software tools which may be used for developing user interfaces; *see also user interface management system*

user interface management system (UIMS)

any set of software tools which may be used for managing the user interface; *see also user interface development system*

user response time

the period required for or taken by a user to enter a command or reply to a display prompt

user-computer interaction

see human-computer interaction

user-computer interface

see human-computer interface

user-friendly

(sl) designed according to human factors/ergonomic principles with claimed or demonstrated ease of use

utricle

an expanded region in the membranous labyrinth of the vestibular apparatus which contains endolymph and the utricular macula

utricular macula
the mechanoreceptor of the utricle, composed of sensory hair cells and a gelatinous mass with embedded otoliths; *syn.* macula

Comment: involved in static equilibrium

uvula
a muscular tissue descending from the midline of the soft palate

V

vacation
an employee benefit in which a given number of days off from work per year are provided with pay for the employee
Comment: may be certain restrictions as to when the time off may be taken

vacuum
a condition in which air pressure approaches zero

vagina
a tubular structure associated with the female genitalia and located between the bladder and rectum; *adj.* vaginal

vagus nerve
the tenth cranial nerve, having a wide distribution and having motor functions including speech production and swallowing, with sensory functions including pressure/touch in the pharyngeal region, abdominal distention, and nausea; *pl.* vagi; *adj.* vagal

validate
demonstrate that a test, standard, or other device has validity; *n.* validation

validity
the degree to which a test or other measurement device really measures what it was designed to measure; *see criterion-related validity, face validity, construct validity, content validity*

valsalva maneuver
a procedure in which the nostrils are pinched off, the mouth kept closed, and pressure voluntarily increased in the nasopharyngeal region to aid in equalizing pressures within the nasopharynx and middle ear; also valsalva

value [1]
the estimated, appraised, or actual market worth of a product or service

value [2]
the ratio of the cumulative benefits of an item to the cumulative costs of that item

value [3]
see Munsell value

value [4]
the numerical magnitude of some measurable quantity

value added
that difference between the sales income from goods and the costs of materials, supplies, and any outside services used in their production and delivery

value adding time
that portion of the time in the work sequence which increases a product's value by work performed on the product

value analysis
a systematic study to determine costs in each production phase for manufacturing an item, either during the engineering phase of product development or on an already existing product
Comment: generally with the intent to reduce costs by eliminating unnecessary steps

value control
see value engineering

value engineering
the application of engineering techniques toward providing a functional product or service at the lowest cost; *syn.* value control

value index
a dimensionless number which represents the ratio of worth to cost

value of human capital
the present economic worth of an

individual's expected future earnings from a given age due to employment

value set

those social standards commonly accepted by an individual, group, or society; *syn.* values

vapor

a gas, usually pertaining to a substance which is normally either a solid or liquid at room temperature and pressure

vapor pressure

that pressure exerted by the gas phase of a substance when in equilibrium with its liquid or solid source

variability

that quality which leads to obtaining different results under the same or different conditions; *see also* **dispersion**

variable

an entity, often represented by a symbol, which is capable of having any value within a specified set or range

variable cost

a manufacturing or service cost which changes with the quantity produced or the level of services

variable element

a work element whose time for completion varies due to one or more changes in characteristics of the product or service, either across jobs or from one work cycle to another

variable error

a deviation from an obtained value on replication of a psychophysical experiment due to random or unknowingly altered conditions between the measurements

variance [1]

a measure of the dispersion about the mean of a distribution, represented by the second moment of the deviations from a mean value in a normally distributed population; *see also* **standard deviation**

$$\mathrm{var} = \frac{\sum X^2 - \dfrac{(\sum X)^2}{N}}{N-1} \quad (computational)$$

$$\mathrm{var} = \frac{\sum (X_i - \overline{X})^2}{N} \quad (theoretical)$$

variance [2]

a deviation from a standard or otherwise normal or specified value

variance ratio test

see **F test**

varicose vein

an enlarged vein, especially in the leg and visible as blue streaks beneath the skin, generally due to overstretching by an excessive venous pressure for an extended period of time, resulting in valve damage

vasoconstriction

a decrease in the cross-sectional area of a blood vessel

vasodilation

an increase in the cross-sectional area of a blood vessel; also vasodilatation

vasomotor

pertaining to neural control of the arterial smooth muscles in regulating blood flow; *see* **vasoconstriction, vasodilation**

vector [1]

any physical quantity having both magnitude and direction, and which may be combined using certain rules

vector [2]

a matrix composed of a single row or column; a column or row within a matrix

vector [3]

any organism which is capable of carrying microorganisms from an ill person to a healthy person, spreading the disease

vector graphics

see **stroke writing**

veg

a psychophysical scale for heaviness

veiling luminance

a luminance added to an object or display which reduces the contrast and may result in disability glare

veiling reflection

a diffusion of the external lighting impinging on a surface or display which tends to result in disability glare

vein

a vessel in the cardiovascular system which carries blood back toward the heart; *adj.* venous

velar [1]

see velum

velar [2]

articulated with the tongue on or near the velum

velocity ($\vec{v}$)

a vector representing the rate of change of position or displacement of an object with time

$$\vec{v} = \frac{d\vec{x}}{dt}$$

velocity control

see rate control

velum

see palatine velum; adj. velar

velvet

a fabric having a short pile of silk, cotton, or other material on a closely woven backing

velveteen

a fabric resembling velvet which has a short pile and a cotton filling

Venn diagram

a graphical means of illustrating the relationships between sets in set theory, usually using a square or circle to represent a set

venous valve

a structure within the veins of the legs which permits only unidirectional flow of blood back toward the heart

ventilate [1]

circulate fresh air and remove stale/contaminated air within a closed space; *n.* ventilation

ventilate [2]

oxygenate and remove carbon dioxide from the blood in the lungs via diffusion in the alveoli; *n.* ventilation

ventilation rate

the amount of gases and water vapor exchanged between the atmosphere and the lungs per unit time

ventilometer

an instrument used for measuring the various lung capacities and volumes as a function of time

ventral

pertaining to the underside of an object, or the side opposite the back of an organism

verbal

pertaining to words, often referring to spoken words

verification by analysis

the use of techniques such as interpolation or extrapolation of pre-existing data, references, simulation, or modeling to illustrate that a system or subsystem complies with requirements or specifications

verification by demonstration

the use of more qualititative techniques such as observing performance, maintainability/servicing, or other appropriate aspects to verify that a system complies with requirements or specifications

verification by inspection

a technique for confirming that a product meets design specifications/requirements through the use of vision or simple measuring devices and reviewing supporting documentation

verification by similarity

the use of techniques such as a comparison with the design, manufacturing, and quality control standards of a previously existing product which has met the same or greater criteria to indicate that the item under consideration will likely meet the necessary requirements and specifications

verification by test
a technique for confirming that a product meets specifications/requirements through the use of measurements taken during and/or following the controlled application of appropriate functional and environmental stimuli, often with the use of sophisticated lab test equipment, recorded data, and some analysis of that test data

vernier
a smaller scale, in association with a larger one, which is divided to permit interpolation between the divisions of the larger scale

vernier acuity
a measure of the alignment judging ability of the eyes, determined by the reported existence or degree of any lack of alignment or co-linearity of two parallel lines, with one placed vertically above the other in a plane

vertebra
one of the bones which form part of the spine; *pl.* vertebrae; *adj.* vertebral

vertebral column
see ***spine***

vertex
the highest point on the top of the head with the head oriented in the Frankfort plane; *syn.* crown

vertex plane
that horizontal plane which intersects the top of the head and is parallel to the floor or other reference surface when in an erect posture

vertical grip reach, sitting
the vertical distance above the sitting surface which an individual can reach while gripping a pointer held in a clenched fist perpendicular to the longitudinal axis of the arm
　Comment: measured with the individual sitting erect with the shoul-

der rotated upward 180° or as close to that as possible, and the elbow and wrist fully extended

vertical grip reach, standing
the vertical distance above the floor which an individual can reach while gripping a pointer held in a clenched fist perpendicular to the longitudinal axis of the arm
　Comment: measured with the individual standing erect with the shoulder rotated upward 180° or as close to that as possible, and the elbow and wrist fully extended

vertical job enlargement
see ***job enrichment***

vertical leg room
see ***knee well height***

vertical reach
the vertical distance from the floor or other reference surface to the tip of the middle finger (digit III) when one arm, wrist, the hand, and the fingers are extended vertically
　Comment: measured with the individual standing erect

vertical reach, sitting
the vertical distance from the sitting surface to the tip of the middle finger (digit III) when one arm, wrist, hand, and the fingers are extended vertically
　Comment: measured with the individual sitting erect

vertical scroll
view text or alphanumeric information on a screen by advancing vertically line-by-line under operator control to examine the information in a file above or below that currently visible; *opp.* horizontal scroll

vertical trunk circumference, sitting
an off-vertical surface loop distance around the torso, from the crotch, passing diagonally up and over the midshoulder, down through the small of the back, to return to the crotch passing over the posterior protuberance of the buttock
　Comment: measured with the indi-

vidual sitting erect

vertical trunk circumference, standing
an off-vertical surface loop distance around the torso, from the crotch, passing diagonally up and over the midshoulder, down through the small of the back, to return to the crotch passing over the posterior protuberance of the buttock; also vertical trunk circumference
Comment: measured with the individual standing erect with the body weight equally distributed on both feet

vertigo
a condition in which objects in the environment appear either to revolve around the individual or the individual appears to rotate

very high frequency (VHF)
that portion of the electromagnetic spectrum consisting of radiation frequencies between 30 MHz and 300 MHz

very low density lipoprotein
a substance having the function of carrying triglycerides from the liver through the blood vessels to the cells, forming low density lipoprotein after the release of the triglycerides

very low frequency (VLF)
that portion of the electromagnetic spectrum consisting of radiation frequencies between 10 KHz and 30 KHz

vesicant
any substance which produces blistering

vestibular
see vestibule

vestibular apparatus
that portion of the vestibular system physically located within the inner ear for transduction of mechanical stimuli

vestibular membrane
a delicate membrane separating the cohlear duct from the scala vestibuli in the cochlea; *syn.* Reissner's membrane

vestibular nerve
that branch of the vestibulocochlear

nerve which carries equilibrium information from the inner ear to the brain

vestibular nystagmus
that nystagmus produced either from stimulation of the labyrinth or a diseased/damaged vestibular system; *syn.* labyrinthine nystagmus

vestibular sense
that sense mediated by the vestibular system, generally involving head position, motion, or orientation with respect to gravity

vestibular system
those structures consisting of the otoliths, semicircular canals, vestibular nerve, and related sensory structures of the inner ear which transduce head motion, orientation, and equilibrium/balance and convey that information to the brain

vestibule
a cavity within the osseous labyrinth of the inner ear which interconnects the cochlea and the semicircular canals; *adj.* vestibular

vestibulo-ocular reflex (VOR)
those reflex eye movements during a head movement which are intended to maintain a stationary image on the retina

vestibulocochlear nerve
the eighth cranial nerve, having a sensory function and which combines the vestibular and auditory branches; *see auditory nerve, vestibular nerve*

vibrating conveyor
see oscillating conveyor

vibration
any mechanical oscillation of a structure or system

vibration isolation
the separation of structures with the use of one or more vibration isolators or other means

vibration isolator
a resilient support which separates a mechanical system from the structure on which it is mounted and greatly or

essentially completely reduces the transfer of vibrational energy from one to the other

vibration sense

the sensing of vibration via repeated activation of the touch/pressure receptors in the skin; *syn.* vibratory sense, vibrotactile sense

vibration syndrome

any sign or symptom associated with the use of vibrating tools or equipment, ranging from numbness, blanching, or tingling to a recognized disease; *see **Raynaud's phenomenon, beat knee***

vibration white finger (VWF)

*see **Raynaud's phenomenon***

Vibroacoustic Payload Environment Prediction System (VAPEPS)

a computer modeling program and database for predicting vibroacoustic levels within vehicles

vibroacoustics

the field of study dealing with the combined or interrelated effects of sound and vibration on systems and/or people

vibrotactile sense

*see **vibration sense***

vibrotactile stimulation

the application of a mechanical vibration by any means to cause displacement of the touch/pressure receptors in the skin

video display terminal (VDT)

a computer workstation having a monitor capable of processing one or more types of television signals or other displays, and having one or more computer input devices for interaction; *see also **visual display terminal**, **video display unit***

video display unit (VDU)

a monitor capable of processing one or more types of television format signals for viewing, with no interactive controls available other than simple channel selection, picture adjustments, volume, etc.; *see also **visual display unit**, **video display terminal***

videography

the use of video electronic media for the acquisition, presentation, and/or study of moving visual images

view

focus one's visual attention on; *ger.* viewing

vigilance

a state in which an individual sustains a high level of attention in an attempt to detect a signal, a change in signal, or a particular activity

violet

that hue typically perceived when a normal retina is stimulated with electromagnetic radiation with wavelengths between approximately 390 nm to 450 nm

virgin wool

that wool fresh from clipping of sheep and used for the first time

virtual environment

a computer-generated, three-dimensional environmental simulation in which the user is able to sense and interact with that environment via some set of transducers or computer input devices; *syn.* virtual presence, artificial reality, virtual reality, virtual image display

virtual image

an image in which light appears to emanate from one or more points, as behind a mirror or as presented in a computer-driven image generator, but in reality does not

virtual image display

*see **virtual environment***

virtual presence

*see **virtual environment***

virtual reality

*see **virtual environment***

virtual workspace

a type of virtual environment which provides an analog to an individual's actual workspace

virus [1]

a computer code, generally written with

the malicious intent to interact with system software to overwrite files, present a message, or other such activity in a computer, possibly effectively rendering the files or system useless; *adj.* viral

virus [2]

any of a group of biological entities which generally have a protein shell encasing nucleic acids and which are dependent on other organisms to support their reproduction; *adj.* viral

viscera

see viscus

viscosity

a measure of that resistance which a liquid substance exhibits to flow over itself

viscous

having a high viscosity

viscus

an organ within one of the bodily cavities; *pl.* viscera; *adj.* visceral

visibility [1]

a measure of the capability of being seen through a combination of factors such as luminous intensity, contrasts, intervening conditions between the observer and object(s), object size, and distance from the observer

visibility [2]

the recognizability of an individual or group to another individual, group, or the public at large through repeated appearances, media coverage, or other means

visibility limit

see visual range

visibility meter

a photometer which measures the amount of reduction in light intensity required to bring the visibility of objects to their threshold value

visibility reference function

a graphical curve or mathematical equation providing the relationship between an individual's ability to detect an object against a background and the background luminance

visible

capable of being seen with the eye

visible spectrum

that range of wavelengths/frequencies of the electromagnetic spectrum which the eye is capable of normally detecting, approximately 380 nm to 750 nm wavelength

vision

the capacity for seeing; the act of seeing; *adj.* visual; *syn.* sight

vision test

any eye test for measuring visual acuity or color sensitivity; *see Snellen test, Ishihara test, Stilling test*

visor

any device which may be placed over or in front of the eyes to shield or protect them from intense light radiation or glare, blast effects, wind, blown grit or dust, or any other noxious physical agent

visual acuity

a measure of the ability to resolve distinct objects or fine detail with the eye; *see Snellen acuity, vernier acuity, stereoscopic acuity; see also resolution acuity*

visual adaptation

a change in visual sensitivity over prolonged viewing of a particular intensity, color, or other aspect of light; *see dark adaptation, light adaptation, color adaptation*

visual angle

the angle subtended at the nodal point of the eye by the height or width of an object in the visual field

visual axis

an imaginary line, internal to the eye, projected from the point being fixated, through the lens, to the center of the fovea; *syn.* visual line; *see also optical axis*

> *Comment:* the visual axis is separated from the optical axis by about 4°

visual clutter

having too much patternless detail in

the visual field or on a display such that a sensory overload condition exists; *see also visual noise*

visual coding

any type of coding which relies wholly or primarily on the visual modality; *see symbol, shape coding, color coding, size coding*

visual colorimetry

the use of the human eyes to judge relative similarities and differences in hue; *see also visual photometry*

visual comfort

having an adequately lighted visual environment, without glare, and with pleasing hue(s); *opp.* visual discomfort

visual comfort probability (VCP)

an estimate of the probability that a given observer will rate a given lighting system as equal to or more comfortable than the visual comfort-discomfort boundary under given conditions

visual correspondence

the use of a visual display which is slaved to the position of the operator's head

visual display terminal (VDT)

a computer or other workstation having a monitor capable of processing/presenting information for viewing in one or more formats and having one or more input devices for interaction; *see also video display terminal, visual display unit*

visual display unit (VDU)

a monitor capable of presenting one or more types of signals for viewing, but with no interactive capability; *see also video display unit, visual display terminal*

visual environment

that external physical and psychological volume having characteristics generated by all of the following: (a) the luminous environment, (b) the structure of the volume, and (c) any objects within that volume

visual field

that part of the visual environment which can be seen by the eye(s) at any given instant with the head and eyes stationary; *see also field of view, central visual field, peripheral visual field, monocular visual field, binocular visual field*

visual field defect

any impairment where an individual cannot see one or more regions within the visual field

visual flight rules (VFR)

a set of regulations which dictate that an aircraft flight may only be made under conditions in which navigation can be safely accomplished using visual references to the earth's horizon, with sufficient visibility, and below certain altitudes; *opp.* instrument flight rules

visual line

see visual axis

visual noise

an array of elements or images on a display or in the visual field which appear to or are intended to have no pattern; *see also visual clutter*

visual perception

the process of, or the product from, interpreting visual stimuli

visual persistence

see afterimage

visual photometer

any device for judging the equality of brightness of two surfaces using the eyes, rather than instrumentation hardware; *see also physical photometer*

visual photometry

the study or process of using the eye instead of a photoelectric device as the sensing element for brightness differences; *see also visual colorimetry*

visual pigment

any of the chemicals involved in transduction of light energy to chemical energy in the retina

visual position constancy

the tendency for the visual field to appear stable as the observer moves his head or eyes, due to the vestibulo-ocular reflex

visual purple
   see rhodopsin

visual range
   the maximum distance at which the contrast between one or more distinct objects and their background enables them to be distinguished by an observer under given environmental conditions; *syn.* visibility limit

visual space
   the integrated sum of all possible visual fields from a given body location, given only those bodily movements which would normally be available to an individual under specified conditions

visual strain
   see eyestrain

visual surround
   all portions of the visual space except that pertaining to the task at hand

visual task
   any task or portion of a task which requires vision and the perception/integration of visual stimuli for its performance

visualization
   the ability to create a mental visual image, either of some original object or an imagined object

vital capacity
   the maximum volume of air which can be expelled from the lungs after a maximal inspiration; *syn.* lung vital capacity, respiratory capacity

vital statistics
   that organized personal data pertaining to health, deaths, births, marriages, and divorces which is a matter of public record

vitamin
   one of a group of organic compounds not normally produced by the body but which is required in small quantities for normal body health and metabolism

vitreous humor
   the transparent, gelatinous substance which fills the posterior cavity of the eyeball; *syn.* vitreum, vitrina

vocabulary
   the number of words or terms readily available to an individual or computer for on-line use

vocal
   pertaining to certain speech organs or structures; conveyed using the voice

vocal cord
   one of two ligaments within the larynx which, on movement, are involved in speech or other sound production; also vocal chord; *syn.* vocal ligament

vocal fold
   a thin mucous membrane covering the vocal cord

vocal ligament
   see vocal cord

vocal tract
   the combination of passageways and enclosing structures which are involved in the mechanical production of speech sounds

vocalis
   a medial portion of the thyroarytenoid muscle which attaches to the vocal cord

vocational aptitude test
   any examination for determining which occupation an individual is best suited

vocational guidance
   the use of results from one or more vocational aptitude tests, interviews, trend forecasts, or other measures to counsel an individual in job selection

vocoder
   a device which produces voice-like sounds, and which, through the appropriate combinations of sounds, can synthesize speech

voice
   carry out the purposeful vibration of the vocal cords to produce a phoneme; *ger.* voicing

voice recognition [1]
   the human attribution of the vocal

sounds from an individual to that individual

voice recognition [2]

the use of a computer to compare the spectra of selected spoken words by one or more specified individuals with those spectra of words previously established in memory to perform a match; *see speech recognition, word recognition, speaker identification, speaker verification*

voice-activated (VOX)

pertaining to a piece of equipment having the capability to initiate certain operations in response to sounds in the frequency range of the human voice; *syn.* voice-operated

voice-operated

see voice-activated

voiced sound

any sound occurring as quantities of subglottal air are forced against tightened vocal cords, causing them to open and close at certain intervals and air to resonate in vocal cavities; *opp.* unvoiced sound

voiceprint

a selected spectral density function of an individual's speech; also voice print

vola

the palm of the hand or the sole of the foot; *syn.* volar surface; *adj.* volar

volar

see vola

volitional movement

see active movement

volt (V)

a unit of electrical potential or electromotive force; that potential difference between two points of a conductor carrying a constant current of one amp when the power dissipated between those two points is one watt

volt-ampere (VA)

the SI unit of apparent electrical power; the mathematical product of voltage and amperage, either available to or used by a system; also volt-amp

volume [1]

the amount of three dimensional space which an object occupies

volume [2]

the loudness of an audio output

volume control

a potentiometer or other device which changes the loudness of an audio output from an amplifier

volume display

any display which indicates relative volume of an audio output or input; *syn.* volume indicator

volume indicator

see volume display

volume velocity (U_v)

the flow rate of a medium due to a sound wave through a cross-sectional or surface area

voluntary

being under willful control; *syn.* volitional

voluntary informed consent

a voluntary agreement by a potential experimental subject to allow himself to be exposed to the conditions of a test; *syn.* informed consent

voluntary muscle

a muscle which is normally controllable by the individual without any highly specialized training; *opp.* involuntary muscle

voluntary standard

any standard which is complied with on a voluntary basis, without any legal requirement or consensual agreement to do so

vomer

a facial/skull bone in the mid-sagittal plane which forms part of the nasal septum

vomit

(v) forcibly eject swallowed material from the stomach

vomitus

that material ejected from the gastrointestinal tract during the process of vomiting

Von Frey filament
  any of a set of fibers of various lengths and thicknesses which is calibrated to exert a given force when pressed on the skin; *syn.* Von Frey hair; *see also* **hair esthesiometer**

Von Frey hair
  see **Von Frey filament**

wage

the monetary compensation for services provided to an employer

wage incentive

a financial reward reflected in a worker's wages for greater than normal performance

wage incentive plan

an incentive plan for determining additional wages to be paid an individual or group based on exceeding standard or normal performance, and having the intent to increase output; *syn.* incentive wage system

wage rate

the hourly or other time-based wage described in monetary terms; *syn.* rate; *see also* **piece rate**

waist [1]

that level above the hip and below the thorax at which the torso has a minimum breadth when viewed from the front

Comment: if such a level is not apparent, as in pregnant or obese individuals, use the level at which the belt is worn

waist [2]

the level of omphalion

waist back length

the surface distance, along the spine, from the waist level to cervicale; *syn.* waist back

Comment: measured with the individual standing erect and the weight balanced equally between both feet

waist breadth

the horizontal linear distance across the torso at the waist

Comment: measured with the individual standing erect

waist circumference

the surface distance around the torso at the waist height

Comment: measured without tissue compression with the individual standing erect, the weight equally balanced between both feet, and the waist muscles relaxed

waist circumference, sitting

the surface distance around the torso at that level represented by the waist height

Comment: measured without tissue compression, with the individual sitting erect

waist depth

the horizontal linear distance from the back to the front of the torso at the waist

Comment: measured with the individual standing erect, the weight evenly distributed on both feet, and the torso muscles relaxed

waist depth, sitting

the waist depth in a sitting individual

Comment: measured with the individual sitting erect and the torso muscles relaxed

waist front from cervicale

the surface distance from cervicale to the waist level, passing along the base of the neck to the neck-shoulder intersection and following the contour of the anterior body to the waist level in the midsagittal plane

Comment: measured with the individual standing erect and the torso muscles relaxed

waist front length

the surface distance in the midsagittal plane from the waist level to supra-

sternale; *syn.* waist front
> *Comment:* measured with the individual standing erect and the torso muscles relaxed

waist height
the vertical distance from the floor or other reference surface to the waist
> *Comment:* measured with the individual standing erect

waist height, omphalion
see **omphalion height**

waist height, sitting
the vertical distance from the upper sitting surface to the waist level
> *Comment:* measured with the individual sitting erect

waiting line
see **queue**

waiting line theory
see **queuing theory**

waiting time
see **delay time**

walk
a type of gait in which each foot is normally placed sequentially along a line parallel to the direction of travel, with at least one foot being on the ground at all times; *see also* **gait**

walking ventilation
a measure of the amount and content of air expired while performing a mild exercise, usually consisting of walking on a level treadmill
> *Comment:* measured after a few minutes walking at a slow-to-moderate pace

wander
an apparent rapid shift of the target position from its mean on a radar screen; *syn.* target glint, target scintillation, scintillation

warm color
a red or yellow color; a color which seems to be brighter than another for a given intensity; *opp.* cool color

warmup [1]
the brief process of an individual stretching muscles, doing practice problems, or other as appropriate to the situation prior to participating in some form of exertion or performance test; also warm-up

warmup [2]
the process of the functioning parts of an electrical or electromechanical system becoming operational after the application of power

warning [1]
a statement either attached to or otherwise accompanying some product which provides information about the safe use of the product; *see also* **warning label**

warning [2]
a signal that a hazardous situation exists and that immediate corrective action is required to avert possible loss of life and property
> *Comment:* usually annunciated via some audible and/or visible means

warning [3]
a statement issued by local authorities, or weather/other authorized officials indicating that a specified threat is highly probable in a certain region within a specified period of time; *see also* **watch**

warning label
a label containing text, iconic, and/or graphic information attached to some product indicating some type of potential hazard; *see* **warning**

warp
a twisting effect due to uneven stresses

washerwoman's itch
a form of dermatitis appearing on the hands, generally consisting of various skin eruptions from fungal infections or contact dermatitis, and due to having one's hands in water a great deal of the time

watch [1]
a period of duty consisting of vigilance and monitoring for possible hazards, normally performed by a qualified individual or group at some location

watch [2]

a statement issued by local authorities, weather/other authorized officials indicating that a specific threat is possible in a certain region within a specified period of time; *see also warning*

watchkeeping

the process of maintaining a watch

watchmaker's cramp [1]

an occupational disease consisting of painful contractions of the muscles in the hand

watchmaker's cramp [2]

a spasm of the extra-ocular skeletal muscles in or near the orbit due to the containment of a jeweler's lens over the eye with those muscles

water balance

see fluid balance

water purification

the use of any of a variety of processes to remove or neutralize one or more impurities in water

water repellent

(adj) having a tendency for water to flow off a material, rather than be absorbed; also water repellant

water repellent

(n) any chemical or other substance applied to clothing or other materials to resist water wetting; also water repellant

water treatment

any type of processing of water to enable it meet certain standards or desired characteristics, especially including purification for drinking purposes

waterfall illusion

a motion aftereffect in which a stationary background or horizontal lines of graphics or text appear to move in the opposite direction after the individual has viewed scrolling lines on a display

waterproof

impermeable to water

watt (W, w)

a unit of power in the SI/MKS system; equal to the production of energy at the rate of one Joule per second

watt-second

see joule

wave

a disturbance propagated through a medium such that the value of the disturbance is a function of time and/or position

wave number

the reciprocal of the wavelength

waveform

a pictorial or graphical representation of a wave

wavefront

a locus of points which act as a continuous surface having the same phase

wavelength (λ)

the distance between one point on a cycle of a periodic waveform and the corresponding point at the same phase of an adjacent cycle of the same wave parallel to the direction of propagation

weak color

see pastel

wear

(n) that deterioration of a surface due to relative motion between it and another surface

wear-out phase

that period of time occurring after a system has performed much of its useful life and components begin to fail due to aging or other factors; *see life characteristic curve*

weber (Wb)

that unit of magnetic flux in the MKS system which produces in one turn of a circuit a potential of one volt as the flux is reduced linearly in one second to zero

Weber fraction

see Weber ratio

Weber ratio

a measure of the relative discriminability between the just noticeable difference in a stimulus and the original or another stimulus intensity, equal to the constant in Weber's Law;

syn. Weber fraction

$$k = \frac{\Delta I_s}{I_s}$$

where:
k = Weber ratio
ΔI_s = just noticeable difference change in stimulus intensity
I_s = original stimulus intensity

Weber test
a method for determining unilateral hearing loss in which the handle of a vibrating tuning fork is placed against the forehead; also Weber's test; *see also* ***tuning fork test***
> *Comment:* a person with normal hearing hears the sound from the midline (equally in both ears); a conductive hearing impaired individual hears the sound coming from the side with the affected ear; a central hearing loss impaired individual hears the sound better in the normal ear

Weber's law
a rule which states that the Weber ratio remains constant for a given sensory parameter over the normal sensory range

Weber-Fechner law
a psychophysical rule which attempts to describe the relationship between the degree of response or sensation strength of a sense organ to the intensity of the stimulus as a logarithmic function, e.g., below; *syn.* Fechner's law

$$A = k \log B \quad where \; B = \frac{I_s}{I_0}$$

where:
A = the magnitude of the sensation
k = Weber ratio
I_s = the stimulus intensity presented
I_0 = absolute threshold for that stimulus

week
a period of seven consecutive days
> *Comment:* the starting and ending

days may vary according to religious, legal, or other criteria

weekend
the consecutive days within a week during which most workers normally do not work, typically Saturday and Sunday in a standard workweek

weight [1]
*see **body weight***

weight [2]
a value, based on previous information, which is assigned to some score or variable to change the outcome of results

weight [3]
the force with which the mass of a person or object is attracted to the earth or other body, according to Newton's second law; *see also **body weight***

$$\vec{W} = m\vec{g}$$

weight as reported
an individual's body weight as stated by the individual himself, without instrumental confirmation

weight of lift
a guideline for the maximum number of pounds which an individual could be expected to lift under given circumstances

weight velocity
the rate at which body weight increases during physical maturation

weight-height tables
a data matrix providing the average weight and height of individuals in a population at various ages

weighted mean skin temperature $(\bar{T}_s)$
a weighted measure using the proportions of the body surface represented by the various body segments such as the arms, trunk, head, etc. which is intended to represent the average temperature of the skin over its total body surface; *see also **mean skin temperature***

weighting network
an electrical network, whether incorporated via hardware directly into

sound level meters or vibration measuring equipment or incorporated via computer software and data processing, such that they conform to a specified weighting curve; see *A scale, B scale, C scale*

weightlessness

a condition in which no gravitational or other accelerating force can be consciously detected by the observer, and in which an individual or object may remain suspended indefinitely in air, subject only to air movement and other forces; see *microgravity*

welder's flashburn

a form of actinic keratoconjunctivitis caused by observing a welding operation without the use of ultraviolet filtering glasses

wet and dry bulb thermometer

see *psychrometer*

wet bulb globe temperature (WBGT)

a measure of heat stress represented by the synthesis of three weighted variables: the dry bulb temperature, the wet bulb temperature, and the radiant temperature according to the following relationship; see *heat stress index*

$$WBGT = 0.7\, WBT \times 0.1\, DBT \times 0.2\, GT$$

where:
WBT = wet bulb temperature
DBT = dry bulb temperature
GT = globe temperature

wet bulb globe temperature index

a measure of heat stress based on the wet bulb globe temperature and the precautions which should be taken for each range; see *heat stress index*

wet bulb temperature (WBT)

that temperature obtained with an air current passing a calibrated liquid thermometer whose bulb is enclosed by a wet gauze; that temperature obtained from the instrumentation equivalent; *syn.* psychrometric wet bulb temperature, natural wet bulb temperature

wet bulb temperature index

a measure of heat stress based on the wet bulb temperature and the precautions which should be taken for various ranges of readings; see *heat stress index*

wet globe temperature

a measure of heat stress using a Botsball and a thermometer; see *heat stress index*

wet Kata thermometer

a measure of heat stress obtained by a device similar to a wet bulb thermometer, having a silk sleeve along the length of an alcohol-filled tube and graduations marked only at 95° and 100°F, which is used to measure the rate of cooling in hot environments and determining the relaxation allowance required under those conditions; see *heat stress index*

wet strength

a measure of the tensile strength of a material when saturated with water

wet suit

an insulated, water-repellent garment designed to protect an individual exposed to cold water by retaining body heat

what-you-see-is-what-you-get (WYSIWYG)

a display in which the hardcopy is expected to be as the display appears

whiplash

a rapid, severe neck hyperextension followed by a rapid hyperflexion from a posterior impulse imparted to the body below the neck level

Comment: can result in spine, spinal cord, or brain injury

whisker

a short-term growth of facial hair on the sides of the face and/or near the chin

whisper

a manner of speaking very softly, involving the transmission of speech sounds produced by the passage of air through the glottis without vibrating the vocal cords

whistleblower

an individual who reports what he considers to be wrongdoing within an organization

white [1]

pertaining to a broad spectrum of some energy form, consisting of approximately equal intensity levels across the frequency band or spectrum; *see white noise*

white [2]

an appropriate mixture of frequencies/ wavelengths of the electromagnetic spectrum within the visual range which are perceived by the eye as the achromatic color white

white blood cell

see leukocyte

white collar

pertaining to that type of work usually done by management or office personnel, as opposed to factory or production line work; *opp.* blue collar

white finger

see Raynaud's phenomenon; syn. vibration white finger

white muscle

skeletal (striated) muscle tissue having a pale appearance in the fresh or living state, with more myofibrils and less sarcoplasm and myoglobin than red muscle, and which exhibits short latency and rapid response to stimulation; *syn.* fast twitch muscle; *opp.* red muscle

white noise

that sound or electromagnetic noise whose power density spectrum is approximately flat at all frequencies over the normal audio or other frequency range of interest; *syn.* white sound

white point

see achromatic point

white sound

see white noise

whitener

see brightener

whiteout

an atmospheric condition, found most frequently in the arctic region, where the horizon is non-discernable, objects do not cast shadows, and only nearby dark objects are visible

whole learning

a learning situation in which all of the material to be learned is presented and processed as a single unit; *see also part learning*

whole-body

pertaining to the entire body as a unit, usually in terms of some effects on the body

wide band

containing a broad spectrum of frequencies; also wideband; *opp.* narrow band

wide band analysis

a type of frequency analysis in which intensity level measurements are made over an octave or third octave

width

that distance representing the side-to-side dimension of an object

Wilcoxon matched-pairs signed-ranks test

a non-parametric statistical test using the rankings of difference magnitudes between related samples; *syn.* Wilcoxon signed-ranks test, signed-ranks test; *see also sign test*

Wilcoxon signed-ranks test

see Wilcoxon matched-pairs signed-ranks test

wild value

(sl); *see outlier*

willful misconduct

the deliberate non-compliance with rules or regulations

wind

that motion of a portion of the atmosphere relative to the earth's surface

wind cheater

see windbreaker

wind noise

that noise due to airflow around,

through, or otherwise in relation to a vehicle or other object

windblast

the effect of the exposure of a person or object to air when either the air or the person/object is moving

windbreaker

a light garment for covering the torso to maintain body heat by protect the body from minor windchill effects or other environmental conditions which cause a only slight degree of thermal discomfort due to cold; *syn.* wind cheater

windchill

the sensation of an effective reduction in air temperature due to wind velocity and water vaporization; also wind chill

windchill index (WCI)

an experimentally-derived index which combines the effects of wind and temperature in an attempt to describe the severity of a cold environment as a single number

$$WCI = \frac{(10.4 + 6.69\sqrt{V} - .447V) \times (91.4° - t_{DB})}{1.8}$$

where:
V = wind velocity in mph
t_{DB} = dry bulb temperature in °F

windchill temperature (WCT)

an equivalent ambient temperature value for body cooling which would be experienced if there were no wind

window [1]

a rectangular, or approximately rectangular, independent display structure or partition serving a specific function

window [2]

a restricted or selected portion of a data set or time period

window [3]

any structure containing a transparent material through which light may enter a structure or an observer may see outside a structure

windpipe

see **trachea**

wink [1]

a time division equal to 1/2000 or 0.0005 minute or about 0.03 second

wink [2]

the brief conscious closing of one eyelid

wink counter

a clock which is designed to have a face with 100 graduations and two hands, the small one of which revolves twice each minute, the large 20 times per minute

wire-frame

a model or image which displays only the edges, corners, and connecting structures of an object

wool

a natural fiber, normally obtained from sheep

word processing

the use of a computer for text processing, and possibly including some simple graphics, within a range of user-specified formats

word processor

any software application which performs word processing functions

word recognition

a capability in which an artificial system can compare the auditory signal generated by certain words and respond if a particular match is found; *syn.* discrete word recognition, isolated word recognition

word wrap

the automated displacement of a word at the end of a line of text to begin the following line of text when that word would extend the original line beyond set margins

work [1]

the scalar product of the force and distance through which an object is moved

$$W = \vec{F} \cdot \vec{d}$$

work [2]

the physical, physiological, and/or mental effort expended in performing a task; *see also* **positive work, negative work**

work [3]

any task or activity which is considered occupational, not recreational or the activities of daily living, which occurs at an appropriate workplace, has at least a somewhat structured environment, and for which one hopes to be or is typically compensated, either monetarily or through trade

work aid

any device, system, data, and/or information which enables or enhances worker performance in the work environment; *syn.* job aid, job performance aid

work content

that physical labor performed in carrying out tasks, plus the rest and relaxation permitted to recover from fatigue

$$WC = basic\ time + relaxation\ allowance$$
$$+ additional\ work\ allowance$$

work curve

a graphical, longitudinal record of the mental and/or physical work output of an individual or group within a series of specified units of time

work cycle

one complete period, including all work elements used within that period, of any repetitive pattern, process, or operation required to complete a job or task

work cycle time

that period of time required or used to complete one work cycle; *syn.* overall cycle time; *see* **standard time**

work design

the structuring of the complete working environment, including personnel, workstation and workplace layouts, equipment, supplies, procedures, and all their interrelationships; *see also* **job design**

work distribution chart

a chart illustrating all indirect worker activities to be carried out by a work unit and the individuals responsible for carrying out those activities

work efficiency

the relative work output for a given amount of energy used

$$E_w = \frac{external\ work\ performed}{physiological\ energy\ expended}$$

work element

see **therblig**

work environment

the total physical, physiological, social, and psychological environment within which a worker performs his tasks; *syn.* working conditions, working environment

work equipment

the complete set of tools, machinery, jigs, and other devices available for use by a worker in a given work environment

Work Factor®

a predetermined motion time system

work flow chart

any symbolic flow chart illustrating how work is to be performed

work humanization

the process of making jobs more easily performed, ideally with greater efficiency and less effort; *syn.* humanization of work

Comment: usually by the application of human factors techniques

work injury

see **occupational injury**

work measurement [1]

the use of any of a variety of methodologies to determine how much time is required by a qualified worker to produce a certain output in a specified job; *see* **time study, time and motion study, activity sampling**

work measurement [2]
see physiological work measurement

work metabolism
the physiological energy consumption in excess of resting metabolism attributed to the performance of a specific task

work pace
the rate at which a task or activity is done, whether externally- or self-paced

work physiology
the study or consideration of the body's metabolic responses and costs to involvement in physical effort in the work environment; *syn.* occupational physiology

work psychology
the study or consideration of the cognitive or mental aspects of the work environment

work sampling
see activity sampling

work sampling study
see activity sampling

work simplification
the planned improvement of the work environment with the goal of enabling workers to produce more with greater efficiency and less effort

work specification
some form of written documentation which provides details of job procedures, the duties and responsibilities of the person executing the job, the work place layout, and the tools and equipment to be used

work standardization
the process of setting up uniform working environments across workers or groups in terms of tools and equipment used, procedures, and any other factors affecting performance

work standards method
the use of standard times, output, and related measures in management's evaluation of a worker

work strain
see occupational strain

work stress
see occupational stress

work stressor
see occupational stressor

work study
the use of any motion and time study techniques to systematically analyze work methods and procedures dealing with those factors involving efficiency and economy
> *Comment:* normally used with the intent of optimizing the use of all resources in a given task

work surface
any surface or plane which represents the principle area within which motion occurs at the workplace, which supports the tools required for a worker to perform his job, and for which illumination intensity and other environmental variables are generally specified; also worksurface, working surface; *syn.* plane of work, workplane

work system
an integrated group of one or more machines and/or workers for coordinated activities in the output of some product or service

work task
a specific job function or set of reponsibilities assigned to one or more workers

work therapy
see occupational therapy

work tolerance
the length of time for which a worker can perform effectively at some task without a rest period and/or the onset of unnecessary discomfort, illness or injury

work triangle
see working triangle

work unit [1]
a group of workers who function as a team or group

work unit [2]
any unitary amount of quantifiable work output

work-rest cycle
a single sequence of a repetitive set involving activity followed by rest; a single sequence of a repetitive set involving heavy physical activity followed by lighter activity and/or rest; also work rest cycle

work-rest ratio
the ratio of activity time to non-activity time in a work-rest cycle; the ratio of heavy physical activity to lighter activity and rest in a work-rest cycle

Workability
a test battery for measuring quantitatively the residual capabilities of individuals with physical disabilities

workday
the distribution and/or number of hours at work during a single 24-hour period

worker accommodation
a job design aspect which considers the capabilities of the worker and the demands on him in a given job; *syn.* accommodation of workers
Comment: may include job aids, platforms, chairs, tools

Workers' Compensation
an insurance system which provides for payment to employees or their families in the event of an occupational illness, injury, or fatality resulting in the loss of wages, regardless of any negligence; *syn.* workman's compensation, workmen's compensation
Comment: usually required by state law and financed by a tax on employers

worker-type flow process chart
a flow process chart which indicates what an employee normally does or is expected to do during some process; *syn.* man-type flow process chart, person-type flow process chart

working area
that region of the workplace within which a worker moves about in the course of performing his normal tasks

working conditions
see work environment

working environment
see work environment

working memory
an intermediate duration, generally of seconds to minutes, form of memory which is transferred or encoded from sensory memory and capable of manipulation; *syn.* short-term memory

working rule
see shop rule

working standard
a physical standard, prepared from secondary standards, which is available for common use

working triangle
the concept that the stove, refrigerator, and sink make up the three corners of a triangle in the kitchen and that the sum of the triangle legs should be within certain limits for greatest efficiency/productivity; also work triangle

workload
an indicator of the level of total mental and/or physical effort required to carry out one or more tasks at a specific performance level; *see mental workload, physical workload*

workman's compensation
see workers' compensation

workmen's compensation
see workers' compensation

workpiece hazards
the capacity for injury to occur from handling the item being worked or from incidental processing of that piece such as flying chips, sparks, hot metal, etc.

workplace
that smaller region within a worksite at which an individual or group normally works; also work place; *syn.* workspace

workplace design
the process of developing a workplace, including accommodations and locations for the machines, worker(s), tools,

and other devices; *see workplace layout, motion efficiency principles, prerequisites of biomechanical work tolerance*

workplace layout
the description and/or physical arrangement of the workplace, including provisions for worker(s), materials, furniture, tools, equipment, movement, maintenance, the external environment, and any necessary interactions for performing a certain task or job; also work place layout; *syn.* workspace layout; *see also workplace design*

workplane
see work surface

worksample test
a brief examination given to a prospective employee to determine his mastery of the skill(s) required for a particular job; *syn.* trade test

worksite
the physical plant, facility, or other location operated or controlled by an employer and at which an individual or group works

workspace
see workplace

workspace layout
see workplace layout

workstation
a single location within a workplace at which instrumentation or equipment is located and at which a worker might remain for extended periods of time to perform control, monitoring, processing, or other functions; also work station

workweek
the number of days or the pattern of days a worker is expected to perform at the workplace within a calendar week; also work week

worry
a state of anxiety due to a feared or expected outcome from some anticipated event

worsted
a fabric made from long combed wool fibers

worth
the lowest product cost which will consistently perform the required function

wound
a loss of tissue or the interruption of normal anatomical relationships due to mechanical injury; *see also injury*
Comment: typically due to weapons as opposed to machinery or tools

wraparound grasp
a technique for gripping where an object is held against the palm with all four fingers wrapped around it and the thumb overlays the index finger

wrist [1]
the collection of carpal bones and other tissues which form the junction of the forearm and the hand

wrist [2]
that structure on a robotic or teleoperator arm which serves an analogous function to the human wrist, permitting flexion, extension, and rotation

wrist breadth
the linear distance perpendicular to the forearm longitudinal axis between the radial and ulnar styloid prominences of the wrist
Comment: measured with the flesh compressed and hand digits extended

wrist circumference, distal
the surface distance around the wrist just distal to the styloid processes of the radius and ulna and proximal to the hand
Comment: measured with minimal tissue compression and the musculature of the arm and hand relaxed

wrist circumference, styloid level
the surface distance around the wrist at the level of the tip of the styloid process of the radius

Comment: measured with minimal tissue compression and the arm and hand musculature relaxed

wrist height
the vertical distance from the floor or other reference surface to stylion; *syn.* stylion height
Comment: measured with the individual standing erect and the arm hanging naturally at the side

wrist-finger speed
the ability to make rapid, simple, repetitive movements of the fingers, hand(s), and wrist(s), with little concern for accuracy and eye-hand coordination aspects

wristwatch study
a cumulative timing study using a standard wristwatch

written comprehension
see written verbal comprehension

written expression
see written verbal expression

written standard practice
an outline of the methods for a worker to use in some operation, often also including the tools and equipment used and a workplace layout diagram; *see* **standard practice sheet**
Comment: may be in hardcopy or digital form

written verbal comprehension
the ability to understand written language; *syn.* written comprehension

written verbal expression
the ability to put words or concepts into written language for communication purposes; *syn.* written expression

$\bar{x}$ – chart
a chart showing the arithmetic mean with upper and lower control limits for dimensions in quality control where sampled values may be plotted; also x-bar chart

x axis [1]
the horizontal axis having a left-to-right extent on a two- or three-dimensional graph in the rectangular coordinate system; *see also* **y axis, z axis, abscissa, ordinate**

x axis [2]
the horizontal axis having a forward-to-back extent in a vehicular coordinate system; *see also* **y axis, z axis**

X-Y controller
any device having the ability to control a cursor in the screen X and Y dimensions

X-Y plotter
a computer-driven graphics printing device which plots two-dimensional figures

X-Y-Z controller
any device having the ability to control a cursor in the screen X and Y dimensions while also producing apparent movement in the Z dimension

xiphoid process
the most inferior segment of the sternum

Y

y axis [1]

the vertical axis on a two- or three-dimensional graph in the rectangular coordinate system; *see also x axis, z axis, abscissa, ordinate*

y axis [2]

the horizontal axis having a left-to-right extent in a vehicular coordinate system; *see also x axis, z axis*

yaw

(n) a rotation or oscillation about the vertical (z) axis

yaw axis

the vertical axis through an aircraft, spacecraft, ship, or other vehicle capable of motion in three dimensions about which it may yaw; *syn.* z axis, yawing axis

yawing axis

see yaw axis

yawing moment

a torque or force which tends to cause yaw

yellow marrow

a yellowish-colored marrow, consisting of fat cells and a few blood cells

yellow spot

see macula lutea

yes-no design

an experimental design, used for testing thresholds, in which a subject is required to state whether (yes) or not (no) he believes a stimulus was presented during a given trial

yield strength

that stress level at which a material deviates beyond a certain proportionality between stress and strain; *syn.* strength

Young's modulus

the ratio of tension to strain in a material within the elastic range for that material

Young's three-component theory

see Young-Helmholtz theory

Young-Helmholtz theory

a theory of vision which proposed that the eye contains three types of receptors — one sensitive primarily to red, one primarily to green, and one primarily to blue-violet; *syn.* Young's three component theory, three-component theory

Z

z axis [1]

the horizontal axis having the extent of depth and perpendicular to the x and y axes in the rectangular coordinate system; *see also x axis, y axis, abscissa, ordinate*

z axis [2]

the vertical axis in a vehicular system; *see also yaw axis*

z score

see standard score

Zeitgeber

see entraining agent

zero defects

the concept of perfect product quality, without any flaws

zero fault tolerant

having no redundancy; pertaining to a condition in which a single fault in a system will cause that system or the function performed by it to fail

zero gravity

see microgravity; syn. zero-g

Comment: strictly, a non-existent condition

zero level

see threshold of audibility

zero-crossing

the transition of a time-varying signal through the baseline level, or zero amplitude

zero-order control

see position control

zinc chills

see metal fume fever

zipper

a fastener for clothing which interlaces two separated strips of material

zone

a partitioned area of a display used for some specific purpose or function

zoning

the designation of certain sections of land for specified types of uses, usually within a city having a land use plan

zoom

transform a portion of the field of view into a close-up image, either on a display or with a camera lens system

zygoma

see zygomatic bone

zygomatic arch

the projecting bony arch forming the most lateral portion of the face which extends horizontally along the side of the head, from beneath the eye orbit to near otobasion superior, and generally including parts of both the zygomatic bone and the temporal bone; occasional *syn.* cheekbone

zygomatic bone

a lateral facial bone forming part of the orbit; *syn.* cheekbone, zygoma

BIBLIOGRAPHY

Applied Ergonomics Handbook
London: Butterworths Scientific., 1974.

Code of Federal Regulations
Washington, D.C.: U.S. Government Printing Office., 1990.

CSP Refresher Guide (Vol. 1-3)
Des Plaines, IL: American Society of Safety Engineers., 1991.

AGARD (1955)
Anthropometry and Human Engineering. London: Butterworths Scientific.

AGARD (1989)
Human Performance Assessment Methods (AGARDograph No. 308). Essex, England: Specialized Printing Services.

AGARD (1990a)
Speech Analysis and Synthesis and Man-machine Speech Communications for Air Operations (AGARD lecture series No. 170). Essex, England: Specialized Printing Services.

AGARD (1990b)
Occupant Crash Protection in Military Air Transport (AGARDograph No. 306). Essex, England: Specialized Printing Services.

AGARD (1990c)
High G Physiological Protection Training (AGARDograph No. 322). Essex, England: Specialized Printing Services.

Alexander, D. C. (1986)
Practice and Management of Industrial Ergonomics. Englewood Cliffs, NJ: Prentice Hall.

ANSI (1982)
Industrial Engineering Terminology (Standard Z94). Norcross, GA: Institute of Industrial Engineers.

ANSI/HFS (1988)
American National Standard for Human Factors Engineering of Visual Display Terminal Workstations (Standard 100-1988). Santa Monica, CA: Human Factors Society.

ANSI/IIE (1990)
Industrial Engineering Terminology (revised edition). Norcross, GA: Industrial Engineering and Management Press.

ASHRAE (1965–66)
Guide and Data Book: Fundamentals and Equipment. New York: ASHRAE.

ASHRAE (1989)
ASHRAE Handbook — Fundamentals. Atlanta: ASHRAE.

Åstrand, P.-O. and Rodahl, K. (1977)
Textbook of Work Physiology: Physiological Bases of Exercise (2nd ed). New York: McGraw-Hill.

Barham, J. N. and Wooten, E. P. (1973)
Structural Kinesiology. New York: Macmillan.

Battinelli, T. (1989)
Physique and Fitness: The influence of Body Build on Physical Performance. New York: Human Sciences Press.

Behnke, A. R. and Wilmore, J. H. (1974)
Evaluation and Regulation of Body Build

and Composition. Englewood Cliffs, NJ: Prentice-Hall.

Bendat, J. S. and Piersol, A. G. (1971)
Random Data: Analysis and Measurement Procedures. New York: Wiley-Interscience.

Berchem-Simon, O. (Ed.) (1982)
Ergonomics Glossary: Terms Commonly Used in Ergonomics. Utrecht/Antwerp: Bohn, Scheltema, and Holkema.

Beyer, W. H. (Ed.) (1980)
Handbook of Mathematical Sciences (5th ed.). Boca Raton, FL: CRC Press.

Blake, M. P. and Mitchell, W. S. (Eds.) (1972)
Vibration and Acoustic Measurement Handbook. New York: Spartan Books.

Blitz, J. (1964)
Elements of Acoustics. Washington: Butterworths.

Blum, M. L. and Naylor, J. C. (1968)
Industrial Psychology: Its Theoretical and Social Foundations. New York: Harper and Row.

Board of Certified Safety Professionals (1991)
Examination Information (6th ed.). Savoy, IL: BCSP.

Boff, K. R. and Lincoln, J. E. (Eds.) (1988)
Engineering Data Compendium: Human Perception and Performance. Wright Patterson AFB, OH: AAMRL.

Borden, G. J. and Harris, K. S. (1984)
Speech Science Primer (2nd ed.). Baltimore, MD: Williams and Wilkins.

Bruning, J. L. and Kintz, B. L. (1977)
Computational Handbook of Statistics (2nd ed.). Glenview, IL: Scott, Foresman and Co.

Burger, E. and Korzak, G. (1986)
Dictionary of Robot Technology. Amsterdam: Elsevier.

Carlisle, K. E. (1986)
Analyzing Jobs and Tasks. Englewood Cliffs, NJ: Educational Technology Publications.

Carlsoo, S. (transl. W. P. Michael) (1972)
How Man Moves: Kinesiological Studies And Methods. London: Heinemann.

Carpenter, M. B. (1978)
Core Text of Neuroanatomy (2nd ed.). Baltimore, MD: Williams and Wilkins.

Carterette, E. C. and Friedman, M. P. (Eds.) (1975)
Seeing (Handbook of Perception, Vol. V). New York: Academic Press.

Chaffin, D. B. and Andersson, G. B. J. (1984)
Occupational Biomechanics. New York: John Wiley & Sons.

Chestnut, H. (1967)
Systems Engineering Methods. New York: John Wiley & Sons.

Churchill, E., Churchill, T. and Kikta, P. (1977)
The AMRL Anthropometric Data Bank Library (AMRL-TR-77-1) (Vols. 1–4). Wright-Patterson AFB, OH: USAF.

Chusid, J. G. (1976)
Correlative Neuroanatomy and Functional Neurology (16th ed.). Los Altos, CA: Lange.

Cleland, D. I. and Kerzner, H. (1985)
A Project Management Dictionary of Terms. New York: Van Nostrand Reinhold.

Cleland, D. I. and King, W. R. (Eds.) (1988)
Project Management Handbook (2nd ed.). New York: Van Nostrand Reinhold.

Cochran, G. Van B. (1982)
A Primer of Orthopedic Biomechanics. New York: Churchill Livingstone.

Cotman, C. W. and McGaugh, J. L. (1980)
Behavioral Neuroscience: An Introduction. New York: Academic Press.

Davis, M. (1972)
Understanding Body Movement: An Annotated Bibliography. Bloomington, IN: Indiana University Press.

Dempster, W. T. (1955)
Space Requirements of the Seated Operator (WADC TR 55-159). WPAFB, OH: USAF.

Department of Defense (U.S.) (1989)
Human Engineering Design Criteria for Military Systems, Equipment, and Facilities (MIL-STD 1472D). Hunstville, AL: Redstone Arsenal.

Diffrient, N., Tilley, A. R., and Bardagjy, J. C. (1974)
Humanscale. Cambridge, MA: MIT Press.

Easterby, R., Kroemer, K. H. E., and Chaffin, D. B. (1982)
Anthropometry and Biomechanics: Theory and Application. New York: Plenum Press.

Eastman Kodak Company, Human Factors Section (1983)
Ergonomic Design for People at Work (Vol. 1). Belmont, CA: Lifetime Learning Publications.

Eastman Kodak Company, The Ergonomics Group (1986)
Ergonomic Design for People at Work (Vol. 2). New York: Van Nostrand Reinhold.

Edwards, A. L. (1979)
Multiple Regression and the Analysis of Variance and Covariance. San Francisco, CA: W. H. Freeman.

Elzey, F. F. (1971)
A Programmed Introduction to Statistics (2nd ed.). Belmont, CA: Brooks/Cole.

Ernst and Young Quality Improvement Consulting Group (1990)
Total Quality: An Executive's Guide for the 1990s. Homewood, IL: Dow Jones-Irwin.

Fleishman, E. A. and Quaintance, M. K. (1984)
Taxonomies of Human Performance: The Description of Human Tasks. Orlando: Academic Press.

Fox, E. L. (1979)
Sports Physiology. Philadelphia: W. B. Saunders.

General Services Administration (1984)
Federal Standard 595a: Colors. Washington, D.C.: GSA.

Gilmore, W. E., Gertman, D. I., and Blackman, H. S. (1989)
The User–Computer Interface in Process Control: A Human Factors Engineering Handbook. Boston: Academic Press.

Glaser, E. M. and Ruchkin, D. S. (1976)
Principles of Neurobiological Signal Analysis. New York: Academic Press.

Gorsuch, R. L. (1983)
Factor Analysis (2nd ed.). Hillsdale, NJ: Erlbaum Associates.

Gowitzke, B. A. and Milner, M. (1980)
Understanding the Scientific Bases of Human Movement (2nd ed.). Baltimore, MD: Williams and Wilkins.

Grandjean, E. (Ed.) (1988)
Fitting the Task to the Man: A Textbook of Occupational Ergonomics (4th ed.). London: Taylor and Francis.

Grossman, S. P. (1967)
A Textbook of Physiological Psychology. New York: John Wiley & Sons.

Guilford, J. P. (1965)
Fundamental Statistics in Psychology and Education (4th ed.). New York: McGraw-Hill.

Guyton, A. C. (1971)
Textbook of Medical Physiology (4th ed.). Philadelphia: W. B. Saunders.

Hardy, J. D., Gagge, A. P. and Stolwijk, J. A. J. (Eds.) (1970)
Physiological and Behavioral Temperature Regulation. Springfield, IL: Thomas.

Harper, H. A. (1965)
Review of Physiological Chemistry (10th ed.). Los Altos, CA: Lange Medical Publications.

Harriman, P. L. (1968)
Handbook of Psychological Terms. Totowa, NJ: Littlefield, Adams and Co.

Harris, R. J. (1975)
A Primer of Multivariate Statistics. New York: Academic Press.

Hay, J. G. (1985)
The Biomechanics of Sports Techniques (3rd ed.). Englewood Cliffs, NJ: Prentice-Hall.

Hays, W. L. (1963)
Statistics. New York: Holt, Rinehart and Winston.

Helander, M. (Ed.) (1988)
Handbook of Human–Computer Interaction. Amsterdam: North-Holland.

Hinsie, L. E. and Campbell, R. J. (1970)
Psychiatric Dictionary (4th ed.). London: Oxford University Press.

Hole, J. W., Jr. (1981)
Human Anatomy and Physiology (2nd ed.). Dubuque, IA: Wm. C. Brown.

Hollies, N. R. S. and Goldman, R. F. (Eds.) (1977)
Clothing Comfort. Ann Arbor: Ann Arbor Science Publishers.

Hopkinson, R. G. and Collins, J. B. (1970)
The Ergonomics of Lighting. London: Macdonald Technical and Scientific.

Hunt, R. W. G. (1987)
Measuring Colour. New York: Halstead Press.

Hunter, R. S. and Harold, R. W. (1987)
The Measurement of Appearance (2nd ed.). New York: John Wiley & Sons.

Hutchins, D. (1985)
Quality Circles Handbook. New York: Nichols Publishing.

IBM (1984)
Human Factors of Workstations with Visual Displays (3rd ed.). San Jose: IBM.

Information Handling Services (1990/1991)
Index and Directory of Industry Standards. Englewood, CO: Information Handling Services.

International Labor Office (1974)
Introduction to Work Study (Revised ed.). Geneva: ILO.

Jacob, S. W., Francone, C. A., and Lossow, W. J. (1978)
Structure and Function in Man (4th ed.).

Philadelphia: W. B. Saunders.

Johnsen, E. G. and Corliss, W. R. (1971)
Human Factors Applications in Teleoperator Design and Operation. New York: Wiley-Interscience.

Jones, F. W. (1929)
Measurements and Landmarks in Physical Anthropology. Honolulu: Bernice P. Bishop Museum.

Kahn, J. (1987)
Principles and Practice of Electrotherapy. New York: Churchill Livingstone.

Kahneman, D. (1973)
Attention and Effort. Englewood Cliffs, NJ: Prentice Hall.

Kamenetz, H. L. (1983)
Dictionary of Rehabilitation Medicine. New York: Springer.

Kantowitz, B. H. and Sorkin, R. D. (1983)
Human Factors: Understanding People–System Relationships. New York: John Wiley & Sons.

Kaufman, J. E. (Ed.) (1984)
IES Lighting Handbook (Reference Volume). New York: Illuminating Engineering Society of North America.

Kenedi, R. M. (Ed.) (1964)
Biomechanics and Related Bio-engineering Topics. Oxford: Pergamon Press.

Kerlinger, F. N. (1973)
Foundations of Behavioral Research (2nd ed.). New York: Holt, Rinehart and Winston.

Kerlinger, F. N. and Pedhazur, E. J. (1973)
Multiple Regression in Behavioral Research. New York: Holt, Rinehart, and Winston.

Kling, J. W. and Riggs, L. A. (1971)
Experimental Psychology (3rd ed.). New York: Holt, Rinehart and Winston.

Kroemer, K. H. E. (1970)
Human Strength: Terminology, Measurement, and Interpretation of Data (AMRL-TR-69-9). WPAFB: USAF.

Kroemer, K. H. E. (1982)
Ergonomics of VDT Workplaces (Ergonomics Guide). American Industrial Hygiene Association.

Kroemer, K. H. E. (1991)
Ergonomics. *Encyclopedia of Human Biology* (Vol. 3), 473–480. New York: Academic Press.

Kroemer, K. H. E., Kroemer, H. J., and Kroemer-Elbert, K. E. (1986)
Engineering Physiology: Physiological Bases of Human Factors/Ergonomics. Amsterdam: Elsevier.

Kroemer, K. H. E., Marras, W. S., McGlothlin, J. D., McIntyre, D. R., and Nordin, M. (1990)
On the measurement of human strength. *Int. J. of Ind. Ergonomics*, 6:199–210.

Kroemer, K. H. E., Snook, S. H., Meadows, S. K., and Deutsch, S. (Eds.) (1988)
Ergonomic Models of Anthropometry, Human Biomechanics, and Operator–Equipment Interfaces. Washington, D.C.: National Academy Press.

Laird, C. E. and Rozier, C. K. (1979)
Toward understanding the terminology of exercise mechanics. *Physical Therapy*, 59:287–292.

Lapedes, D. N. (Ed.) (1974)
Dictionary of Scientific and Technical Terms. New York: McGraw Hill.

Larkin, J. A. (1969)
Work Study. New York: McGraw-Hill.

Lasker, G. W. (1961)
Physical Anthropology (2nd. ed.). New York: Holt, Rinehart and Winston.

Lewis, D. (1966)
Quantitative Methods in Psychology. Iowa City: University of Iowa.

Lieberman, P., and Blumstein, S. E. (1988)
Speech Physiology, Speech Perception, and Acoustic Phonetics. New York: Cambridge University Press.

Lohman, T. G., Roche, A. F., and

Martorell, R. (1988)
Anthropometric Standardization Reference Manual. Champaign, IL: Human Kinetics Books.

Lohr, J. B. and Wisniewski, A. A. (1987)
Movement Disorders: A Neuropsychiatric Approach. New York: Guilford Press.

Luce, G. G. (1971)
Biological Rhythms in Human and Animal Physiology. New York: Dover.

Marconnet, P. and Komi, P. V. (Eds.) (1987)
Muscular Function in Exercise and Training. Basel: Karger.

Maxwell, A. E. (1977)
Multivariate Analysis in Behavioral Research. London: Chapman and Hall.

Maynard, H. B. (Ed) (1971)
Industrial Engineering Handbook (3rd ed). New York: McGraw-Hill.

McCormick, E. J. and Sanders, M. S. (1982)
Human Factors in Engineering and Design (5th ed.). New York: McGraw-Hill.

Mechtly, E. A. (1969)
The International System of Units: Physical Constants and Conversion Factors (SP-7012) (Revised). Washington, D.C.: NASA.

Medland, A. J. and Burnett, P. (1986)
CAD/CAM in Practice: A Manager's Guide to Understanding and Using CAD/CAM. New York: John Wiley & Sons.

Miller, M. A., Drakontides, A. B., and Leavell, L. C. (1977)
Anatomy and Physiology (17th ed.). New York: Macmillan.

Miner, R. W. (Ed.) (1955)
Dynamic Anthropometry. New York: New York Academy of Sciences.

Moore, F. D., Olesen, K. H., McMurrey, J. D., Parker, H. V., Ball, M. R., Boyden, C. M. (1963)
The Body Cell Mass and its Supporting Environment. Philadelphia: Saunders.

Mountcastle, V. B. (1980)
Medical Physiology (14th ed.). St. Louis: Mosby.

Myers, J. L. (1972)
Fundamentals of Experimental Design (2nd ed.). Boston: Allyn and Bacon.

NASA (1969)
Databook for Human Factors Engineers. NASA Ames Research Center.

NASA (1973)
Bioastronautics Data Book (SP-3006) (2nd ed.). Washington, D.C.: NASA.

NASA (1987)
Guidelines for Noise and Vibration Levels for the Space Station (Contractor Report 178310). Hampton, VA: NASA Langley Research Center.

NASA (1988)
Space Station Freedom Program Human–Computer Interface Guide (USE 1000, Version 2.1). Houston: NASA JSC.

NASA (1989)
Man-Systems Integration Standard (NASA-STD 3000, Vol. II). Houston: Johnson Space Center.

National Society for Performance and Instruction (1986)
Introduction to Performance Technology (Vol. 1). Washington, D. C.: NSPI.

Newburgh, L. H. (1949)
Physiology of Heat Regulation and the Science of Clothing. Philadelphia: Saunders.

Niebel, B. W. (1976)
Motion and Time Study (6th ed.). Homewood, IL: Richard D. Irwin.

Nieman, D. C. (1990)
Fitness and Sports Medicine: An Introduction. Palo Alto, CA: Bull Publishing Co..

Nimeroff, I. (Ed.) (1972)
Selected NBS Papers on Colorimetry (NBS Special Publ. 300, Vol. 9). Washington, D.C.: U.S. Government Printing Office.

NIOSH (1981)
Work Practices Guide for Manual Lifting. Cincinnati, OH: USDHHS.

NIOSH (1987)
A Guide to Safety in Confined Spaces. Cincinnati, OH: USDHHS.

Nunnally, J. C. (1978)
Psychometric Theory (2nd ed.). New York: McGraw-Hill.

Osol, A. et al. (Eds.) (1972)
Gould Medical Dictionary (3rd ed.). New York: McGraw-Hill.

Pact, V., Sirotkin-Roses, M., and Beatus, J. (1984)
The Muscle Testing Handbook. Boston: Little Brown and Co..

Pelsma, K. H. (Ed.) (1987)
Ergonomics Sourcebook: A Guide to Human Factors Information. Lawrence, Kansas: The Report Store.

Plutchik, R. (1968)
Foundations of Experimental Research. New York: Harper and Row.

Pollock, M. L., Wilmore, J. H., and Fox, S. M., III (1984)
Exercise in Health and Disease. Philadelphia: Saunders.

Roche, A. F. (Ed.) (1985)
Body-Composition Assessment in Youth and Adults. Columbus, OH: Ross Laboratories.

Roebuck, J. A., Kroemer, K. H. E., and Thomson, W. G. (1975)
Engineering Anthropometry Methods. New York: John Wiley & Sons.

Ross, A. O. (1976)
Psychological Aspects of Learning Disabilities and Reading Disorders. New York: McGraw-Hill.

Ruch, T. and Patton, H. D. (1979)
Physiology and Biophysics (20th ed). Philadelphia: W. B. Saunders.

Salvendy, G. (Ed.) (1982)
Handbook of Industrial Engineering. New York: John Wiley & Sons.

Salvendy, G. (Ed.) (1987)
Handbook of Human Factors. New York: John Wiley & Sons.

Sanders, M. S. and McCormick, E. J. (1987)
Human Factors in Engineering and Design (6th ed.). New York: McGraw-Hill.

Schmidt, R. F. (Ed.) (1978)
Fundamentals of Neurophysiology (2nd ed.). New York: Springer-Verlag.

Siegel, S. (1956)
Nonparametric Statistics for the Behavioral Sciences. New York: McGraw-Hill.

Singleton, W. T. (Ed.) (1982)
The Body at Work: Biological Ergonomics. London: Cambridge University Press.

Smith, N. J. and Stanitski, C. L. (1987)
Sports Medicine: A Practical Guide. Philadelphia: W. B. Saunders.

Society of Automotive Engineers (1986)
Human Tolerance to Impact Conditions as Related to Motor Vehicle Design (J885). Warrendale, PA: SAE.

Society of Automotive Engineers (1988)
Aerospace Glossary for Human Factors Engineers (ARP 4107). Warrendale, PA: SAE.

Strauss, R. H. (1979)
Sports Medicine and Physiology. Philadelphia: W. B. Saunders.

Strong, P. (1971)
Biophysical Measurements. Beaverton, OR: Tektronix.

Suslick, K. S. (Ed.) (1988)
Ultrasound: Its Chemical, Physical, and Biological Effects. New York: VCH Publishers.

Thorndike, R. M. (1978)
Correlational Procedures for Research. New York: Gardner Press.

Tichauer, E. R. (1978)
The Biomechanical Basis of Ergonomics. New York: John Wiley & Sons.

Toglia, J. U. (1976)
Electronystagmography: Technical Aspects and Atlas. Springfield, IL: Thomas.

Tortora, G. J. (1980)
Principles of Human Anatomy (2nd ed). New York: Harper and Row.

U.S. Congress, Office of Technology Assessment (1991)
Biological Rhythms: Implications for the Worker. Washington, D.C: U.S. Government Printing Office.

U.S. Department of Defense (1980)
Military Handbook: Anthropometry of U.S. Military Personnel (DOD-HDBK-743). Washington, D. C.: DoD.

U.S. Department of Defense Human Factors Engineering Technical Advisory Group (1986)
Workload Assessment: Techniques and Tools (A workshop at the Human Factors Society Annual Meeting).

University of Surrey, Materials Handling Research Unit (1980)
Force Limits in Manual Work. Surrey, England: IPC Science and Technology Press.

Van Cott, H. P. and Kinkade, R. G. (1972)
Human Engineering Guide to Equipment Design (Rev. ed.). Washington, D.C.: U.S. Government Printing Office.

Wasserman, G. S. (1978)
Color Vision: An Historical Introduction. New York: John Wiley & Sons.

Webb Associates (1978)
Anthropometric Source Book. Vol. II: A Handbook of Anthropometric Data (NASA RP 1024). Ohio: Yellow Springs.

Webster (Guralnik) (1986)
New World Dictionary of the American Language (2nd college ed.). New York: Simon and Shuster.

Weidner, R. T. and Sells, R. L. (1968)
Elementary Modern Physics (2nd ed.). Boston: Allyn and Bacon.

Winer, B. J. (1971)
Statistical Principles in Experimental Design (2nd ed.). New York: McGraw-Hill.

Winter, D. A. (1979)
Biomechanics of Human Movement. New York: John Wiley & Sons.

APPENDIX A
ABBREVIATIONS, ACRONYMS OF TERMS

$\vec{\alpha}$	acceleration	4M	Micro-Matic Methods and Measurement
α	alpha	$\vec{a}$	instantaneous acceleration
α	angular acceleration	$\bar{a}$	average acceleration
α_m	acoustic absorption coefficient	°	degree
α_r	reflection coefficient	Å	Angstrom
β	beta	A	ampere
β_F	spectral fluorescent radiance factor	A	assemble
		ABO	aviator's breathing oxygen
β_R	spectral reflected radiance factor	abt	abtesla
		AC	alternating current
γ^2	coherence	AD	avoidable delay
λ	wavelength	ADA	Americans with Disabilities Act of 1990
λ_c	complementary wavelength		
λ_d	dominant wavelength	ADAM	Advanced Dynamic Anthropomorphic Manikin
μ	micro-		
μ	coefficient of static friction	ADC	analog-to-digital converter
μ	coefficient of friction	ADL	activities of daily living
μ	coefficient of sliding friction	AE	above elbow
π	pi	AGSM	anti-g straining maneuver
ρ	rho	AI	speech articulation index
ρ	reflectance	AIR	aircraft accident incidence rate
ρ	Spearman rank-order correlation coefficient		
		AIS	Abbreviated Injury Scale
τ	transmittance	AK	above knee
τ	torque	AL	action limit
τ	Kendall rank-order correlation coefficient	ALARA	as low as reasonably achievable
		ALR	action limit ratio
χ^2	chi square	AMI	Available Motions Inventory
ω	angular frequency		
ω	angular velocity	AMI	airspeed/mach indicator
Φ	luminous flux	ANCOVA	analysis of covariance
Ω	ohm	ANOVA	analysis of variance
$\mho$	mho		

ANS	autonomic nervous system	CAR	Computer Assessment of Reach
asb	apostilb		
ATB	articulated total body model	CAT	computerized axial tomography
ATC	Air Traffic Controller		
ATP	adenosine triphosphate	CAT	computer assisted tomography
$\vec{B}$	magnetic field		
b	bit	cd	candela
B	bel	CD	change direction
B	byte	CDF	cumulative distribution function
BARS	behavior-anchored rating scale		
		CET	corrected effective temperature
BD	balancing delay		
BE	below elbow	cff	critical flicker frequency
BEV	barrier equivalent velocity	cff	critical fusion frequency
BFO	blood forming organs	CFR	Code of Federal Regulations
BI	Barthel Index	CGS	Centimeter-Gram-Second system
BIA	bioelectrical impedance analysis		
		CHI	Comfort Health Index
bit	binary digit	CHI	computer-human interface
BIT	Built-in Test	CHP	Chemical Hygiene Plan
BITE	Built-in Test Equipment	CIM	computer-integrated manufacturing
BK	below knee		
BMI	body mass index	CIR	Crash Injury Research project
BMR	basal metabolic rate		
BMTS	Basic Motion Time Study	C_m	compliance
BTPS	body temperature and pressure, saturated conditions	CNS	central nervous system
		co	cardiac output
BTU	British Thermal Unit	CO	carbon monoxide
BW	bandwidth	CO2	carbon dioxide
c	specific heat	COMBIMAN	Computer Biomechanical Man
C	coulomb		
C	capacitance	CORELAP	Computerized Relationship Layout Planning
C	Celsius		
C	Centigrade	cp	candlepower
C	Munsell chroma	CPM	Critical Path Method
C&W	caution and warning	cps	cycle per second
CA	chronological age	CR	critical ratio
CAE	cost of accidents per employee	CRAFT	Computerized Relative Allocation of Facilities Technique
CAI	computer-aided instruction		
cal	calorie	CRI	CIE color rendering index
Cal	Calorie	CRP	corneo-retinal potential
CAL-3D CVS	CAL-3D crash victim simulator	CRT	choice reaction time
		CRT	cathode ray tube
CAP	Control Assessment Protocol	CSA	compressed spectral array
		CSDG	Crash Survival Design Guide
CAPE	Computerized Accommodated Percentage Evaluation		
		CSF	cerebrospinal fluid
		CSG	constructive solid geometry

CSP	Certified Safety Professional	EMG	electromyogram
CTD	cumulative trauma disorder	EMG	electromyography
CTI	Clerical Task Inventory	EMI	electromagnetic interference
CTS	carpal tunnel syndrome		ence
CU	coefficient of utilization	EMS	electrical muscle stimulation
CWW	compressed workweek		tion
d	coefficient of determination	EMU	extravehicular mobility unit
DA	disassemble	ENG	electronystagmogram
DA	developmental age	EOG	electrooculogram
DAT	Differential Aptitude Test	EP	evoked potential
db	database	EPP	end-plate potential
dB	decibel	ERG	electroretinography
dB(A)	A-weighted sound pressure level	ERISA	Employee Retirement Income Security Act
dB(B)	B-weighted sound pressure level	ERV	expiratory reserve volume
		ESR	electrical skin resistance
dB(C)	C-weighted sound pressure level	ET	effective temperature
		EVA	extravehicular activity
DC	direct current	F	farad
df	degree of freedom	F	Fahrenheit
DL	difference limen	FAST	Functional Analysis Systems Technique
DNL	day-night average sound level		
		fc	footcandle
DO	do	f_{cl}	clothing area factor
DOD-STD	Department of Defense Standard	FEV	forced expiratory volume
		fff	flicker fusion frequency
DQ	developmental quotient	FFM	fat-free mass
DR	defensive response	FFW	fat-free weight
DRI	dynamic response index model	fL	footlambert
		FLSA	Fair Labor Standards Act
DSK	Dvorak keyboard	FMEA	failure mode and effects analysis
E	experimenter		
E	irradiance	FRC	functional residual capacity
E	index of forecasting efficiency	FRP	floor reference point
		ft	foot
E	illuminance	ft-lb	foot-pound
E	examine	$\vec{g}$	gravitational field
EAS	Employee Aptitude Survey	g	gram
ECG	electrocardiogram	$g, g_{x,y, or z}$	gravity
ECW	extracellular water	G	grasp
EEG	electroencephalography	G	giga-
EEG	electroencephalogram	g-LOC	gravity-induced loss of consciousness
EEO	Equal Employment Opportunity		
		GATB	General Aptitude Test Battery
EGG	electrogoniogram		
EGG	electrogoniography	GFI	ground fault interrupter
EHF	extremely high frequency	GIGO	garbage in, garbage out
EKG	electrokardiogram	GIS	OSHA General Industry Standard
ELF	extremely low frequency		
EMF	electromagnetic field	gm	gram

GMT	Greenwich Mean Time	IEMG	integrated electromyogram
GOMS	goals, operators, methods, and selection rules	IFR	instrument flight rules
		i_m	permeability index
GSR	galvanic skin response	INM	Integrated Noise Model
GUI	Graphical User Interface	IQ	intelligence quotient
Gy	gray	IR	infrared
H	information	ISCC-NBS	Inter-Society Color Council
H/C	handcontroller		– National Bureau of Stan-
H	henry		dards color system
H	hold	I_t	earnings profile
HANES	Health And Nutrition Ex-	I_T	total thermal insulation
	amination Survey		value of clothing
HANS ™	Head And Neck Support	IVA	intravehicular activity
HAZMAT	hazardous material	J	joule
H_b	body heat content	J	impulse
Hb	hemoglobin	JND	just noticeable difference
HbO2	oxyhemoglobin	JROM	joint range of motion
h_c	convective heat transfer co-	JSI	Job Severity Index
	efficient	k	coefficient of alienation
H_C	hue composition	K, k	kilo-
HCI	human-computer interac-	K	Kelvin
	tion	k^2	coefficient of non-determi-
HCI	human-computer interface		nation
HDTV	high-definition television	KE	kinetic energy
HEP	human error probability	kg	kilogram
HF	high frequency	kgf	kilogram force
HIC	Head Injury Criterion	l	liter
HMD	helmet-mounted display	L_A	A-weighted sound pressure
HMD	head-mounted display		level
HMI	human-machine interface	L_{dn}	day-night average sound
hp	horsepower		level
HPD	hearing protection device	L	radiance
hr	hour	L	angular momentum
HSI	heat stress index	L	lambert
HUD	head-up display	LAN	Local Area Network
Hz	Hertz	LASER	Light Amplification by
HZE	high-energy heavy ion		Stimulated Emission of Ra-
I	acoustic intensity		diation
I	inspect	lbf	pound force
I	rotational inertia	lbm	pound mass
I/O	input/output	LBM	lean body mass
I_a	thermal insulation value of	LBO	lamp burnout
	air	LCD	liquid crystal display
IADL	instrumental activities of	LD_{50}	median lethal dose
	daily living	LDD	luminaire dirt depreciation
IAP	intra-abdominal pressure	LDL	low density lipoprotein
IC	inspiratory capacity	LED	light-emitting diode
I_{cl}	thermal insulation value of	LEF	lighting effectiveness factor
	clothing	lel	lower explosive limit
ICW	intracellular water	L_{eq}	equivalent sound level

LET	linear energy transfer	MRP	material requirements planning
LF	low frequency		
lfl	lower flammable limit	MS	multiple sclerosis
L_I	sound intensity level	MSD	Master Standard Data
LLD	lamp lumen depreciation	MTA	Motion Time Analysis
LLF	light loss factor	MTBF	mean time between failures
lm	lumen	MTM	Methods Time Measurement
lm-hr	lumen-hour		
lm-sec	lumen-second	MTTF	mean time to failure
LOS	line of sight	MVC	maximal voluntary contraction
L_P	sound pressure level		
LP	linear programming	MVV	maximum voluntary ventilation
lx	lux		
m	meter	n	nano-
m	slope	n	index of refraction
m	milli-	N	newton
M	radiant exitance	N-m	newton-meter
M	move	N_2	nitrogen
M	mega-	NASA-STD	National Aeronautics and Space Administration Standard
MAC	maximum allowable concentration		
MANOVA	multivariate analysis of variance	NCC	noise criterion curve
		NCS	Natural Color System
MANPRINT	Manpower and Personnel Integration	NGT	nominal group technique
		NIPTS	noise-induced permanent threshold shift
MAP	Michigan Anthropometric Processor		
		NITTS	noise-induced temporary threshold shift
MAP	Manufacturing Automation Protocol		
		NMR	Nuclear Magnetic Resonance
MDW	measured daywork		
ME	mechanical efficiency	NO_2	nitrogen dioxide
MF	medium frequency	Np	néper
mg-h	milligram-hour	NRC	noise reduction coefficient
MIL-STD	Military Standard	NRR	noise reduction rating
mired	microreciprocal degree	nt	nit
mirek	microreciprocal kelvin	NVG	night vision goggles
MK^{-1}	reciprocal megakelvin	O	observer
MKS	Meter-Kilogram-Second system	O_2	oxygen
		O_3	ozone
MMH	manual materials handling	OCR	optical character recognition
MOST	Maynard Operation Sequence Technique	OD	optical density
		OJT	on-the-job training
MPC	maximum permissible concentration	OKN	optokinetic nystagmus
		OT	occupational therapist
MPC	man process chart	oz	ounce
MPD	maximum permissible dose	$\vec{p}$	linear momentum
MPL	maximum permissible limit	p	probability
MRI	Magnetic Resonance Imaging	p	pico-
		p	acoustic pressure

P	operator performance	QA	quality assurance
P	position index	QC	quality circles
P	position	QC	quality control
p-p	peak-to-peak amplitude	QF	quality factor
P4SR	predicted four-hour sweat rate	Q_i	quartile
Pa	pascal	r	Pearson product-moment correlation coefficient
PAQ	Position Analysis Questionnaire	R	roentgen
PB	phonetically balanced	R	rest for overcoming fatigue
PCM	pulse code modulation	R	reach
PDF	probability density function	R	thermal resistance value
PDF	probability distribution function	R	multiple correlation coefficient
P_e	excitation purity	R	electrical resistance
PE	potential energy	R&D	research and development
PEL	permissible exposure limit	r^2	coefficient of determination
PERT	Program Evaluation and Review Technique	r^2	generality
PFD	primary flight display	RA	relaxation allowance
ph	phot	rad	radiation absorbed dose
PL	plan	RBC	red blood cell
PLAID	Panel Layout And Integrated Design	RBE	relative biological effectiveness
PMTS	predetermined motion time system	RC	room criterion curve
Pn	plan	rd	rutherford
PNL	perceived noise level	RDA	Recommended Daily Allowance
PNS	peripheral nervous system	REAT	real ear attenuation at threshold
PP	pre-position	rem	roentgen equivalent man
PPE	personal protective equipment	REM	rapid eye movement
PRE	progressive resistance exercises	rep	roentgen equivalent physical
PROME-THEUS	Program for European Traffic with Highest Efficiency and Unprecedented Safety	rf	radio frequency
		RH	relative humidity
		RJP	realistic job preview
		RL	release load
PSE	point of subjective equality	R_m	mechanical resistance
psig	gauge pressure	RMI	repetitive motion injury
PSIL	preferred speech interference level	rms	root mean square
		RNA	ribonucleic acid
PT	physical therapy	ROC	receiver operating characteristic
PTS	permanent threshold shift	ROM	range of motion
PTT	part-task trainer	ROW	right-of-way
PWC	physical work capacity	r_{pb}	point biserial r
PWL	sound power level	RPE	Rating of Perceived Effort
Q	light quantity	RQ	respiratory quotient
Q	quality factor	RSDD	room surface dirt depreciation
Q	semi-interquartile range		

RSI	repetitive strain injury	STP	standard temperature and pressure
RT	reaction time		
RTD	repetitive trauma disorder	STPD	standard temperature and pressure, dry
s, sec	second		
S/N	signal-to-noise ratio	STRES	Standardized Tests for Research with Environmental Stressors
SAD	seasonal affective disorder		
SAMMIE	System for Aiding Man-Machine Interaction Evaluation		
		surfactant	surface active agent
		Sv	sievert
SAS	space adaptation syndrome	SWAT	Subjective Workload Assessment Technique
sb	stilb		
SCBA	self-contained breathing apparatus	$\overline{T}_b$	mean body temperature
		$\overline{T}_s$	weighted mean skin temperature
SCOT	Standard Colors of Textiles		
SCR	skin conductance response	T	reverberation time
SDO	Standard Deviate Observer	T	temperature
SEA	statistical energy analysis	T	tesla
SEM	standard error of the mean	T	talbot
SERDS	Standard Ergonomic Reference Data System	T	transmittance
		TACAN	Tactical Air Navigation
SEU	speech enhancement unit	TBF	total body fat
Sh	search	tbsp	tablespoon
SHF	superhigh frequency	TBT	total bottom time
SHM	simple harmonic motion	TBW	total body water
SI	Severity Index	T_c	color temperature
SI	Système International d'Unites	td	troland
		T_d	distribution temperature
SIC	Standard Industrial Classification	TE	transport empty
		TENS	transcutaneous electrical nerve stimulation
SIL	speech interference level		
SMAC	spacecraft maximum allowable concentration	THERP	Technique for Human Error Rate Prediction
SNR	signal-to-noise ratio	TKA line	trochanter-knee-ankle line
SO$_2$	sulfur dioxide	TL	transport loaded
SPE	solar particle event	TLV	threshold limit value
SPEX	specularly reflected light excluded	TLV-C	threshold limit value - ceiling
SPH	sound protective helmet	TLV-STEL	threshold limit value - short term exposure limit
SPINC	specularly reflected light included		
		TLV-TWA	threshold limit value - time-weighted average
SPL	sound pressure level		
sr	steradian	TLX	Task Load Index
SRP	seat reference point	TMJ	temporo-mandibular joint disorder
SRR	skin resistance response		
SRT	speech reception threshold	TMU	time measurement unit
St	select	TOBEC	total body electrical conductivity
stanine	standard nine scale		
STC	sound transmission class	TOC	total organic carbon
STEL	short-term exposure limit	TQ	Total Quality
STI	speech transmission index	TQM	Total Quality Management

TTS	temporary threshold shift	VAPEPS	Vibroacoustic Payload Environment Prediction System
TWA	time-weighted average		
U	use		
UD	unavoidable delay	VCP	visual comfort probability
uel	upper explosive limit	VDT	video display terminal
ufl	upper flammable limit	VDT	visual display terminal
UHF	ultrahigh frequency	VDU	visual display unit
UIDS	user interface development system	VDU	video display unit
		VFR	visual flight rules
UIMS	user interface management system	VHF	very high frequency
		VIR	marine vessel accident incidence rate
U_v	volume velocity		
UV	ultraviolet	VLF	very low frequency
UV-A	ultraviolet A	VOR	vestibulo-ocular reflex
UV-B	ultraviolet B	VOX	voice-activated
UV-C	ultraviolet C	VWF	vibration white finger
$\vec{v}$	velocity	W	sound power level
$V(\lambda)$	photopic spectral luminous efficiency function	W	watt
		W	Kendall's coefficient of concordance
$V'(\lambda)$, $V_{\lambda'}$	scotopic spectral luminous efficiency function	Wb	weber
$\dot{V}_{CO_2}$	carbon dioxide production rate	WBGT	wet bulb globe temperature
		WBT	wet bulb temperature
$\dot{V}_E$	pulmonary ventilation	WCI	windchill index
$\dot{V}_{O_2\,max}$	maximal aerobic capacity	WCT	windchill temperature
		WYSIWYG	what-you-see-is-what-you-get
$\dot{V}_{O_2}$	oxygen consumption rate		
V	volt	X_m	mechanical reactance
V_λ	luminosity function	Z	electrical impedance
VA	volt-ampere	Z	impedance
		Z_m	mechanical impedance

APPENDIX B1
ORGANIZATIONS AND
THEIR ACRONYMS

Accreditation Board for Engineering and Technology (ABET)

Acoustical Society of America (ASA)

Advisory Group for Aerospace Research and Development (AGARD)

Aerospace Human Factors Association (AsHA)

Aerospace Medical Association (AsMA)

Aerospace Medical Panel (AMP)

Air Force Institute of Pathology (U.S.) (AFIP)

Air Force Office of Scientific Research (U.S.) (AFOSR)

Alcohol, Drug Abuse, and Mental Health Administration (ADAMHA)

American Academy of Orthopedic Surgeons (AAOS)

American Academy of Physical Medicine and Rehabilitation (AAPMR)

American Alliance for Health, Physical Education, Recreation and Dance (AAHPERD)

American Association for Automotive Medicine (AAAM)

American Association for the Advancement of Science (AAAS)

American Association of Electromyography and Electrodiagnosis (AAEE)

American Association of Retired Persons (AARP)

American Association of Textile Chemists and Colorists (AATCC)

American Board of Health Physics (ABHP)

American Board of Industrial Hygiene

American College of Occupational and Environmental Medicine (ACOEM)

American College of Sports Medicine (ACSM)

American Conference of Government Industrial Hygienists (ACGIH)

American Congress of Rehabilitation Medicine (ACRM)

American Dental Association (ADA)

American Hospital Formulary Service (AHFS)

American Industrial Hygiene Association (AIHA)

American Institute for Mining Engineers (AIME)

American Institute of Architects (AIA)

American Institute of Medical Climatology (AIMC)

American Institute of Plant Engineers (AIPE)

American Medical Association (AMA)

American National Standards Institute (ANSI)

American Occupational Therapy Association (AOTA)

American Physical Therapy Association (APTA)

American Physiological Society (APS)

American Production and Inventory Control Society (APICS)

American Psychological Association (APA)

American Psychological Society (APS)

American Public Transit Association (APTA)

American Rehabilitation Foundation (ARF)

American Society for Engineering Education (ASEE)

American Society for Quality Control (ASQC)

American Society for Testing and Materials (ASTM)

American Society of Biomechanics

American Society of Civil Engineers (ASCE)

American Society of Heating, Refrigeration, and Air Conditioning Engineers (ASHRAE)

American Society of Mechanical Engineers (ASME)

American Society of Safety Engineers (ASSE)

American Society on Aging (ASA)

American Speech and Hearing Association (ASHA)

Armstrong Aerospace Medical Research Laboratory (USAF) (AAMRL)

Army Aeromedical Research Laboratory (U.S.) (AARL)

Army Institute of Environmental Medicine (ARIEM)

Army Medical Biomechanical Research Laboratory (AMBRL)

Army Research Institute (ARI)

Association Colombiana de Ergonomia

Association for Quality and Participation (AQP)

Association for the Advancement of Medical Instrumentation (AAMI)

Association for the Advancement of Rehabilitation Technology (RESNA)

Association Mexicana de Ergonomia

Association of Aviation Psychologists (AAP)

Association of Driver Educators for the Disabled (ADED)

Association of Ergonomics Societies of Yugoslavia

Back Pain Association

Belgian Ergonomics Society

Bioelectromagnetics Society (BEMS)

Biofeedback Society of America (BSA)

Biophysical Society

Board of Certification in Professional Ergonomics (BCPE)

Board of Certified Safety Professionals (BCSP)

Brazilian Association of Ergonomics

British Occupational Hygiene Society (BOHS)

British Psychological Society (BPS)

British Standards Institution (BSI)

Bureau des temps élémentaires

Bureau of Labor Standards (U.S.) (BLS)

Bureau of Labor Statistics (U.S.) (BLS)

Bureau of the Census

Canadian Centre for Occupational Health and Safety (CCOHS)

Centers for Disease Control (CDC)

Civil Aeromedical Institute (CAMI)

Civil Aeronautics Board (CAB)

Commission Internationale de l'Eclairage (CIE)

Committee on Hearing, Bioacoustics, and Biomechanics (CHABA)

Conference of European Posts and Telecommunications Administrations (CEPT)

Consumer Product Safety Commission (CPSC)

Crew System Ergonomic Information Analysis Center (CSERIAC)

Defence and Civil Institute of Environmental Medicine (Canada) (DCIEM)

Department of Commerce (U.S.)

Department of Defense (U.S.) (DOD)

Department of Energy (U.S.) (DOE)

Department of Health and Human Services (U.S.) (DHHS)

Department of Labor (U.S.)

Department of Transportation (U.S.) (DOT)

Deutsches Institut für Normung (DIN)

Educational Testing Service (ETS)

Electronics Industries Association (EIA)

Employment Standards Administration

Environmental Protection Agency (EPA)

Equal Employment Opportunity Commission (EEOC)

Ergonomics Association of Southern Africa

Ergonomics Society

Ergonomics Society of Australia

European Association for Work and Organizational Psychology (EAWOP)

European Computer Manufacturers Association (ECMA)

European Organization for Quality Control (EOQC)

Federal Aviation Administration (FAA)

Federal Highway Administration (FHWA)

Federal Railroad Administration (FRA)

Federal Transit Administration (FTA)

Food and Drug Administration (FDA)

General Aviation Manufacturers Association (GAMA)

General Aviation Safety Panel (GASP)

General Services Administration (GSA)

Gerontological Society of America

Gesellschaft für Arbeitswissenschaft

High Speed Rail Association (HSRA)

Human Ergology Society

Human Factors Association of Canada (HFAC)

Human Factors Society (HFS)

Illuminating Engineering Society (IES)

Industrial Designers Society of America (IDSA)

Institute for Consumer Ergonomics (ICE)

Institute of Aviation Medicine (RAF) (IAM)

Institute of Electrical and Electronics Engineers (IEEE)

Institute of Environmental Sciences (IES)

Institute of Industrial Engineers (IIE)

Institute of Management Sciences

Institute of Production Engineers

Instrument Society of America (ISA)

Intelligent Vehicle Highway Society of America (IVHSA)

Inter-Society Color Council (ISCC)

International Association of Lighting Designers

International Civil Aviation Organization (ICAO)

International Commission on Radiological Protection (ICRP)

International Computer Integrated Manufacturing Society

International Council on Health, Physical Education and Recreation (ICHPER)

International Dance - Exercise Association (IDEA)

International Design for Extreme Environments Association (IDEEA)

International Ergonomics Association (IEA)

International Foundation for Industrial Ergonomics and Safety Research (IFIESR)

International Institute for Production Engineering Research (CIRP)

International Institute for Safety in Transportation (IST)

International Labor Office/Organization (ILO)

International Material Management Society

International Olympic Committee (IOC)

International Organization for Standardization (ISO)

International Radiation Commission (IRC)

International Radiation Protection Association (IRPA)

International Society of Biomechanics (ISB)

International Society of Electrophysiological Kinesiology (ISEK)

International Telegraph and Telephone Consultative Committee (CCITT)

Israel Ergonomics Society

Japan Ergonomics Research Society

Marine Safety Council (MSC)

Marine Technology Society (MTS)

Maritime Administration (MARAD)

Method Time Measurement Association for Standards and Research (MTMASR)

Mine Safety and Health Administration (MSHA)

National Academy of Science (NAS)

National Advisory Council for Aeronautics (NACA)

National Aeronautics and Space Administration (NASA)

National Center for Health Statistics

National Clearinghouse on Technology and Aging

National Commission for Employment Policy

National Computer Graphics Association (NCGA)

National Council of Acoustical Consultants (NCAC)

National Electrical Manufacturer's Association (NEMA)

National Fire Protection Association (NFPA)

National Health Service (NHS)

National Hearing Conservation Association (NHCA)

National Highway Traffic Safety Administration (NHTSA)

National Institute for Occupational Safety and Health (NIOSH)

National Institute for Standards and Technology (NIST)

National Institutes of Health (NIH)

National Labor Relations Board (NLRB)

National Research Council (NRC)

National Safety Council (NSC)

National Sanitation Foundation

National Science Foundation (NSF)

National Service Robot Association (NSRA)

National Society for Performance and Instruction

National Transportation Safety Board (NTSB)

Naval Air Development Center (NADC)

Naval Air Material Center

Naval Biodynamics Laboratory (NBDL)

Nederlandse Vereniging Voor Ergonomie

New Zealand Ergonomics Society

Nordic Ergonomics Society (NES)

North Atlantic Treaty Organization (NATO)

Nuclear Regulatory Commission (NRC)

Occupational Safety and Health Administration (OSHA)

Occupational Safety and Health Review Commission

Öesterreich Arbeitsgemeinschaft für Ergonomie

Office of Federal Contract Compliance Programs (OFCCP)

Office of Naval Research (ONR)

Office of Technology Assessment (OTA)

Operations Research Society of America

Optical Society of America (OSA)

Polish Ergonomics Society

Professional Aeromedical Transport Association (PATA)

Profit Sharing Research Foundation

Psychonomic Society

Robot Institute of America

Royal Society for the Prevention of Accidents (RoSPA)

SAFE Association

Safety Equipment Institute (SEI)

School of Aerospace Medicine (USAF) (SAM)

Select Committee on Aging

Societa Italiana di Ergonomia

Societé d'Ergonomie de Langue Française (SELF)

Society for Information Display (SID)

Society for Photo-optical Engineering (SPIE)

Society of American Value Engineers (SAVE)

Society of Automotive Engineers (SAE)

Society of Experimental Test Pilots (SETP)

Society of Manufacturing Engineers (SME)

South East Asian Ergonomics Society (SEAES)

System Safety Society

Technical Association of the Pulp and Paper Industry (TAPPI)

Transportation Safety Equipment Institute

Underwriter's Laboratories (UL)

United States Olympic Committee (USOC)

Veteran's Administration (VA)

World Health Organization (WHO)

APPENDIX B2
ORGANIZATION ACRONYMS AND
THEIR NAMES

AAAM   American Association for Automotive Medicine

AAAS   American Association for the Advancement of Science

AAEE   American Association of Electromyography and Electrodiagnosis

AAHPERD   American Alliance for Health, Physical Education, Recreation and Dance

AAMI   Association for the Advancement of Medical Instrumentation

AAMRL   Armstrong Aerospace Medical Research Laboratory (USAF)

AAOS   American Academy of Orthopedic Surgeons

AAP   Association of Aviation Psychologists

AAPMR   American Academy of Physical Medicine and Rehabilitation

AARL   Army Aeromedical Research Laboratory (U.S.)

AARP   American Association of Retired Persons

AATCC   American Association of Textile Chemists and Colorists

ABET   Accreditation Board for Engineering and Technology

ABHP   American Board of Health Physics

ACGIH   American Conference of Government Industrial Hygienists

ACOEM   American College of Occupational and Environmental Medicine

ACRM   American Congress of Rehabilitation Medicine

ACSM   American College of Sports Medicine

ADA   American Dental Association

ADAMHA   Alcohol, Drug Abuse, and Mental Health Administration

ADED   Association of Driver Educators for the Disabled

AFIP   Air Force Institute of Pathology (U.S.)

AFOSR   Air Force Office of Scientific Research (U.S.)

AGARD   Advisory Group for Aerospace Research and Development

AHFS   American Hospital Formulary Service

AIA   American Institute of Architects

AIHA   American Industrial Hygiene Association

AIMC   American Institute of Medical Climatology

AIME   American Institute for Mining Engineers

AIPE	American Institute of Plant Engineers	ASQC	American Society for Quality Control
AMA	American Medical Association	ASSE	American Society of Safety Engineers
AMBRL	Army Medical Biomechanical Research Laboratory	ASTM	American Society for Testing and Materials
AMP	Aerospace Medical Panel	BCPE	Board of Certification in Professional Ergonomics
ANSI	American National Standards Institute	BCSP	Board of Certified Safety Professionals
AOTA	American Occupational Therapy Association	BEMS	Bioelectromagnetics Society
APA	American Psychological Association	BLS	Bureau of Labor Statistics (U.S.)
APICS	American Production and Inventory Control Society	BLS	Bureau of Labor Standards (U.S.)
APS	American Psychological Society	BOHS	British Occupational Hygiene Society
APS	American Physiological Society	BPS	British Psychological Society
APTA	American Physical Therapy Association	BSA	Biofeedback Society of America
APTA	American Public Transit Association	BSI	British Standards Institution
AQP	Association for Quality and Participation	CAB	Civil Aeronautics Board
		CAMI	Civil Aeromedical Institute
ARF	American Rehabilitation Foundation	CCITT	International Telegraph and Telephone Consultative Committee
ARI	Army Research Institute	CCOHS	Canadian Centre for Occupational Health and Safety
ARIEM	Army Institute of Environmental Medicine	CDC	Centers for Disease Control
ASA	American Society on Aging	CEPT	Conference of European Posts and Telecommunications Administrations
ASA	Acoustical Society of America	CHABA	Committee on Hearing, Bioacoustics, and Biomechanics
ASCE	American Society of Civil Engineers		
ASEE	American Society for Engineering Education	CIE	Commission Internationale de l'Eclairage
AsHA	Aerospace Human Factors Association	CIRP	International Institute for Production Engineering Research
ASHA	American Speech and Hearing Association	CPSC	Consumer Product Safety Commission
ASHRAE	American Society of Heating, Refrigeration, and Air Conditioning Engineers	CSERIAC	Crew System Ergonomic Information Analysis Center
AsMA	Aerospace Medical Association	DCIEM	Defence and Civil Institute of Environmental Medicine (Canada)
ASME	American Society of Mechanical Engineers		

DHHS	Department of Health and Human Services (U.S.)	ICHPER	International Council on Health, Physical Education and Recreation
DIN	Deutsches Institut für Normung	ICRP	International Commission on Radiological Protection
DOD	Department of Defense (U.S.)	IDEA	International Dance - Exercise Association
DOE	Department of Energy (U.S.)		
DOT	Department of Transportation (U.S.)	IDEEA	International Design for Extreme Environments Association
EAWOP	European Association for Work and Organizational Psychology	IDSA	Industrial Designers Society of America
ECMA	European Computer Manufacturers Association	IEA	International Ergonomics Association
EEOC	Equal Employment Opportunity Commission	IEEE	Institute of Electrical and Electronics Engineers
EIA	Electronics Industries Association	IES	Institute of Environmental Sciences
EOQC	European Organization for Quality Control	IES	Illuminating Engineering Society
EPA	Environmental Protection Agency	IFIESR	International Foundation for Industrial Ergonomics and Safety Research
ETS	Educational Testing Service		
FAA	Federal Aviation Administration	IIE	Institute of Industrial Engineers
FDA	Food and Drug Administration	ILO	International Labor Office / Organization
FHWA	Federal Highway Administration	IOC	International Olympic Committee
FRA	Federal Railroad Administration	IRC	International Radiation Commission
FTA	Federal Transit Administration	IRPA	International Radiation Protection Association
GAMA	General Aviation Manufacturers Association	ISA	Instrument Society of America
GASP	General Aviation Safety Panel	ISB	International Society of Biomechanics
GSA	General Services Administration	ISCC	Inter-Society Color Council
		ISEK	International Society of Electrophysiological Kinesiology
HFAC	Human Factors Association of Canada		
HFS	Human Factors Society	ISO	International Organization for Standardization
HSRA	High Speed Rail Association		
IAM	Institute of Aviation Medicine (RAF)	IST	International Institute for Safety in Transportation
ICAO	International Civil Aviation Organization	IVHSA	Intelligent Vehicle Highway Society of America
ICE	Institute for Consumer Ergonomics	MARAD	Maritime Administration
		MSC	Marine Safety Council

MSHA	Mine Safety and Health Administration	NSRA	National Service Robot Association
MTMASR	Method Time Measurement Association for Standards and Research	NTSB	National Transportation Safety Board
MTS	Marine Technology Society	OFCCP	Office of Federal Contract Compliance Programs
NACA	National Advisory Council for Aeronautics	ONR	Office of Naval Research
		OSA	Optical Society of America
NADC	Naval Air Development Center	OSHA	Occupational Safety and Health Administration
NAS	National Academy of Science	OTA	Office of Technology Assessment
NASA	National Aeronautics and Space Administration	PATA	Professional Aeromedical Transport Association
NATO	North Atlantic Treaty Organization	RESNA	Association for the Advancement of Rehabilitation Technology
NBDL	Naval Biodynamics Laboratory	RoSPA	Royal Society for the Prevention of Accidents
NCAC	National Council of Acoustical Consultants	SAE	Society of Automotive Engineers
NCGA	National Computer Graphics Association	SAM	School of Aerospace Medicine (USAF)
NEMA	National Electrical Manufacturer's Association	SAVE	Society of American Value Engineers
NES	Nordic Ergonomics Society	SEAES	South East Asian Ergonomics Society
NFPA	National Fire Protection Association	SEI	Safety Equipment Institute
NHCA	National Hearing Conservation Association	SELF	Societé d'Ergonomie de Langue Française
NHS	National Health Service		
NHTSA	National Highway Traffic Safety Administration	SETP	Society of Experimental Test Pilots
NIH	National Institutes of Health	SID	Society for Information Display
NIOSH	National Institute for Occupational Safety and Health	SME	Society of Manufacturing Engineers
NIST	National Institute for Standards and Technology	SPIE	Society for Photo-optical Engineering
NLRB	National Labor Relations Board	TAPPI	Technical Association of the Pulp and Paper Industry
NRC	Nuclear Regulatory Commission	UL	Underwriter's Laboratories
NRC	National Research Council	USOC	United States Olympic Committee
NSC	National Safety Council	VA	Veteran's Administration
NSF	National Science Foundation	WHO	World Health Organization